Other Titles in This Series

(Continued in the back of this publication)

Dynamical Systems and Probabilistic Methods in Partial Differential Equations

Lectures in
APPLIED
MATHEMATICS

Volume 31

Dynamical Systems and Probabilistic Methods in Partial Differential Equations

1994 Summer Seminar
on Dynamical Systems and
Probabilistic Methods for Nonlinear Waves
June 20–July 1, 1994
MSRI, Berkeley, CA

Percy Deift
C. David Levermore
C. Eugene Wayne
Editors

American Mathematical Society
Providence, Rhode Island

The Proceedings of the 1994 AMS-SIAM Summer Seminar on Dynamical Systems and Probabilistic Methods for Nonlinear Waves were prepared by the American Mathematical Society with support from the National Science Foundation, grant DMS-9318637; the office of Naval Research, grant N00014-94-0700; and the Department of Energy, grant DE-FG02-94ER25208.A000.

1991 *Mathematics Subject Classification*. Primary 58F40;
Secondary 34C35, 35Q99.

Library of Congress Cataloging-in-Publication Data

Summer Seminar on Dynamical Systems and Probabilistic Methods for Nonlinear Waves (1994 : Berkeley, Calif.)
  Dynamical systems and probabilistic methods in partial differential equations : 1994 Summer Seminar on Dynamical Systems and Probabilistic Methods for Nonlinear Waves, June 20–July 1, 1994, MSRI, Berkeley, CA / Percy Deift, C. David Levermore, C. Eugene Wayne, editors.
    p. cm. — (Lectures in applied mathematics, ISSN 0075-8485 ; v. 31)
  Includes bibliographical references.
  ISBN 0-8218-0368-9 (alk. paper)
  1. Differentiable dynamical systems—Congresses. 2. Differential equations. Partial—Congresses. I. Deift, Percy, 1945– . II. Levermore, C. D. III. Wayne, C. Eugene, 1956– . IV. Title. V. Series: Lectures in applied mathematics (American Mathematical Society) ; v. 31.
  QA614.8.S86 1994
  515′.353—dc20   95-44661
                                                                                  CIP

---

10 9 8 7 6 5 4 3 2   01 00 99 98

Contents

Preface

This volume contains some of the lectures presented at the 1994 AMS/SIAM Summer Seminar, held June 20–July 1 at the Mathematical Sciences Research Institute in Berkeley. It was the intent of the organizers of the summer seminar to introduce the participants to as many of the interesting and active applications of dynamical systems to problems in applied mathematics as the time constraints of the workshop allowed. Consequently, this book covers a great deal of ground. Nonetheless, the pedagogical orientation of the lectures has been retained in this volume, and as such, we hope that it will serve as an ideal introduction to these varied and interesting topics.

While the focus of the workshop was quite broad, several organizing principles emerged. The first was the increasing role of dynamical systems theory in our understanding of partial differential equations. The first three contributions of the present volume are devoted to this theme. In particular, all of these lecturers stressed the importance that the geometrical structures present in the phase spaces of these systems have for our understanding of their dynamics. A second theme was the central importance of certain prototypical partial differential equations. These equations, which include the complex Ginzburg-Landau, nonlinear Schrödinger and Korteweg-de Vries equations, arise in many different contexts and hence have an importance that transcends their apparently special form. In this book, two sets of lectures explore this phenomenon in greater detail for the complex Ginzburg-Landau equation, one examining in detail the sorts of phenomena that can arise in this equation, and the other focussed on showing rigorously how a knowledge of the behavior of the solutions of the Ginzburg-Landau equation implies information about a host of more complicated systems. In addition to their ubiquity, the nonlinear Schrödinger and Korteweg-de Vries equation share the additional remarkable property of being completely integrable. The meaning and consequences of complete integrability are explored in the lectures of section 2. Finally, the last set of lectures looks specifically at problems in fluid mechanics and turbulence. More specifically, it examines the extent to which one can determine the limits of popular physically motivated heuristic theories of fluids like the renormalization group and the Kolmogorov scaling law.

Percy Deift
C. David Levermore
C. Eugene Wayne

Section I

Dynamical Systems and PDE's

An Introduction to KAM Theory

C. EUGENE WAYNE

1. Introduction

Over the past thirty years, the Kolmogorov-Arnold-Moser (KAM) theory has played an important role in increasing our understanding of the behavior of non-integrable Hamiltonian systems. I hope to illustrate in these lectures that the central ideas of the theory are, in fact, quite simple. With this in mind, I will concentrate on two examples and will forego generality for concreteness and (I hope) clarity. The results and methods which I will present are well-known to experts in the field but I hope that by collecting and presenting them in as simple a context as possible I can make them somewhat more approachable to newcomers than they are often considered to be.

The outline of the lectures is as follows. After a short historical introduction, I will explain in detail one of the simplest situations where the KAM techniques are used – the case of diffeomorphisms of a circle. I will then go on to discuss the theory in its original context, that of nearly-integrable Hamiltonian systems.

The problem which the KAM theory was developed to solve first arose in celestial mechanics. More than 300 years ago, Newton wrote down the differential equations satisfied by a system of massive bodies interacting through gravitational forces. If there are only two bodies, these equations can be explicitly solved and one finds that the bodies revolve on Keplerian ellipses about their center of mass. If one considers a third body (the "three-body-problem"), no exact solution exists – even if, as in the solar system, two of the bodies are much lighter then the third. In this case, however, one observes that the mutual gravitational force between these two "planets" is much weaker than that between either planet and the sun. Under these circumstances one can try to solve the problem perturbatively, first ignoring the interactions between the planets. This gives an **integrable** system, or one which can be solved explicitly, with

1991 *Mathematics Subject Classification.* Primary 58F05; Secondary 34C20 70H05.
The author was supported in part by NSF Grant #9203359

each planet revolving around the sun oblivious of the other's existence. One can then try to systematically include the interaction between the planets in a perturbative fashion. Physicists and astronomers used this method extensively throughout the nineteenth century, developing series expansions for the solutions of these equations in the small parameter represented by the ratio of the mass of the planet to the mass of the sun. However, the convergence of these series was never established – not even when the King of Sweden offered a very substantial prize to anyone who succeeded in doing so. The difficulty in establishing the convergence of these series comes from the fact that the terms in the series have **small denominators** which we shall consider in some detail later in these lectures. One can obtain some physical insight into the origin of these convergence problems in the following way. As one learns in an elementary course in differential equations, a harmonic oscillator has a certain natural frequency at which it oscillates. If one subjects such an oscillator to an external force of the same frequency as the natural frequency of the oscillator, one has **resonance** effects and the motion of the oscillator becomes unbounded. Indeed, if one has a typical nonlinear oscillator, then whenever the perturbing force has a frequency that is a rational multiple of the natural frequency of the oscillator, one will have resonances, because the nonlinearity will generate oscillations of all multiples of the basic driving frequency.

In a similar way, one planet exerts a periodic force on the motion of a second, and if the orbital periods of the two are commensurate, this can lead to resonance and instability. Even if the two periods are not exactly commensurate, but only approximately so the effects lead to convergence problems in the perturbation theory.

It was not until 1954 that A. N. Kolmogorov [8] in an address to the ICM in Amsterdam suggested a way in which these problems could be overcome. His suggestions contained two ideas which are central to all applications of the KAM techniques. These two basic ideas are:

- Linearize the problem about an approximate solution and solve the linearized problem – it is at this point that one must deal with the small denominators.
- Inductively improve the approximate solution by using the solution of the linearized problem as the basis of a Newton's method argument.

These ideas were then fleshed out, extended, and applied in numerous other contexts by V. Arnold and J. Moser, ([1], [9]) over the next ten years or so, leading to what we now know as the KAM theory.

As I said above, we will consider the details of this procedure in two cases. The first, the problem of showing that diffeomorphisms of a circle are conjugate to rotations, was chosen for its simplicity – the main ideas are visible with fewer technical difficulties than appear in other applications. We will then look at the KAM theory in its original setting of small perturbations of integrable

Hamiltonian systems. I'll attempt to parallel the discussion of the case of circle diffeomorphisms as closely as possible in order to keep our focus on the main ideas of the theory and ignore as much as possible the additional technical complications which arise in this context.

Acknowledgments: It is a pleasure to thank Percy Deift and Andrew Török for many helpful comments about these notes.

2. Circle Diffeomorphisms

Let us begin by discussing one of the simplest examples in which one encounters small denominators, and for which the KAM theory provides a solution. It may not be apparent for the moment what this problem has to do with the problems of celestial mechanics discussed in the introduction, but almost all of the difficulties encountered in that problem also appear in this context but in ways which are less obscured by technical difficulties – this is, if you like, our warm-up exercise.

We will consider orientation preserving diffeomorphisms of the circle, or equivalently, their lifts to the real line:

$$\phi : R^1 \to R^1$$

$$\phi(x) = x + \tilde{\eta}(x) \text{ with } \tilde{\eta}(x+1) = \tilde{\eta}(x) \text{ and } \tilde{\eta}'(x) > -1 \ .$$

We wish to consider ϕ as a dynamical system, and study the behavior of its "orbits" – *i.e.* we want to understand the behavior of the sequences of points $\{\phi^{(n)}(x)(\mathrm{mod}1)\}_{n=0}^{\infty}$, where $\phi^{(n)}$ means the n-fold composition of ϕ with itself. Typical questions of interest are whether or not these orbits are periodic, or dense in the circle.

The simplest such diffeomorphism is a rotation $R_\alpha(x) = x + \alpha$. Note that we understand "everything" about its dynamics. For instance, if α is rational, all the orbits of R_α are periodic, and none are dense. However, we would like to study more complicated dynamical systems than this. Thus we will suppose that

$$(1) \qquad\qquad \phi(x) = x + \alpha + \eta(x) \ ,$$

where as before, $\eta(x+1) = \eta(x)$ and $\eta'(x) > -1$. As I said in the introduction, I will not attempt to consider the most general case, but rather will focus on simplicity of exposition. Thus I will consider only **analytic** diffeomorphisms. Define the strips $S_\sigma = \{z \in C \mid |Imz| < \sigma\}$. Then I will assume that

$$\eta \in B_\sigma = \{\eta \mid \eta(z) \text{ is analytic on } S_\sigma,$$

$$\eta(x+1) = \eta(x) \text{ and } \sup_{|Imz|<\sigma} |\eta(z)| \equiv \|\eta\|_\sigma < \infty\} \ .$$

Note that one can assume that $\sigma < 1$, without loss of generality.

Our goal in this section will be to understand the dynamics of $\phi(x) = x + \alpha + \eta(x)$ when η has small norm. One way to do this is to show that the dynamics of ϕ are "like" the dynamics of a system we understand – for instance, suppose that we could find a change of variables which transformed ϕ into a pure rotation. Then since we understand the dynamics of the rotation, we would also understand those of ϕ. If we express this change of variables as $x = H(\xi)$, where $H(\xi+1) = 1 + H(\xi)$ preserves the periodicity of ϕ, then we want to find H such that

$$H^{-1} \circ \phi \circ H(\xi) = R_\rho(\xi) \ ,$$

or equivalently

(2) $$\phi \circ H(\xi) = H \circ R_\rho(\xi) \ .$$

Such a change of variables is said to conjugate ϕ to the rotation R_ρ.

REMARK 2.1. *The relationship between this problem and the celestial mechanics questions discussed in the introduction now becomes more clear. In that case we wanted to understand the extent to which the motion of the solar system when we included the effects of the gravitational interaction between the various planets was "like" that of the simple Kepler system.*

In order to answer this question we need to introduce an important characteristic of circle diffeomorphisms, the **rotation number**

DEFINITION 2.1. *The rotation number of ϕ is*

$$\rho(\phi) = \lim_{n \to \infty} \frac{\phi^{(n)}(x) - x}{n} \ .$$

REMARK 2.2. *It is a standard result of dynamical systems theory that for any homeomorphism of the circle the limit on the right hand side of this equation exists and is independent of x. (See [6], p. 296.)*

REMARK 2.3. *Note that from the definition of the rotation number, it follows immediately that for any homeomorphism H, the map $\tilde{\phi} = H^{-1} \circ \phi \circ H$ has the same rotation number as ϕ. (Since $\tilde{\phi}^{(n)} = H^{-1} \circ \phi^{(n)} \circ H$, and the initial and final factors of H and H^{-1} have no effect on the limit.)*

As a final remark about the rotation number we note that if $\phi(x) = x + \alpha + \eta(x)$, then an easy induction argument shows that $\rho(\phi) = \alpha + \lim_{n \to \infty} \frac{1}{n} \sum_{j=0}^{n-1} \eta \circ \phi^{(j)}(x)$. In particular, if $\alpha = \rho$, we have $\lim_{n \to \infty} \frac{1}{n} \sum_{j=0}^{n-1} \eta \circ \phi^{(j)}(x) = 0$, so we have proved:

LEMMA 2.1. *If $\phi(x) = x + \rho + \eta(x)$ has rotation number ρ, then there exists some x_0 such that $\eta(x_0) = 0$.*

We must next ask about the properties we wish the change of variables H to have. If we only demand that H be a homeomorphism, then **Denjoy's Theorem** ([6] p. 301) says that if the rotation number of ϕ is irrational, we can always find an H which conjugates ϕ to a rotation. However, if we want more detailed information about the dynamics it makes sense to ask that H have additional smoothness. In fact, it is natural to ask that H be as smooth as the diffeomorphism itself – in this case, analytic. (There will, in general, be some loss of smoothness even in this case. We will find, for example, that while there exists an analytic conjugacy function, H, its domain of analyticity will be somewhat smaller than that of ϕ.) Surprisingly, the techniques which Denjoy used fail completely in this case, and the answer was not known until the late fifties when Arnold applied KAM techniques to answer the question in the case when

η is small. Even more surprisingly, in order to even state Arnold's theorem, we have to discuss a little number theory.

Any irrational number can be approximated arbitrarily well by rational numbers, and in fact, **Dirichlet's Theorem** even gives us an estimate of how good this approximation is. More precisely, it says that given any irrational number ρ, there exist infinitely many pairs of integers (m, n) such that $|\rho - (m/n)| < 1/n^2$. On the other hand, most irrational numbers can't be approximated much better than this.

DEFINITION 2.2. *The real number ρ is of type (K, ν) if there exist positive numbers K and ν such that $|\rho - (m/n)| > K|n|^{-\nu}$, for all pairs of integers (m, n).*

PROPOSITION 2.1. *For every $\nu > 2$, almost every irrational number ρ is of type (K, ν) for some $K > 0$.*

Proof: The proof is not difficult, but would take us a bit out of our way. The details can be found in [**3**], page 116, for example. Note also, that we can assume without loss of generality that $K \leq 1$, since if ρ is of type $(\tilde{K}, \nu)$ for some $\tilde{K} > 1$, it is also of type $(1, \nu)$.

THEOREM 2.1 (ARNOLD'S THEOREM [**1**]). *Suppose that ρ is of type (K, ν). There exists $\epsilon(K, \nu, \sigma) > 0$ such that if $\phi(x) = x + \rho + \eta(x)$ has rotation number ρ, and $\|\eta\|_\sigma < \epsilon(K, \nu, \sigma)$, then there exists an analytic and invertible change of variables $H(x)$ which conjugates ϕ to R_ρ.*

As mentioned above, Arnold's proof of this theorem used the KAM theory. The proof can be broken into two main parts – an analysis of a linearized equation, and a Newton's method iteration step. These same two steps will reappear in the next section when we discuss nearly integrable Hamiltonian systems, and they are characteristic of almost all applications of the KAM theory.

REMARK 2.4. *It may seem that by assuming that the diffeomorphism is of the form $\phi(x) = x + \rho + \eta(x)$, where ρ is the rotation number of ϕ, we are considering a less general situation than that described above in which we allowed ϕ to have the form $x + \alpha + \eta(x)$. As we shall see below, there is no real loss of generality in this restriction.*

Step 1: Analysis of the Linearized equation

Note that since $\|\eta\|_\sigma$ is small, the diffeomorphism ϕ is "close" to the pure rotation R_ρ. Thus, we might hope that if a change of variables H which satisfies (2) exists is would be close to the identity *i.e.* $H(x) = x + h(x)$, where h is "small". If we make this assumption and substitute this form of H in (2), we find that h should satisfy the equation

$$(3) \qquad\qquad h(x + \rho) - h(x) = \eta(x + h(x))$$

If we now expand both sides of this equation, retaining only terms of first order in the (presumably) small quantities h and η, we find:

$$(4) \qquad\qquad h(x + \rho) - h(x) = \eta(x)$$

Since all the functions in this equation are periodic, and the equation is linear in the unknown function h, we can immediately write down a (formal) solution for the coefficients in the Fourier series of h. If $\hat{\eta}(n)$ is the nth Fourier coefficient of η, then the nth Fourier coefficient of h is

$$(5) \qquad\qquad \hat{h}(n) = \frac{\hat{\eta}(n)}{e^{2\pi i n \rho} - 1} , \qquad n \neq 0 .$$

In just a moment, we will address whether or not the function

$$h(x) = \sum_{n \neq 0} \hat{h}(n) e^{2\pi i n x} = \sum_{n \neq 0} \frac{\hat{\eta}(n)}{e^{2\pi i n \rho} - 1} e^{2\pi i n x}$$

makes any sense, however, we first note that even if (5) defines a well-behaved function, it will not solve (4) but rather:

$$(6) \qquad h(x + \rho) - h(x) = \eta(x) - \int_0^1 \eta(x)dx = \eta(x) - \hat{\eta}(0) .$$

This is because the zeroth Fourier coefficient of h drops out of (4). The fact that h does not solve (4) will complicate the estimates below. The problems with showing that (5) converges arise due to the presence of the factors of $e^{2\pi i n \rho} - 1$ in the denominator of the summands, and these are the (in)famous **small denominators** which plagued celestial mechanics in the last century and which the KAM theory finally overcame. We first note that if ρ is rational, there is little hope that the sum defining h will converge since the denominators in this sum will vanish for the infinitely many n for which $\rho n = m$ for some $m \in Z$. Thus, we can only hope for success if $\rho \notin Q$. If ρ is irrational, the denominator will still be large whenever $n\rho \approx m$. However, by assuming that ρ is of type (K, ν), we have some control over how close to zero the denominator can become. In fact, the following lemma immediately allows us to estimate $h(x)$.

LEMMA 2.2. *If ρ is of type (K, ν), then*

$$|e^{2\pi i n \rho} - 1| = |e^{2\pi i m}(e^{2\pi i (\rho n - m)} - 1)| \geq 4K|n|^{-(\nu-1)} \text{ if } n \neq 0 .$$

Proof: Since ρ is of type (K, ν), we know that $|\rho n - m| \geq K|n|^{-(\nu-1)}$ and the lemma follows by writing $|e^{2\pi i(\rho n - m)} - 1| = 2|\sin(\pi(\rho n - m))|$, and then using the fact that $|\sin(\pi x)| \geq 2|x|$, for $|x| \leq 1/2$.

$\square$

The other fact which we must use to estimate h is the fact that since $\eta \in B_\sigma$, Cauchy's theorem immediately gives an estimate on its Fourier coefficients of the form $|\hat{\eta}(n)| \leq \|\eta\|_\sigma e^{-2\pi\sigma|n|}$. Combining this remark with Lemma 2.2, we see that if $|Imz| \leq \sigma - \delta$, one has

$$|h(z)| = |\sum_{n \neq 0} \frac{\hat{\eta}(n)e^{2\pi inz}}{e^{2\pi in\rho} - 1}| \leq \sum_{n \neq 0} \frac{|n|^{(\nu-1)}}{4K} \|\eta\|_\sigma e^{-2\pi\sigma|n|} e^{2\pi|n|(\sigma-\delta)}$$

$$\leq \frac{\Gamma(\nu)}{K(2\pi\delta)^\nu} \|\eta\|_\sigma ,$$

where $\Gamma(\nu) = \int_0^\infty x^{\nu-1}e^{-x}dx$, and we have assumed that $2\pi(K+1)\delta \leq 4\pi\delta < 1$. Thus we have proven,

PROPOSITION 2.2. *If ρ is of type (K, ν), and $\eta \in B_\sigma$, then $h(x)$, defined by (5) is an element of $B_{\sigma-\delta}$ for any $\delta > 0$, and if $4\pi\delta < 1$, we have the estimate:*

$$\|h\|_{\sigma-\delta} \leq \frac{\Gamma(\nu)}{K(2\pi\delta)^\nu} \|\eta\|_\sigma .$$

REMARK 2.5. *Note that we do **not** get an estimate on h in B_σ – we lose some analyticity. This is why we can't use an ordinary Implicit Function Theorem to solve (2). Indeed, if we were to attempt to proceed with an ordinary iterative method to solve (2), we would find that we gradually lost **all** of the analyticity of our approximate solution. This is where the second "big idea" of the KAM theory enters the picture, namely:*

Step2: Newton's Method in Banach Space

Recall that Newton's method says that if you want to find the roots of some nonlinear equation, you should take an approximate solution and then use a linear approximation to the function whose roots you seek to improve your approximation to the solution. You then use this improved approximation as your new starting point and iterate this procedure. In the present circumstance, we regard ϕ as an approximation to R_ρ and then use the linear approximation, $H(x) = x + h(x)$, to the conjugating function to improve that approximation. Recall that if $h(x)$ had solved (2) exactly, then $H^{-1} \circ \phi \circ H = R_\rho$. Our hope is that if we use the $H(x)$ that comes from solving (6), $H^{-1} \circ \phi \circ H$ will be closer to R_ρ than ϕ was and then we can iterate this process.

The first thing we must do is check that H is invertible. Since $H(z) = z + h(z)$, H will be invertible with analytic inverse on any domain on which $\|h'\| < 1$. Cauchy's theorem and Proposition 2.2 imply that $\|h'\|_{\sigma-2\delta} \leq \frac{2\pi\Gamma(\nu)}{K(2\pi\delta)^{\nu+1}}\|\eta\|_\sigma$, so we conclude

LEMMA 2.3. *If $2\pi\Gamma(\nu)\|\eta\|_\sigma < K(2\pi\delta)^{(\nu+1)}$ and $4\pi\delta < 1$, then $H(z)$ has an analytic inverse on the image of $S_{\sigma-2\delta}$.*

REMARK 2.6. *Note that if we combine the inequalities in Lemma 2.3 and Proposition 2.2, we find $\|h\|_{\sigma-\delta} < \delta$. Thus, if $z \in S_{\sigma-2\delta}$, $H(z) \in S_{\sigma-\delta}$. Furthermore, H maps the real axis into itself, and the images of the lines $Im z = \pm(\sigma - 2\delta)$ lie outside the strip $S_{\sigma-3\delta}$. With this information it is easy to show that $Range(H|_{S_{\sigma-2\delta}}) \supset S_{\sigma-3\delta}$, so that $H^{-1}(z)$ is defined for all $z \in S_{\sigma-3\delta}$.*

In addition to knowing that the inverse exists, we need an estimate on its properties which the following proposition provides.

PROPOSITION 2.3. *If*

$$2\pi\Gamma(\nu)\|\eta\|_\sigma < K(2\pi\delta)^{(\nu+1)} \text{ and } 4\pi\delta < 1 \ ,$$

then $H^{-1}(z) = z - h(z) + g(z)$, where

$$\|g\|_{\sigma-4\delta} \leq \frac{2\pi\Gamma(\nu)^2}{K^2(2\pi\delta)^{(2\nu+1)}}\|\eta\|_\sigma^2 \ .$$

Proof: If we define $g(z)$ by $g(z) = H^{-1}(z) - z + h(z)$, then we see that

$$z = H^{-1} \circ H(z) = z + h(z) - h(z + h(z)) + g(z + h(z)) \ .$$

Thus, $g(z+h(z)) = h(z+h(z)) - h(z) = \int_0^1 h'(z+sh(z))h(z)ds$. Setting $\xi = H(z)$, this becomes

$$g(\xi) = \int_0^1 h'(H^{-1}(\xi) + sh(H^{-1}(\xi)))h(H^{-1}(\xi))ds \ .$$

In a fashion similar to that in the remark just above, $Range(H|_{S_{\sigma-3\delta}}) \supset S_{\sigma-4\delta}$, so if $\xi \in S_{\sigma-4\delta}$, $H^{-1}(\xi) \in S_{\sigma-3\delta}$, and hence $H^{-1}(\xi) + sh(H^{-1}(\xi)) \in S_{\sigma-2\delta}$, so applying the estimates on h and h' from above we obtain

$$\|g\|_{\sigma-4\delta} \leq 2\pi\left(\frac{\Gamma(\nu)}{K}\right)^2 \frac{\|\eta\|_\sigma^2}{(2\pi\delta)^{2\nu+1}}$$

$\square$

Let us now examine the transformed diffeomorphism $\tilde{\phi}(x) = H^{-1} \circ \phi \circ H(x)$. Since h is only an approximate solution of (2), $\tilde{\phi}$ will not be exactly a rotation, but since h did solve the **linearized** equation (6), we hope that $\tilde{\phi}$ will differ from a rotation only by terms that are of second order in the small quantities h and η. Using the form of H^{-1} given by Proposition 2.3, we find

$$
\begin{aligned}
\tilde{\phi}(x) &= H^{-1} \circ \phi \circ H(x) = H^{-1}(x + h(x) + \rho + \eta(x + h(x))) \\
&= x + h(x) + \rho + \eta(x + h(x)) - h(x + \rho + h(x) + \eta(x + h(x))) + \\
&\quad + g(x + h(x) + \rho + \eta(x + h(x))) \\
&= x + \rho + \{h(x) - h(x + \rho) + \eta(x)\} + \{\eta(x + h(x)) - \eta(x)\} + \\
&\quad + \{h(x + \rho) - h(x + \rho + h(x) + \eta(x + h(x)))\} + \\
&\quad + g(x + h(x) + \rho + \eta(x + h(x))) \ .
\end{aligned}
$$

We first note that because h solves (6), the first expression in braces in the second to last line of this sequence of equalities is equal to $\hat{\eta}_0$. The next expression in braces equals $\int_0^1 \eta'(x + sh(x))h(x)ds$. If $2\pi\Gamma(\nu)\|\eta\|_\sigma < K(2\pi\delta)^{\nu+1}$, $4\pi\delta < 1$, and $x \in S_{\sigma - 4\delta}$, we can bound the norm of this expression on $B_{\sigma - 4\delta}$ by $2\pi\|\eta\|_\sigma \frac{\Gamma(\nu)}{K(2\pi\delta)^{\nu+1}}\|\eta\|_\sigma$. Similarly, the quantity in braces in the next to last line may be rewritten as $\int_0^1 h'(x + \rho + s(h(x) + \eta(x + h(x))))(h(x) + \eta(x + h(x))ds$. Once again, assuming that the conditions on $\|\eta\|_\sigma$ and δ described above hold, and that $x \in S_{\sigma - 4\delta}$, then we can bound the norm of this expression on $B_{\sigma - 4\delta}$ by

$$
\frac{2\pi\Gamma(\nu)}{K(2\pi\delta)^{\nu+1}}\Big\{\frac{\Gamma(\nu)}{K(2\pi\delta)^{\nu}}\|\eta\|_\sigma + \|\eta\|_\sigma\Big\}\|\eta\|_\sigma < \frac{4\pi(\Gamma(\nu))^2}{K^2(2\pi\delta)^{2\nu+1}}\|\eta\|_\sigma^2 \ ,
$$

where the last inequality used the fact that $2\pi\delta K < 1$. Finally, if $|Im x| < \sigma - 6\delta$, then $|Im(x + h(x) + \rho + \eta(x + h(x)))| < \sigma - 4\delta$, (since $\|\eta\|_\sigma < \delta$), so that the last term in this expression is bounded by Proposition 2.3.

Define $\tilde{\eta}(x)$ by $\tilde{\phi}(x) = x + \rho + \tilde{\eta}(x)$. By Remark 2.3, $\tilde{\phi}$ has rotation number ρ, so by Lemma 2.1 there exists x_0 such that $\tilde{\eta}(x_0) = 0$. Combining this remark with the expression for $\tilde{\phi}$ just above, we find

$$
\begin{aligned}
\hat{\eta}(0) &= -\{\eta(x_0 + h(x_0)) - \eta(x_0)\} \\
&\quad - \{h(x_0 + \rho) - h(x_0 + \rho + h(x_0) + \eta(x_0 + h(x_0)))\} \\
&\quad - g(x_0 + h(x_0) + \rho + \eta(x_0 + h(x_0))) \ .
\end{aligned}
$$

In the previous paragraph we bounded the norm of each of the expressions on the right hand side of this equality, so we conclude that

$$
|\hat{\eta}(0)| \leq 2\pi\|\eta\|_\sigma^2 \frac{\Gamma(\nu)}{K(2\pi\delta)^{\nu+1}} + \frac{4\pi(\Gamma(\nu))^2}{K^2(2\pi\delta)^{2\nu+1}}\|\eta\|_\sigma^2 + \frac{2\pi\Gamma(\nu)^2}{K^2(2\pi\delta)^{2\nu+1}}\|\eta\|_\sigma^2 \ .
$$

Combining this estimate with the bounds above, we have proven,

PROPOSITION 2.4. *If* $2\pi\Gamma(\nu)\|\eta\|_\sigma < K(2\pi\delta)^{(\nu+1)}$, *and* $4\pi\delta < 1$, *then* $\tilde{\phi}(x) = H^{-1} \circ \phi \circ H(x) = x + \rho + \tilde{\eta}(x)$, *where*

$$
\|\tilde{\eta}\|_{\sigma - 6\delta} \leq \frac{16\pi(\Gamma(\nu))^2}{K^2(2\pi\delta)^{2\nu+1}}\|\eta\|_\sigma^2 \ .
$$

REMARK 2.7. *The important thing to note about the estimate of $\tilde{\eta}$ is that in spite of the mess, it is second order in the small quantity $\|\eta\|_\sigma$ as we had hoped. That is, there exists a constant $C(K, \delta, \nu)$ such that $\|\tilde{\eta}\|_{\sigma-6\delta} \leq C(K, \delta, \nu)\|\eta\|_\sigma^2$. This is what makes the Newton's method argument work.*

The proof of Arnold's Theorem is completed by inductively repeating the above procedure. The principal point which we must check is that we don't lose all of our domain of analyticity as we go through the argument – note that $\tilde{\phi}$ is analytic on a narrower strip than was our original diffeomorphism ϕ. The essential reason that there is a nonvanishing domain of analyticity at the completion of the argument is that the amount by which the analyticity strip shrinks at the nth step in the induction will be proportional to the amount by which our diffeomorphism differs from a rotation at the nth iterative step, and thanks to the extremely fast convergence of Newton's method, this is very small.

The Inductive Argument

Let $\phi_0(x) \equiv \phi(x)$, be our original diffeomorphism, and set $\eta_0(x) = \eta(x)$. Define $\phi_1(x) = H_0^{-1} \circ \phi_0 \circ H_0(x)$, and by induction, $\phi_{n+1}(x) = H_n^{-1} \circ \phi_n \circ H_n(x) = x + \rho + \eta_{n+1}(x)$ where $H_n(x) = x + h_n(x)$, and h_n solves

$$h_n(x + \rho) - h_n(x) = \eta_n(x) - \hat{\eta}(0) \ .$$

Also define the sequence of inductive constants:

- $\delta_n = \frac{\sigma}{36(1+n^2)}$, $n \geq 0$ (Note that this insures that $4\pi\delta_0 < 1$.)
- $\sigma_0 = \sigma$, and $\sigma_{n+1} = \sigma_n - 6\delta_n$, if $n \geq 0$.
- $\epsilon_0 = \|\eta\|_\sigma$, and $\epsilon_n = \epsilon_0^{(3/2)^n}$, if $n \geq 0$.

Note that $\sigma^* = \lim_{n \to \infty} \sigma_n > \sigma/2$. We now have:

LEMMA 2.4 (INDUCTIVE LEMMA). *If*

$$\epsilon_0 < \left(\frac{K}{16\pi\Gamma(\nu)} (\frac{\sigma}{36})^{(\nu+1)} \right)^8 \ ,$$

then $\phi_{n+1}(x) = x + \rho + \eta_{n+1}(x)$, with $\eta_{n+1} \in B_{\sigma_{n+1}}$, and $\|\eta_{n+1}\|_{\sigma_{n+1}} \leq \epsilon_{n+1}$. Furthermore, $H_n(x) = x + h_n(x)$ satisfies

$$\|h_n\|_{\sigma_n - \delta_n} \leq \frac{\Gamma(\nu)\epsilon_n}{K(2\pi\delta_n)^\nu} \ ,$$

while $H_n^{-1}(x) = x - h_n(x) + g_n(x)$, where

$$\|g_n\|_{\sigma-4\delta_n} \leq \frac{2\pi\Gamma(\nu)^2\epsilon_n^2}{K^2(2\pi\delta_n)^{2\nu+1}} \ .$$

Proof: Note that Proposition 2.2 and Proposition 2.3 imply that the estimates on h_n and g_n hold for $n = 0$. The estimate on $\|\eta_1\|_{\sigma_1}$ follows by noting that from Proposition 2.4,

$$\|\eta_1\|_{\sigma_1} \leq \frac{16\pi(\Gamma(\nu))^2}{K^2(2\pi\delta)^{2\nu+1}} \|\eta\|_\sigma^2$$

and the hypothesis on the inductive constants in the Inductive Lemma was chosen so that this last expression is less than $\epsilon_0^{(3/2)} = \epsilon_1$. This completes the first induction step.

Now suppose that the induction holds for $n = 0, 1, \ldots, N-1$, so that we know that $\|\eta_N\|_{\sigma_N} \leq \epsilon_N$. To prove that it holds for $n = N$, first choose h_N to solve $h_N(x + \rho) - h_N(x) = \eta_N(x) - \hat{\eta}_N(0)$. By Proposition 2.2, and the inductive estimate on η_N, we will have $\|h_N\|_{\sigma_N - \delta_N} \leq \frac{\Gamma(\nu)\epsilon_N}{K(2\pi\delta_N)^\nu}$, while Proposition 2.3 implies that $H_N^{-1}(x) = x - h_N(x) + g_N(x)$, with $\|g_N\|_{\sigma_N - 4\delta_N} \leq \frac{2\pi\Gamma(\nu)^2\epsilon_N^2}{K^2(2\pi\delta_N)^{2\nu+1}}$. Finally, if we define $\phi_{N+1} = H_N^{-1} \circ \phi_N \circ H_N = x + \rho + \eta_{N+1}$, with η_{N+1} defined in analogy with $\tilde{\eta}$ in Proposition 2.4, then we see that

$$\|\eta_{N+1}\|_{\sigma_{N+1}} \leq \frac{16\pi\Gamma(\nu)^2}{K^2(2\pi\delta_N)^{2\nu+1}}\epsilon_N^2 \ .$$

Once again, if use the hypotheses on the inductive constants we see that this expression is bounded by $\epsilon_N^{(3/2)} = \epsilon_{N+1}$, which completes the proof of the lemma.

$\square$

With the aid of the Inductive Lemma it is easy to complete the proof of Arnold's Theorem. Define

$$\mathcal{H}_N(x) = H_0 \circ H_1 \circ \cdots \circ H_N(x)$$
$$= x + h_N(x) + h_{N-1}(x + h_n(x) + h_{N-2}(x + h_N(x) + h_{N-1}(x + h_N(x))))$$
$$+ \cdots + h_0(x + h_1(x + \ldots) \ldots)$$

By the Induction Lemma, $\mathcal{H}_N$ is analytic on $S_{\sigma_N - 2\delta_N}$ and $\mathcal{H}_N(z) - z$ is bounded by $\sum_{n=0}^{\infty} \frac{\Gamma(\nu)\epsilon_n}{K(2\pi\delta_n)^\nu} \equiv \Delta$. (This sum converges as a consequence of the hypotheses on the induction constants in the Induction Lemma.) Furthermore,

$$\mathcal{H}_{N+1}(z) - \mathcal{H}_N(z) = \mathcal{H}_N \circ H_N(z) - \mathcal{H}_N(z) = \int_0^1 \mathcal{H}_N'(z + th_N(z))h_N(z)ds \ ,$$

so $\|\mathcal{H}_{N+1} - \mathcal{H}_N\|_{\sigma_{N+1}} \leq (\frac{4\Delta}{\delta_N} + 1)\frac{\Gamma(\nu)\epsilon_N}{K(2\pi\delta_N)^\nu}$. Note that by the definition of the inductive constants, the right hand side of this inequality converges if summed over N. Thus, $\mathcal{H}_N$ converges uniformly to some limit $\mathcal{H}$ on S_{σ^*}, and $\mathcal{H}$ is analytic. Furthermore, $\mathcal{H}(z) = z + \tilde{h}(z)$, where the estimates on $\mathcal{H}_N(z) - z$, plus Cauchy's Theorem imply that if $\delta^* = \sigma^*/16$, then $\|\tilde{h}'\|_{\sigma^* - \delta^*} \leq \Delta/\delta^* < \delta^*$, again using the definition of the inductive constants. By an argument similar to that following Lemma 2.3, we see that $\mathcal{H}$ is invertible on the image of $S_{\sigma^* - \delta^*}$ and that this image contains $S_{\sigma^* - 2\delta^*}$.

Finally, note that $\phi \circ \mathcal{H}_N(z) = \mathcal{H}_N \circ \phi_N(z) = \mathcal{H}_N(z + \rho + \eta_N(z))$. As $N \to \infty$, we see that

$$\phi \circ \mathcal{H}(z) = \lim_{N \to \infty} \mathcal{H}_N \circ \phi_N(z) = \lim_{N \to \infty} \mathcal{H}_N(z + \rho + \eta_N(z)) = \mathcal{H} \circ R_\rho(z) \ ,$$

for all $z \in S_{\sigma^* - 2\delta^*}$. (The last equality used the inductive estimate on η_N.) Since $\mathcal{H}$ is invertible on this domain, this implies $\mathcal{H}^{-1} \circ \phi \circ \mathcal{H} = R_\rho$, so $\mathcal{H}$ is the diffeomorphism whose existence was asserted in Arnold's Theorem.

□

REMARK 2.8. *Suppose that in Arnold's Theorem we were given a diffeomorphism of the (apparently) more general form*

$$\psi(x) = x + \alpha + \mu(x) \ .$$

but still with rotation number ρ of type (K, ν) (where $\alpha \neq \rho$.) We can rewrite $\psi(x) = x + \rho + (\alpha - \rho + \mu(x)) \equiv x + \rho + \eta(x)$. If $\|\eta\|_\sigma = \|(\alpha - \rho) + \mu\|_\sigma \leq \epsilon(K, \nu, \sigma)$, then Theorem 2.1 implies that ψ is analytically conjugate to R_ρ

REMARK 2.9. *In examples it may be difficult to determine from inspection of the initial diffeomorphism what the rotation number is. In such cases there is often a parameter in the problem which can be varied and which allows one to show that the conjugacy in Arnold's Theorem exists at least for most parameter values. For instance, the following result can be proven by easy modifications of the previous methods: (See, [1], page 271.)*

THEOREM 2.2. *Consider the family of diffeomophisms:*

(7) $$\phi_{\alpha, \epsilon}(x) = x + \alpha + \epsilon \eta(x) \ ,$$

for $\alpha \in [0, 1]$. For every $\delta > 0$, there exists $\epsilon_0 > 0$ such that if $|\epsilon| < \epsilon_0$, there exists a set $\mathcal{A}(\epsilon) \subset [0, 1]$ such that for $\alpha \in \mathcal{A}(\epsilon)$, $\phi_{\epsilon, \alpha}$ is analytically conjugate to a rotation of the circle, and $|Lebesgue\ measure\ (A(\epsilon)) - 1| < \delta$.

REMARK 2.10. *It is not necessary to work with analytic functions. For instance, Moser [10] showed that if the original diffeomorphism is C^k, and if the rotation number is of type (K, ν), then if k is sufficiently large (depending on ν), and the diffeomorphism is a sufficiently small perturbation of a rotation, the diffeomorphism is conjugate to a rotation, with a $C^{k'}$ conjugacy function, for some $1 \leq k' < k$. Note that this theorem is still "local" in that it demands that the diffeomorphism which we start with be "close" to a pure rotation. More recent work of Hermann [7] and Yoccoz [12], has lead to a remarkably complete understanding of the **global** picture of the dynamics of circle diffeomorphisms. For instance (see [12]), one can write the real numbers as a union of two disjoint sets A and B, and prove that any analytic circle diffeomorphism, ϕ, with rotation number $\rho(\phi) \in B$ is analytically conjugate to the rotation $R_{\rho(\phi)}$, while for any $\alpha \in A$, there exists an analytic circle diffeomorphism with rotation number α which is **not** analytically conjugate to R_α.*

3. Nearly Integrable Hamiltonian Systems

In this section we address the KAM theory in its original setting, namely nearly integrable Hamiltonian systems. Recall that a Hamiltonian system (in Euclidean space) is a system of $2N$ differential equations whose form is given by

$$\dot{p}_j = -\frac{\partial H}{\partial q_j} \quad ; \quad j = 1, \ldots, N \; ,$$

$$\dot{q}_j = \frac{\partial H}{\partial p_j} \quad ; \quad j = 1, \ldots, N \; ,$$

for some function $H(p, q)$. (Here $p = (p_1, \ldots, p_N)$ and $q = (q_1, \ldots, q_N)$.)

In general these equations are just as hard to solve as any other system of $2N$ coupled, nonlinear, ordinary differential equations, but in special circumstances (the **integrable** Hamiltonian systems) there exists a special set of variables known as the **action-angle** variables, (I, ϕ), $I \in R^N$ and $\phi \in T^N$, such that in terms of these variables, $H(I, \phi) = h(I)$. Since the Hamiltonian does not depend on the angle variables ϕ, the equations of motion are very simple – they become

$$\dot{I}_j \; = \; -\frac{\partial H}{\partial \phi_j} = 0 \; ; \; j = 1, \ldots, N \; ,$$

$$\dot{\phi}_j \; = \; \frac{\partial H}{\partial I_j} \equiv \omega_j(I) \; ; \; j = 1, \ldots, N \; .$$

We can solve these equations immediately, and we find that $I(t) = I(0)$, and $\phi(t) = \omega(I)t + \phi(0)$. Thus, for an integrable system with bounded trajectories, the action variables I are constants of the motion, while the angle variables ϕ just precess around an N-dimensional torus with angular frequencies ω given by the gradient of the Hamiltonian with respect to the actions. (In particular, if the components of $\omega(I)$ are irrationally related to one another, $\phi(t)$ is a quasi-periodic function.)

REMARK 3.1. *The three-body (or N-body) problem, in which we ignore the mutual interaction between the planets is an integrable system.*

Now suppose that we start with an integrable Hamiltonian $h(I)$ and make a small perturbation which depends on both the action and the angle variables – as in the case of the solar system when we consider the gravitational interactions between the planets. Then the Hamiltonian takes the form:

$$(8) \qquad\qquad H(I, \phi) = h(I) + f(I, \phi) \; .$$

As before we will assume that the Hamiltonian function is analytic in order to avoid complications. More precisely, if we think of $f(I, \phi)$ as a function on $R^N \times R^N$, which is periodic in ϕ, then we assume that there exists some $I^* \in R^N$ such that H can be extended to an analytic function on the set $\mathcal{A}_{\sigma,\rho}(I^*) = \{(I, \phi) \in C^N \times C^N \mid |I - I^*| < \rho \; , \; |Im(\phi_j)| < \sigma \; , \; j = 1, \ldots, N \; \}$. (I will always use the ℓ^1 norm for N-vectors $-i.e.$ $|I| = \sum_{j=1}^{N} |I_j|$.) We define the norm of a function f, analytic on $\mathcal{A}_{\sigma,\rho}(I^*)$ by $\|f\|_{\sigma,\rho} = \sup_{(I,\phi)\in\mathcal{A}_{\sigma,\rho}(I^*)} |f(I,\phi)|$. (As in the previous section, one can assume without loss of generality that $\sigma < 1$.)

In addition, since we are interested in nearly integrable Hamiltonian systems, we will assume that f has small norm. (Just as we assumed that η was small in the previous section.) Furthermore, we can assume that $\int_{T^N} f(I,\phi)d^N\phi = 0$, since if this were nonzero it could be absorbed by redefining $h(I)$.

Question: *Do the trajectories of this perturbed Hamiltonian system still lie on invariant tori, at least for f sufficiently small?*

To state the answer of this question more precisely, we need an analogue of the numbers of type (K,ν) introduced in the previous section.

DEFINITION 3.1. *We say that a vector $\omega \in R^N$ is of type (L,γ) if*

$$|\langle \omega, n \rangle| = |\sum_{j=1}^{N} \omega_j n_j| \geq L|n|^{-\gamma} \, , \text{ for all } n \in Z^N \backslash 0 \, .$$

REMARK 3.2. *Note that if ρ is of type (K,ν), then the vector $(\rho,-1)$ is of type (L,γ) with $K = L$ and $\gamma = \nu - 1$. Also, we again assume without loss of generality that $L \leq 1$.*

Given this remark, and the fact that we know that the numbers of type (K,ν) are a subset of the real line of full Lebesgue measure, the following result (whose proof we omit) is not surprising.

PROPOSITION 3.1. *If $\gamma > N$, almost every $\omega \in R^N$ is of type (L,γ) for some $L < 1$.*

We are now in a position to state the KAM theorem.

THEOREM 3.1 (KAM). *Suppose that $\omega(I^*) \equiv \omega^*$ is of type (L,γ), and that the Hessian matrix $\frac{\partial^2 h}{\partial I^2}$ is invertible at I^*. (And hence on some neighborhood of I^*.) Then there exists $\epsilon_0 > 0$ such that if $\|f\|_{\sigma,\rho} < \epsilon_0$, the Hamiltonian system (8) has a quasi-periodic solution with frequencies $\omega(I^*)$.*

REMARK 3.3. *Although we have claimed in the theorem only that at least one quasi-periodic solution exists in the perturbed hamiltonian system, we will see in the course of the proof that the whole torus, $I = I^*$, survives.*

REMARK 3.4. *One might wonder why we study quasi-periodic orbits rather than the apparently simpler periodic orbits. If one considers values of the action variables for which the frequencies $\omega_j(I)$ are all rationally related, then the integrable hamiltonian will have an invariant torus filled with **periodic** orbits. However, under a typical perturbation, all but finitely many of these periodic orbits will disappear. Hence, the quasi-periodic orbits are, in this sense, more stable than the periodic ones.*

As we will see, the proof follows very closely the outline of the previous section. In particular, we begin with:

Step 1: Analysis of the Linearized equation

The basic idea is to find new variables $(\tilde{I}, \tilde{\phi})$ such that in terms of these new variables (8) will be integrable. However, not just any change of variables is allowed, because most changes of variables will not preserve the Hamiltonian form of the equations of motion. We will admit only those changes of variables which do preserve the Hamiltonian form of the equations. Such transformations are known as **canonical** changes of variables. There is a large literature on canonical transformations, (for a nice introduction see [2]), but pursuing it would take us too far afield. In order to come to the point in as expeditious a fashion as possible, let us just note the following:

PROPOSITION 3.2. *Suppose that there exists a smooth function $\Sigma(\tilde{I}, \phi)$ such that the equations:*

$$I = \frac{\partial \Sigma}{\partial \phi} \ , \ \tilde{\phi} = \frac{\partial \Sigma}{\partial \tilde{I}} \ ,$$

can be inverted to find $(I, \phi) = \Phi(\tilde{I}, \tilde{\phi})$. Then Φ is a canonical transformation, and Σ is called its generating function.

Proof: See [2], section 48.

REMARK 3.5. *Note that $\Sigma(\tilde{I}, \phi) = \langle \tilde{I}, \phi \rangle$ is the generating function for the identity transformation. (Here, $\langle \cdot, \cdot \rangle$ is the inner product in R^N.)*

REMARK 3.6. *There are other ways of generating canonical transformations. In particular, the Lie transform method has proven to be very convenient for computational purposes [5]. However, the generating function method offers a simple and direct way to prove the KAM theorem and for that reason I have chosen it here.*

We would like to find a canonical transformation $(I, \phi) = \Phi(\tilde{I}, \tilde{\phi})$ such that $\tilde{H}(\tilde{I}, \tilde{\phi}) = H \circ \Phi(\tilde{I}, \tilde{\phi}) = \tilde{h}(\tilde{I})$, or

$$(9) \qquad H(\frac{\partial \Sigma}{\partial \phi}(\tilde{I}, \phi), \phi) = \tilde{h}(\tilde{I}) \ .$$

(This, by the way, is the Hamilton-Jacobi equation. In the last century, Jacobi proved the integrability of a number of physical systems by finding solutions of this equation.) In our example, (9) can be written as:

$$(10) \qquad h((\frac{\partial \Sigma}{\partial \phi}(\tilde{I}, \phi)) + f((\frac{\partial \Sigma}{\partial \phi}(\tilde{I}, \phi), \phi) = \tilde{h}(\tilde{I}) \ .$$

Since H is "close" to an integrable Hamiltonian for f small, we can hope that the canonical transformation is "close" to the identity transformation. Using the fact that we know the generating function for the identity transformation, we will look for canonical transformations whose generating functions are of the form $\Sigma(\tilde{I},\phi) = \langle \tilde{I},\phi \rangle + S(\tilde{I},\phi)$, where S is $\mathcal{O}(\|f\|_{\sigma,\rho})$, the amount by which our Hamiltonian differs from an integrable one. If we substitute this form for Σ into (10) and expand, retaining only terms that are formally of first order in the small quantities $\|f\|_{\sigma,\rho}$ and $\|S\|_{\sigma,\rho}$, we obtain the linearized Hamilton-Jacobi equation:

$$(11) \qquad \langle \omega(\tilde{I}), \frac{\partial S}{\partial \phi}(\tilde{I},\phi) \rangle + f(\tilde{I},\phi) = \tilde{h}(\tilde{I}) - h(\tilde{I}) .$$

Once again, we now have a linear equation involving periodic functions, so if we expand $f(I,\phi) = \sum_{n \in Z^N} \hat{f}(I,n)e^{i2\pi\langle n,\phi \rangle}$, we can solve (11) and we find

$$(12) \qquad S(\tilde{I},\phi) = \frac{i}{2\pi} \sum_{n \in Z^N \backslash 0} \frac{\hat{f}(\tilde{I},n)e^{i2\pi\langle n,\phi \rangle}}{\langle \omega(\tilde{I}),n \rangle}$$

REMARK 3.7. *Once again, as in (6), the function S defined (formally) by (12) does not satisfy (11), but rather*

$$(13) \qquad \langle \omega(\tilde{I}), \frac{\partial S}{\partial \phi}(\tilde{I},\phi) \rangle + f(\tilde{I},\phi) = 0 ,$$

and we will be forced to estimate the difference between these two equations below.

Note that once again, we will face small denominators. Indeed, for a dense set of points I, the denominators in (12) will vanish for infinitely many choices of n. This is the reason that many people (including Poincaré) at the end of the last century believed that these series diverged. Nonetheless, the results of Kolmogorov, Arnold and Moser show that "most" (in the sense of Lebesgue measure) points I give rise to a convergent series. Having S be defined only on the complement of a dense set of points $\tilde{I}$ would be a problem, since we would be hard pressed to take the derivatives we need in order to compute the canonical transformation in Proposition 3.2. To proceed, we take advantage of the fact that because of the analyticity of f, the Fourier coefficients $\hat{f}(\tilde{I},n)$ are decaying to zero exponentially fast as $|n|$ becomes large. Thus, if we truncate the sum defining S to consider only $|n| < M$, for some large M we will make only a relatively small error in the solution of (11). On the other hand, since there are now only finitely many terms in the sum defining S, we can find open sets of action-variables on which the generating function is defined. Before stating the precise estimate on S, we introduce a few preliminaries.

First, define $\Omega \geq 1$, such that

$$\max\left((\sup_{|I-I^*|\leq\rho} \|\frac{\partial^2 h}{\partial I^2}\|), (\sup_{|I-I^*|\leq\rho} \|(\frac{\partial^2 h}{\partial I^2})^{-1}\|) \right) < \Omega .$$

(Here, $\|\cdot\|$ is the norm of the matrix considered as an operator from $C^N \to C^N$ with the ℓ^1 norm.) Analogously, define $\tilde{\Omega}$ such that $\sup_{|I-I^*|\leq\rho} \|(\frac{\partial^3 h}{\partial I^3})\| < \tilde{\Omega}$.

(In this case, $\|\cdot\|$ is the norm of $(\frac{\partial^3 h}{\partial I^3})$ considered as a bilinear operator from $C^N \times C^N \to C^N$.) Next note that if we define

$$S^<(\tilde{I}, \phi) = \frac{i}{2\pi} \sum_{\substack{n \in Z^N \setminus 0 \\ |n| \leq M}} \frac{\hat{f}(\tilde{I}, n)e^{i2\pi\langle n,\phi\rangle}}{\langle \omega(\tilde{I}), n\rangle} \,,$$

$S^<$ will no longer be a solution of (13), but rather will solve

$$\langle \omega(\tilde{I}), \frac{\partial S}{\partial \phi}(\tilde{I}, \phi)\rangle + f^<(\tilde{I}, \phi) = 0 \,,$$

where $f^<(\tilde{I}, \phi) \equiv \sum_{|n| \leq M} \hat{f}(\tilde{I}, n)e^{i2\pi\langle n,\phi\rangle}$. Note that we have already discarded all terms that were formally of more than first order in $\|f\|_{\sigma,\rho}$ in order to derive (11). Thus, if in deriving this equation for $S^<$, we change (11) only by amounts of this order, we won't have qualitatively worsened our approximation. We will choose M in order to insure that this is the case.

PROPOSITION 3.3. *Choose* $0 < \delta < \sigma$, *and set* $M = |\log(\|f\|_{\sigma,\rho})|/(\pi\delta)$. *If* $\rho < L/(2\Omega M^{\gamma+1})$ *and* $4\pi\delta < 1$, *then* $S^<$ *is analytic on* $\mathcal{A}_{\sigma-\delta,\rho}(I^*)$, *and*

$$\|S^<\|_{\sigma-\delta,\rho} \leq \left(\frac{8\Gamma(\gamma+1)}{(2\pi\delta)^{\gamma+1}}\right)^N \frac{(2N^\gamma)\|f\|_{\sigma,\rho}}{2\pi L}$$

Proof: Recall that we chose our domain $\mathcal{A}_{\sigma,\rho}(I^*)$ so that it was centered (in the I variables) at a point with $\omega(I^*) = \omega^*$. Now suppose that we choose $|n| < M$, and consider $\langle \omega(I), n\rangle$ for some other point I in our domain. Writing $I = I^* + (I - I^*)$, we see that $|\langle \omega(I), n\rangle - \langle \omega(I^*), n\rangle| \leq \Omega|n|\rho$. If we then use the fact that ω^* is of type (L, γ), we find that for $|n| \leq M$ and all $|I - I^*| < \rho$,

$$|\langle \omega(I), n\rangle| = |\langle \omega^*, n\rangle + (\langle \omega(I), n\rangle - \langle \omega^*, n\rangle)| \geq \frac{L}{|n|^\gamma} - \Omega|n|\rho \geq \frac{L}{2|n|^\gamma} \,,$$

where the last inequality used the hypothesis on ρ and the fact that $|n| \leq M$. If we combine this observation with the fact that $|\hat{f}(\tilde{I}, n)| \leq \|f\|_{\sigma,\rho}e^{-2\pi\sigma|n|}$, by Cauchy's theorem, we find

$$\begin{aligned}
\|S^<\|_{\sigma-\delta,\rho} &\leq \sum_{|n| \leq M} \frac{2|n|^\gamma}{2\pi L}\|f\|_{\sigma,\rho}e^{-2\pi\delta|n|} \\
&\leq \frac{2\|f\|_{\sigma,\rho}}{2\pi L}N^\gamma(1 + 2\sum_{m=0}^{M}m^\gamma e^{-2\pi\delta|m|})^N \\
&\leq \left(\frac{8\Gamma(\gamma+1)}{(2\pi\delta)^{\gamma+1}}\right)^N \frac{2N^\gamma\|f\|_{\sigma,\rho}}{2\pi L} \,.
\end{aligned}$$

In going from the first to second line of this inequality, we used the fact that

$$|n|^\gamma e^{-2\pi\delta|n|} \leq N^\gamma (\max_j |n_j|)^\gamma e^{-2\pi\delta|n|} \leq N^\gamma \prod_{j=1}^{N} \max(1, |n_j|) e^{-2\pi\delta|n_n|} \, ,$$

so that $\sum_{|n|\leq M} |n|^\gamma e^{-2\pi\delta|n|} \leq N^\gamma (1 + 2\sum_{m=1}^{M} m e^{-2\pi\delta m})^N$.

$\square$

Now that we know that the generating function is well-defined, we can proceed to check that the canonical transformation is defined and analytic, just as we did in Proposition 2.3 in the previous section.

PROPOSITION 3.4. *If*

$$\left(\frac{8\Gamma(\gamma+1)}{(2\pi\delta)^{\gamma+1}}\right)^N \frac{16N^{\gamma+1}\|f\|_{\sigma,\rho}}{2\pi\delta\rho L} < 1 \, ,$$

$\rho < L/(2\Omega M^{\gamma+1})$ *and* $4\pi\delta < 1$, *then the equations*

$$(14) \qquad I = \tilde{I} + \frac{\partial S^<}{\partial \phi} \, , \ and \ \tilde{\phi} = \phi + \frac{\partial S^<}{\partial \tilde{I}} \, ,$$

define an analytic and invertible canonical transformation $(I, \phi) = \Phi(\tilde{I}, \tilde{\phi})$ *on the set* $\mathcal{A}_{\sigma-3\delta,\rho/4}$.

Proof: Just as in the proof of Lemma 2.3 we begin by using the analytic inverse function theorem to check that (14) can be inverted. In both of the expressions in this equation, the inverse function theorem can be applied provided $\|\frac{\partial^2 S^<}{\partial I \partial \phi}\|_{\sigma-2\delta,\rho/2} < 1$. This in turn, follows immediately from the estimate in Proposition 3.3 and Cauchy's Theorem.

The remainder of the proposition follows if we check that the transformation is onto the domain $\mathcal{A}_{\sigma-3\delta,\rho/4}$. (This is analogous to the proof of Proposition 2.3.) Note that if $(\tilde{I}, \phi) \in \mathcal{A}_{\sigma-2\delta,\rho/2}$,

$$\|\frac{\partial S^<}{\partial \phi}\|_{\sigma-2\delta,\rho/2} \leq \left(\frac{8\Gamma(\gamma+1)}{(2\pi\delta)^{\gamma+1}}\right)^N \frac{2N^{\gamma+1}\|f\|_{\sigma,\rho}}{2\pi\delta L} < \rho/8 \, ,$$

by the hypothesis of the Proposition, while

$$\|\frac{\partial S^<}{\partial \tilde{I}}\|_{\sigma-\delta,\rho/4} \leq \left(\frac{8\Gamma(\gamma+1)}{(2\pi\delta)^{\gamma+1}}\right)^N \frac{8N^{\gamma+1}\|f\|_{\sigma,\rho}}{2\pi\rho L} < \delta/2 \, ,$$

again by the hypotheses of the Proposition. Thus, $|I - \tilde{I}| < \rho/8$, while $|\phi - \tilde{\phi}| < \delta/2$. This implies that the canonical transformation maps the set $\mathcal{A}_{\sigma-2\delta,\rho/2}$ onto $\mathcal{A}_{\sigma-3\delta,\rho/4}$, and hence that $(I, \phi) = \Phi(\tilde{I}, \tilde{\phi})$ on this set.

$\square$

REMARK 3.8. *For a vector valued function like $\frac{\partial S^<}{\partial \phi}$ on a domain $\mathcal{A}_{\sigma,\rho}$, $\|\frac{\partial S^<}{\partial \phi}\|_{\sigma,\rho} \equiv \sup_{\mathcal{A}_{\sigma,\rho}} |\frac{\partial S^<}{\partial \phi}|$, where we recall that $|\frac{\partial S^<}{\partial \phi}|$ is the ℓ^1 norm of $\frac{\partial S^<}{\partial \phi}$. This is the origin of the extra factor of N in these estimates.*

Step 2: The Newton Step

Now, just as we did in the case of circle diffeomorphisms, where we transformed our original diffeomorphism with the approximate conjugacy function obtained by solving the linearized conjugacy equation, we transform our original Hamiltonian with the approximate canonical transformation, whose generating function is $S^<$, and show that the difference between the transformed Hamiltonian and an integrable Hamiltonian is of second order in the small quantity $\|f\|_{\sigma,\rho}$. As before, we will use this fact as the basis for a Newton's method argument.

PROPOSITION 3.5. *Define $\tilde{H}(\tilde{I}, \tilde{\phi}) = H \circ \Phi(\tilde{I}, \tilde{\phi}) \equiv \tilde{h}(\tilde{I}) + \tilde{f}(\tilde{I}, \tilde{\phi})$. If*

$$\left(\frac{8\Gamma(\gamma + 1)}{(2\pi\delta)^{\gamma+1}} \right)^N \frac{16 N^{\gamma+1} \|f\|_{\sigma,\rho}}{2\pi\delta\rho L} < 1 \, ,$$

$\rho < L/(2\Omega M^{\gamma+1})$ and $4\pi\delta < 1$, then $\tilde{H}$ is analytic on $\mathcal{A}_{\sigma-3\delta,\rho/4}$, and one has the estimates,

$$\|h - \tilde{h}\|_{\sigma-3\delta,\rho/4} \leq (\Omega + 2) \left(\left(\frac{8\Gamma(\gamma+1)}{(2\pi\delta)^{\gamma+1}} \right)^N \frac{2 N^{\gamma+1} \|f\|_{\sigma,\rho}}{2\pi\delta\rho L} \right)^2 \, ,$$

and

$$\|\tilde{f}\|_{\sigma-3\delta,\rho/4} \leq 2(\Omega + 2) \left(\left(\frac{8\Gamma(\gamma+1)}{(2\pi\delta)^{\gamma+1}} \right)^N \frac{2 N^{\gamma+1} \|f\|_{\sigma,\rho}}{2\pi\delta\rho L} \right)^2 \, .$$

REMARK 3.9. *The important thing to note is that $\tilde{f}$, the amount by which our transformed Hamiltonian fails to be integrable is quadratic in the small quantity $\|f\|_{\sigma,\rho}^2$. Just as in Proposition 2.4 in the previous section, this will form the basis of a Newton's method argument, which will allow us to prove the existence of a quasi-periodic solution with frequencies ω^*.*

Proof: Using Taylor's Theorem, we can rewrite

$$
\begin{aligned}
\tilde{H}(\tilde{I}, \tilde{\phi}) &= H(\tilde{I} + \frac{\partial S^<}{\partial \phi}, \phi(\tilde{I}, \tilde{\phi})) \\
&= h(\tilde{I} + \frac{\partial S^<}{\partial \phi}) + f(\tilde{I} + \frac{\partial S^<}{\partial \phi}, \phi(\tilde{I}, \tilde{\phi})) \\
&= h(\tilde{I}) + \langle \omega(\tilde{I}), \frac{\partial S^<}{\partial \phi} \rangle + \int_0^1 \int_0^t \langle (\frac{\partial \omega}{\partial I}(\tilde{I} + v \frac{\partial S^<}{\partial \phi}) \frac{\partial S^<}{\partial \phi}), \frac{\partial S^<}{\partial \phi} \rangle dv dt \\
&\quad + f(\tilde{I}, \phi) + \int_0^1 \langle \frac{\partial f}{\partial I}(\tilde{I} + t \frac{\partial S^<}{\partial \phi}, \phi), \frac{\partial S^<}{\partial \phi} \rangle dt .
\end{aligned}
$$

From the definition of $S^<$, we know that $\langle \omega(\tilde{I}), \frac{\partial S^<}{\partial \phi} \rangle + f(\tilde{I}, \phi) = f^{\geq}(\tilde{I}, \phi) \equiv \sum_{|n| \geq M} \hat{f}(\tilde{I}, n) e^{2\pi i \langle \phi, n \rangle}$. Thus, we can define

$$
\begin{aligned}
\tilde{h}(\tilde{I}) &= h(\tilde{I}) + average\{ \int_0^1 \int_0^t \langle (\frac{\partial \omega}{\partial I}(\tilde{I} + v \frac{\partial S}{\partial \phi}) \frac{\partial S}{\partial \phi}), \frac{\partial S}{\partial \phi} \rangle dv dt \} \\
&\quad + average\{ \int_0^1 \langle \frac{\partial f}{\partial I}(\tilde{I} + t \frac{\partial S}{\partial \phi}, \phi), \frac{\partial S}{\partial \phi} \rangle dt \} + average f(\tilde{I}, \phi(\tilde{I}, \tilde{\phi})) ,
\end{aligned}
$$

where $average\{ g(\tilde{I}, \tilde{\phi}) \} \equiv \int_{T^N} g(\tilde{I}, \tilde{\phi}) d\tilde{\phi}$, and

$$
\begin{aligned}
\tilde{f}(\tilde{I}, \tilde{\phi}) &= f^{\geq}(\tilde{I}, \phi(\tilde{I}, \tilde{\phi})) \\
&\quad + \int_0^1 \int_0^t \langle (\frac{\partial \omega}{\partial I}(\tilde{I} + v \frac{\partial S^<}{\partial \phi}) \frac{\partial S^<}{\partial \phi}), \frac{\partial S^<}{\partial \phi} \rangle dv dt \\
&\quad + \int_0^1 \langle \frac{\partial f}{\partial I}(\tilde{I} + t \frac{\partial S^<}{\partial \phi}, \phi(\tilde{I}, \phi), \frac{\partial S^<}{\partial \phi} \rangle dt \\
&\quad - average\{ \int_0^1 \int_0^t \langle (\frac{\partial \omega}{\partial I}(\tilde{I} + v \frac{\partial S^<}{\partial \phi}) \frac{\partial S^<}{\partial \phi}), \frac{\partial S^<}{\partial \phi} \rangle dv dt \} \\
&\quad - average\{ \int_0^1 \langle \frac{\partial f}{\partial I}(\tilde{I} + t \frac{\partial S^<}{\partial \phi}, \phi), \frac{\partial S^<}{\partial \phi}) \rangle dt \} \\
&\quad - average f(\tilde{I}, \phi(\tilde{I}, \tilde{\phi})).
\end{aligned}
$$

REMARK 3.10. *Subtracting the average of the three quantities in $\tilde{f}$ insures that when we expand $\tilde{f}$ in a Fourier series, there will be no $n = 0$ coefficient – this was used in solving (11).*

Both $\tilde{h}$ and $\tilde{f}$ are easy to estimate using the estimates of Proposition 3.3 and Cauchy's Theorem. For instance,

$$
\| \int_0^1 \langle \frac{\partial f}{\partial I}(\tilde{I} + t \frac{\partial S^<}{\partial \phi}, \phi), \frac{\partial S^<}{\partial \phi} \rangle dt \|_{\sigma - 3\delta, \rho/4}
$$

$$
\leq \frac{2 \|f\|_{\sigma, \rho}}{\rho} \left(\frac{8 \Gamma(\gamma + 1)}{(2\pi \delta)^{\gamma + 1}} \right)^N \frac{2 N^{\gamma + 2} \|f\|_{\sigma, \rho}}{2\pi \delta L} ,
$$

while,

$$\left\| \int_0^1 \int_0^t \langle (\frac{\partial \omega}{\partial I}(\tilde{I} + v\frac{\partial S^<}{\partial \phi}) \frac{\partial S^<}{\partial \phi}), \frac{\partial S^<}{\partial \phi} \rangle dv dt \right\|_{\sigma - 3\delta, \rho/4}$$

$$\leq \Omega \left(\left(\frac{8\Gamma(\gamma+1)}{(2\pi\delta)^{\gamma+1}} \right)^N \frac{2N^{\gamma+1}\|f\|_{\sigma,\rho}}{2\pi\delta L} \right)^2 .$$

Finally, we have the estimate

$$\|f^{\geq}\|_{\sigma-3\delta,\rho/4} \leq \sum_{|n|\geq M} \|f\|_{\sigma,\rho} e^{-2\pi\delta|n|} \leq \|f\|_{\sigma,\rho} e^{-\pi\delta M} \sum_{|n|\geq M} e^{-\pi\delta|n|}$$

$$\leq (\frac{4}{\pi\delta})^N \|f\|_{\sigma,\rho}^2 ,$$

where the last of these inequalities came from using the definition of M in Proposition 3.3.

If we combine these remarks, we immediately obtain the estimates stated in the Proposition.

□

The Induction Argument

The induction follows closely the lines of the induction step in the case of the circle diffeomorphisms. We have to keep track of two more inductive constants – ρ_n to control the size of the domain of the action variables, and M_n to control how we cut off the sum defining $S^<$ at the nth stage of the iteration. Thus, we define our original Hamiltonian $H(I,\phi) = H_0(I,\phi)$ and set $h(I) = h_0(I)$ and $f(I,\phi) = f_0(I,\phi)$. Also define

- $\delta_n = \frac{\sigma}{36(1+n^2)}$, $n \geq 0$.
- $\sigma_0 = \sigma$, and $\sigma_{n+1} = \sigma_n - 4\delta_n$, if $n \geq 0$.
- $\rho_0 \leq \rho$, and $\rho_{n+1} = \rho_n/8$, with ρ_0 chosen to satisfy the hypothesis of the following Lemma.
- $\epsilon_0 = \|f\|_{\sigma,\rho}$, and $\epsilon_n = \epsilon_0^{(3/2)^{(n/\gamma)}}$, if $n \geq 0$.
- $M_n = |\log \epsilon_n|/(\pi\delta_n)$.

We set $H_{n+1} = H_n \circ \Phi_n = h_{n+1} + f_{n+1}$, with $\hat{f}_{n+1}(I,0) = 0$, where Φ_n is the canonical transformation whose generating function $S_n^<$ solves the equation

$$\langle \omega_n(\tilde{I}), \frac{\partial S_n^<}{\partial \phi}(\tilde{I},\phi) \rangle + f_n^<(\tilde{I},\phi) = 0 ,$$

with $f_n^<(\tilde{I},\phi) \equiv \sum_{|n|\leq M_n} \hat{f}_n(\tilde{I},n) e^{i2\pi\langle n,\phi\rangle}$, and $\omega_n(\tilde{I}) = \frac{\partial h_n}{\partial I}(\tilde{I})$. At the nth stage of the iteration we will work on the domain $\mathcal{A}_{\sigma_n,\rho_n}(I_n) = \{(I,\phi) \in C^N \times C^N \mid |I - I_n| < \rho_n , |Im(\phi_j)| < \sigma_n , j = 1,\ldots,N \}$, where I_n is chosen so that $\omega_n(I_n) = \omega^*$, and we define $\Omega_n = \max(1, \sup \|\frac{\partial^2 h_n}{\partial I^2}\|, \|(\frac{\partial^2 h_n}{\partial I^2})^{-1}\|)$, with the supremum in these expressions running over all I with $|I - I_n| < \rho_n$.

We then have

LEMMA 3.1 (KAM INDUCTION LEMMA). *There exists a positive constant c_1 such that if*

$$\epsilon_0 < 2^{-c_1 N(\gamma+1)} \frac{\sigma^{8N(4\gamma+1)} \rho_0^8 L^{16}}{\Gamma(\gamma+1)^{16N} \Omega^8} \ , \ and \ \rho_0 < \frac{2^{-c_1} L}{\Omega M_0^{\gamma+1}}$$

then

- *The generating function $S_n^<$ satisfies*

$$\|S_n^<\|_{\sigma_n-\delta_n,\rho_n} \leq \left(\frac{8\Gamma(\gamma+1)}{(2\pi\delta_n)^{\gamma+1}} \right)^N \frac{2N^\gamma \epsilon_n}{2\pi L} \ .$$

- Φ_n *is defined and analytic on* $\mathcal{A}_{\sigma_n-3\delta_n,\rho_n/4}(I_n)$ *and maps this set into* $\mathcal{A}_{\sigma_n-2\delta_n,\rho_n/2}(I_n)$.
- $\|f_{n+1}\|_{\sigma_{n+1},\rho_{n+1}} \leq \epsilon_{n+1}$.
- $\|h_{n+1} - h_n\|_{\sigma_{n+1},\rho_{n+1}} \leq \epsilon_{n+1}$.
- $|I_{n+1} - I_n| < \rho_n/8$.

Before proving this lemma, we show how the KAM theorem follows from it. If the perturbation f in our Hamiltonian is sufficiently small, the hypotheses of the Induction Lemma will be satisfied, and roughly speaking, the idea is that as $n \to \infty$, $H_n(I,\phi) \to h^\infty(I)$, an integrable system, since $f_n \to 0$. Since all of the orbits of an integrable system are quasiperiodic, this would complete the proof. However, as n becomes larger and larger, the size of the domain in the action variables on which H_n is defined goes to zero. Thus, we must be a little careful with this limit.

Begin by defining $\Psi_n = \Phi_0 \circ \Phi_1 \circ \ldots \Phi_n$. By the induction lemma, $\Psi_n : \mathcal{A}_{\sigma_n-3\delta_n,\rho_n/4}(I_n) \to \mathcal{A}_{\sigma_0,\rho_0}(I_0)$, and $H_n = H_0 \circ \Psi_{n-1}$. In particular, if $(I^n(t),\phi^n(t))$ is a solution of Hamilton's equations with Hamiltonian H_n, then $\Psi_{n-1}(I^n(t),\phi^n(t))$ is a solution of Hamilton's equations with Hamiltonian H_0.

Consider the equations of motion of H_n:

$$\dot{I} = -\frac{\partial f_n}{\partial \phi} \ , \ \dot{\phi} = \omega_n(I) + \frac{\partial f_n}{\partial I} \ .$$

Since $\|\frac{\partial f_n}{\partial I}\|_{\sigma_n,\rho_n/2} \leq 2\epsilon_n N/\rho_n$, and $\|\frac{\partial f_n}{\partial \phi}\|_{\sigma_n-\delta_n,\rho_n} \leq \epsilon_n N/\delta_n$, the trajectory with initial conditions (I_n,ϕ_0) (for any $\phi_0 \in T^N$), will remain in $\mathcal{A}_{\sigma_n-3\delta_n,\rho_n/4}(I_n)$ for all times $|t| \leq T_n = 2^n$, by our hypothesis on ϵ_0, and the definition of the induction constants. Furthermore, if $(I^n(t),\phi^n(t))$ is the solution with these initial conditions, we have

$$\max\left(\sup_{|t| \leq T_n} |I^n(t) - I_n|, \ \sup_{|t| \leq T_n} |\phi^n(t) - (\omega^* t + \phi_0)| \right) \leq 2^{2n+2}\Omega\epsilon_n N/\rho_n\delta_n \ .$$

Noting that the inductive estimates on I_n imply that there exists I^∞ with $\lim_{n\to\infty} I_n = I^\infty$, we see that for t in any compact subset of the real line,

$(I^n(t), \phi^n(t)) \to (I^\infty, \omega^* t + \phi_0)$ (again using the definition of the inductive constants). Using the inductive bounds on the canonical transformation one can readily establish that

$$\|\Psi_n(I, \phi) - (I, \phi)\|_{\sigma_{n+1}, \rho_{n+1}} \leq \sum_{j=0}^{\infty} 2N \left(\frac{8\Gamma(\gamma+1)}{(2\pi\delta_j)^{\gamma+1}} \right)^N \left(\frac{8N^\gamma \epsilon_j}{2\pi\delta_j \rho_j L} \right) \equiv \Delta ,$$

while

$$\|\Psi_n(I, \phi) - \Psi_{n-1}(I, \phi)\|_{\sigma_{n+1}, \rho_{n+1}}$$
$$= \|\Psi_{n-1} \circ \Phi_n(I, \phi) - \Psi_{n-1}(I, \phi)\|_{\sigma_{n+1}, \rho_{n+1}}$$
$$\leq (2N + \frac{16\tilde{\Delta}}{\rho_n \delta_n}) \left(\frac{8\Gamma(\gamma+1)}{(2\pi\delta_n)^{\gamma+1}} \right)^N \left(\frac{8N^\gamma \epsilon_n}{2\pi\delta_n \rho_n L} \right) .$$

Using the definition of the inductive constants, we see that the sum over n of this last expression converges and hence $\lim_{n\to\infty} \Psi_n(I^\infty, \omega^* t + \phi_0) = (I^*(t), \phi^*(t))$ exists and is a quasi-periodic function with frequency ω^*. Similarly,

$$\lim_{n\to\infty} |\Psi_n(I^\infty, \omega^* t + \phi_0) - \Psi_n(I^n(t), \phi^n(t))| = 0,$$

for t in any compact subset of the real line.

Combining these two remarks, find that

$$\lim_{n\to\infty} \Psi_n(I^n(t), \phi^n(t)) = (I^*(t), \phi^*(t)),$$

so $(I^*(t), \phi^*(t))$ is a quasi-periodic solution of Hamilton's equations for the system with Hamiltonian H_0 as claimed.

$\square$

REMARK 3.11. *Note that this argument is independent of the point ϕ_0 that we take on the original torus. Thus it shows that **every** trajectory on the unperturbed torus is preserved.*

Proof: (of Lemma 3.1.) Note that Propositions 3.3, 3.4, and 3.5, plus the assumption on the induction constants imply that we can start the induction, provided $\mathcal{A}_{\sigma_0 - 3\delta_0, \rho_0/4}(I_0) \supset \mathcal{A}_{\sigma_1, \rho_1}(I_1)$. From the definitions of the domains and the inductive constants, we see that this will follow provided $|I_0 - I_1| < \rho_0/8$. To see that this is so we note that $\omega_0(I_0) = \omega_1(I_1)$. Thus, $\omega_0(I_0) - \omega_0(I_1) = \frac{\partial(h_1 - h_0)}{\partial I}(I_1)$. But, $\|\frac{\partial(h_1 - h_0)}{\partial I}(I_1)\|_{\sigma_0 - 3\delta_0, \rho_0/6} \leq 12\epsilon_1/\rho_0$, while

$$\omega_0(I_0) - \omega_0(I_1) = \frac{\partial\omega_0}{\partial I}(I_0)(I_0 - I_1)$$
$$+ \int_0^1 \int_0^t (\frac{\partial^2\omega_0}{\partial I^2}(I_0 + sI_1)(I_1 - I_0))^2 ds dt .$$

Since $\|\left(\frac{\partial\omega_0}{\partial I}\right)^{-1}\| \leq \Omega$ and $\|\frac{\partial^2\omega_0}{\partial I^2}\| \leq \tilde{\Omega}$, this implies that $|I_0 - I_1| < \rho_0/8$ by the definition of the induction constants, provided $\Omega\tilde{\Omega}\rho_0 < 1/2$, which will follow if the constant c_1 in the Lemma is sufficiently large. This completes the first induction step.

Suppose that the induction argument holds for $n = 0, 1, \ldots, K-1$. To prove it for $n = K$ we first note if $S_K^<$ is defined by:

$$S_K^<(\tilde{I}, \phi) = \frac{i}{2\pi} \sum_{\substack{n \in Z^N \setminus 0 \\ |n| \le M_K}} \frac{\hat{f}_K(\tilde{I}, n) e^{i2\pi \langle n, \phi \rangle}}{\langle \omega_K(\tilde{I}), n \rangle} \; ,$$

then by Proposition 3.3, we have

$$\|S_K^<\|_{\rho_K, \sigma_K - \delta_K} \le \left(\frac{8\Gamma(\gamma + 1)}{(2\pi\delta_K)^{\gamma+1}} \right)^N \frac{2N^\gamma \epsilon_K}{2\pi L} \; .$$

Note that the hypothesis in Proposition 3.3 becomes $\rho_K < L/(2\Omega_K M_K^{\gamma+1})$ where,

$$\Omega_K = \max(1, \sup_{|I - I_K| < \rho_K} \|\frac{\partial^2 h_K}{\partial I^2}\|, \sup_{|I - I_K| < \rho_K} \|(\frac{\partial^2 h_K}{\partial I^2})^{-1}\|)$$

$$\le \Omega \max(1 + \sum_{j=1}^K \frac{64N\epsilon_j}{\rho_j^2}, (1 - \sum_{j=1}^K \frac{64\Omega N\epsilon_j}{\rho_j^2})^{-1}) \le 2\Omega \; ,$$

using the definition of the inductive constants. This observation, plus the hypothesis on ρ_0 in the inductive lemma, guarantees that the hypothesis of Proposition 3.3 is satisfied. Thus, by Proposition 3.4, the canonical transformation Φ_K defined by

$$(15) \qquad\qquad I = \tilde{I} + \frac{\partial S_K^<}{\partial \phi} \; , \text{ and } \tilde{\phi} = \phi + \frac{\partial S_K^<}{\partial \tilde{I}} \; ,$$

is analytic and invertible on the set $\mathcal{A}_{\sigma_K - 3\delta_K, \rho_K/4}(I_K)$, and maps this set into $\mathcal{A}_{\sigma_K, \rho_K}(I_K)$.

If we then define f_{K+1} and h_{K+1}, as we defined $\tilde{f}$ and $\tilde{h}$ in Proposition 3.5 we see that

$$\|f_{K+1}\|_{\sigma_K - 3\delta_K, \rho_K/4} \le 2(\Omega_K + 2) \left(\left(\frac{8\Gamma(\gamma + 1)}{(2\pi\delta_K)^{\gamma+1}} \right)^N \frac{2N^{\gamma+1}\epsilon_K}{2\pi\delta_K \rho_K L} \right)^2$$

while

$$\|h_K - h_{K+1}\|_{\sigma_K - 3\delta_K, \rho_K/4} \le (\Omega_K + 2) \left(\left(\frac{8\Gamma(\gamma + 1)}{(2\pi\delta_K)^{\gamma+1}} \right)^N \frac{2N^{\gamma+1}\epsilon_K}{2\pi\delta_K \rho_K L} \right)^2$$

If we use the bound on ϵ_0, and the definitions of the inductive constants, we see that the quantities on the right hand sides of both of these inequalities are less than ϵ_{K+1}. The proof of the inductive lemma will be completed if we can show that $\mathcal{A}_{\sigma_{K+1}, \rho_{K+1}}(I_{K+1}) \subset \mathcal{A}_{\sigma_K - 3\delta_K, \rho_K/4}(I_K)$. This follows in a fashion very similar to the proof that $\mathcal{A}_{\sigma_1, \rho_1}(I_1) \subset \mathcal{A}_{\sigma_0 - 3\delta_0, \rho_0/4}(I_0)$, which we demonstrated above, so we omit the details.

$$\square$$

REMARK 3.12. *From the point of view of applications of this theory it is often convenient to know not just what happens to a single trajectory, but rather the behavior of whole sets of trajectories. Simple modifications of the preceding argument allow one to demonstrate the following variant of the KAM theorem. (See* [4].) *Consider the family of Hamiltonian systems*

$$(16) \qquad\qquad H_\epsilon = h(I) + \epsilon f(I, \phi) \ .$$

Suppose that there exists a bounded set $V \subset R^N$ such that $\frac{\partial^2 h}{\partial I^2}(I)$ is invertible for all $I \in V$, and that for every ϵ in some neighborhood of zero H_ϵ is analytic on a set of the form $\mathcal{A}_{\sigma, \rho}(V) = \{(I, \phi) \in C^N \times C^N \mid |I - \tilde{I}| < \rho$, for some $\tilde{I} \in V$, and $|Im(\phi_j)| < \sigma$, $j = 1, \dots, N\}$.

THEOREM 3.2. *For every $\delta > 0$, there exists $\epsilon_0 > 0$ such that if $|\epsilon| < \epsilon_0$, there exists a set $P_\epsilon \subset V \times T^N$, such that the Lebesgue measure of $(V \times T^N) \backslash P_\epsilon$ is less than δ and for any point $(I_0, \phi_0) \in P_\epsilon$, the trajectory of (16) with initial conditions (I_0, ϕ_0) is quasi-periodic.*

Thus an informal way of stating the KAM theorem is to say that "most" trajectories of a nearly integrable Hamiltonian systems remain quasi-periodic.

REMARK 3.13. *Just as in the case of Arnold's theorem about circle diffeomorphisms, the KAM theorem also remains true when the Hamiltonian is only finitely differentiable, rather than analytic. For a nice exposition of this theory, see* [11].

REFERENCES

1. V. Arnold. Small denominators, 1: Mappings of the circumference onto itself. *AMS Translations*, 46:213–288, 1965 (Russian original published in 1961).
2. V. Arnold. *Mathematical Methods of Classical Mechanics*. Springer-Verlag, New York, 1978.
3. V. Arnold. *Geometrical Methods in the Theory of Ordinary Differential Equations*. Springer-Verlag, New York, 1982.
4. G. Gallavotti. Perturbation theory for classical hamiltonian systems. In J. Fröhlich, editor, *Scaling and Self-Similarity in Physics*, pages 359–246. Birkhäuser, Boston, 1983.
5. W. Gröbner. *Die Lie-Reihen und ihre Anwendungen*. Deutscher Verlag der Wissenschaften, Berlin, 1960.
6. J. Guckenheimer and P. Holmes. *Nonlinear Oscillations, Dynamical Systems, and Bifurcations of Vector-Fields*. Springer-Verlag, New York, 1983.
7. M. R. Herman. Sur la conjugaison différentiable des difféomorphismes du cercle à des rotations. *Publ. Math. I.H.E.S.*, 49:5–234, 1979.
8. A. N. Kolmogorov. On conservation of conditionally periodic motions under small perturbations of the hamiltonian. *Dokl. Akad. Nauk, SSSR*, 98:527–530, 1954.

9. J. Moser. On invariant curves of area-preserving mappings of an annulus. *Nachr. Akad. Wiss., Göttingen, Math. Phys. Kl.*, pages 1–20, 1962.

10. J. Moser. A rapidly convergent interation method, II. *Ann. Scuola Norm. Sup. di Pisa, Ser. III*, 20:499–535, 1966.

11. J. Pöschel. Integrability of hamiltonian systems on Cantor sets. *Comm. Pure. and Appl. Math.*, 35:653–695, 1982.

12. J.-C. Yoccoz. An introduction to small divisors problems. In *From Number Theory to Physics (Les Houches, 1989)*, chapter 14. Springer Verlag, Berlin, 1992.

DEPARTMENT OF MATHEMATICS, THE PENNSYLVANIA STATE UNIVERSITY, UNIVERSITY PARK, PA 16802

E-mail address: wayne@math.psu.edu

Lectures in Applied Mathematics
Volume **31**, 1996

KAM theory in infinite dimensions

WALTER CRAIG

ABSTRACT. Many of the nonlinear evolution equations of mathematical physics can be viewed as infinite dimensional Hamiltonian systems, posed in a function space as a phase space. This includes nonlinear Schrödinger equations, nonlinear wave equations, and nonlinear systems related to the KdV and the water wave problem. There is a recent class of results whose intent is to describe some of the principal features of the phase space for these equations, through extensions of KAM theory to infinite dimensional settings. The first part of this hour talk will consist in a survey of the results of several authors on this theme. The second part will go into depth on some aspects of the construction of invariant tori for nonlinear evolution equations. The talk will illustrate the connection between invariant tori and Anderson localization in the space of normal modes, and will discuss the development of unstable periodic orbits in the presence of resonance.

§1. Introduction

It is natural to feel that we understand ODE in much more detail than PDE, and therefore it is a natural impulse to regard PDE which describe time dependent phenomena as dynamical systems, posed in an infinite dimensional phase space. As well as being an elegant point of view, it is a successful one if it leads to a better understanding of the solutions of the equations. The problems that I will be addressing today are mainly evolution equations which are Hamiltonian systems with infinitely many degrees of freedom. In the last several years there has been a research effort to extend one of the main tools of finite dimensional Hamiltonian mechanics, the KAM theory, to infinite dimensional settings. The main goal is to understand some of the principal features of the phase space in which the Hamiltonian PDE are posed. I will be describing the work of several people in this talk; my own work on the subject has been in collaboration with E. Wayne, who is lecturing on finite dimensional KAM theory in this series of talks.

1991 *Mathematics Subject Classification.* Primary 35L70, 35Q55, 58F39; Secondary 35B15.
The author was supported in part by NSF Grant # DMS 92-08190

I will describe a class of results on the existence of invariant tori for PDE, considered as infinite dimensional Hamiltonian systems. For initial data on these tori, the flow of the Hamiltonian system corresponds to solutions of the PDE which are time periodic and time quasi-periodic, that is they behave regularly in time. These solutions contrast in character with those exhibited in Dave McLaughlin's talk; numerical solutions of a simultaneously damped and driven version of the nonlinear Schrödinger equation, which apparently undergo irregular behavior when tracked over long time intervals. The impression that we have received in the last two talks on the KAM theory is that these invariant tori are somewhat delicate structures, depending on diophantine properties of the frequency vector and concerning the degree of smoothness of the Hamiltonian system. It is true that the analysis by which we construct them is detailed. On the other hand the fact that they exist is very robust, in the sense that they are present for a large class of Hamiltonian systems, and we will see today that there are invariant tori present in the phase space of a large class of PDE. Indeed, if we somehow consider the space of nonlinearities, say, for the nonlinear wave equation, then most - in the sense of an open dense set - do possess invariant tori in their phase space. Failure to have these structures can only occur when there is either (i) uncharacteristically severe linear resonance, or (ii) lack of genuine nonlinearity in the problem. But I am getting ahead of myself, and will start by giving you the examples I have in mind.

§2. Hamiltonian systems

This is a selection of the most basic PDE that come to mind when working on the infinite dimensional extensions of the KAM theory. The first one that comes to mind is the nonlinear wave equation

$$(1) \qquad \partial_t^2 u = \partial_x^2 u - g(x, u) , \qquad 0 \le x \le \pi$$

with either Dirichlet boundary conditions

$$u(0, t) = 0 = u(\pi, t) \qquad \text{(Dir.)}$$

or periodic ones

$$u(x, t) = u(x + \pi, t) \qquad \text{(Per.)} .$$

This is a second order equation, but may be made into a first order Hamiltonian system with the following Hamiltonian function,

$$H(u, p) = \int_0^\pi \tfrac{1}{2} p(x)^2 + \tfrac{1}{2} (\partial_x u(x))^2 + G(x, u) \, dx$$

where

$$\partial_u G(x, u) = g(x, u) = g_1(x) u + g_2(x) u^2 + g_3(x) u^3 + \cdots$$

The canonical equations

$$\partial_t \begin{pmatrix} u \\ p \end{pmatrix} = \begin{pmatrix} 0 & 1 \\ -1 & 0 \end{pmatrix} \begin{pmatrix} \delta_u H(u,p) \\ \delta_p H(u,p) \end{pmatrix}$$

$$= \begin{pmatrix} p \\ \partial_x^2 u - g(x,u) \end{pmatrix}$$

are easily seen to be equivalent to (1). We will be assuming in our later analysis the technical conditions that $g(x,u)$ is analytic in (x,u) and periodic in x, and if we consider Dirichlet boundary conditions (Dir.), that $g(x,u) = -g(-x,-u)$.

The next example is the nonlinear Schrödinger equation,

$$(2) \qquad\qquad i\partial_t \psi = \tfrac{1}{2}\partial_x^2 \psi - Q(x,\psi,\overline{\psi}) \qquad 0 \le x \le 2\pi$$

where we also consider either boundary conditions (Dir.) or (Per.). Letting $G(x,\psi,\overline{\psi})$ be a real valued function, such that $\partial_{\overline{\psi}} G = Q$, this equation has the Hamiltonian

$$H_{nls}(\psi,\overline{\psi}) = \int_0^{2\pi} \tfrac{1}{2}|\partial_x \psi|^2 + G(x,\psi,\overline{\psi})\ dx\ ,$$

and equation (2) can be written as

$$\partial_t \psi = J\delta_{\overline{\psi}} H_{nls}$$

with $J = i$ representing the symplectic form.

The third natural class of examples is equations of KdV type,

$$(3) \qquad\qquad \partial_t q = \tfrac{1}{6}\partial_x^3 q - \partial_x(g(x,q)), \qquad 0 \le x \le 2\pi$$

which have Hamiltonian

$$H_{KdV}(q) = \int_0^{2\pi} \tfrac{1}{12}(\partial_x q)^2 + G(x,q)\ dx$$

where $\partial_q G = g$. The equation (3) can be rewritten

$$\partial_t q = J\delta_q H_{KdV}(q)\ ,$$

with this time $J = -\partial_x$, given by a nonclassical symplectic structure.

Related in origin to the KdV equation is the Boussinesq system given through the Hamiltonian

$$H_B(q,p) = \int_0^{2\pi} \tfrac{1}{2}p^2 + \tfrac{1}{2}q^2 - \tfrac{1}{6}(\partial_x q)^2 + G(x,p,q)\ dx$$

and

$$J = \begin{pmatrix} 0 & \partial_x \\ \partial_x & 0 \end{pmatrix}\ ,$$

which results in the system of PDE

$$(4) \qquad\qquad \partial_t \begin{pmatrix} q \\ p \end{pmatrix} = \partial_x \begin{pmatrix} p + \partial_p G(x,p,q) \\ q + \tfrac{1}{3}\partial_x^2 q + \partial_q G(x,p,q) \end{pmatrix}\ .$$

The system that is given here is ill posed as an initial value problem, and it is not one of the examples that we will pursue in this talk.

Finally I want to mention one of my favorite systems, albeit one of the least tractable, the water wave problem. This is the system that describes the evolution of surface waves in a body of an inviscid incompressible fluid which is in addition taken to be irrotational. It can be written in Hamiltonian form

$$(5) \qquad \partial_t \begin{pmatrix} \eta \\ \xi \end{pmatrix} = \begin{pmatrix} 0 & 1 \\ -1 & 0 \end{pmatrix} \begin{pmatrix} \delta_\eta H_{ww} \\ \delta_\xi H_{ww} \end{pmatrix} ,$$

where the Hamiltonian comes from an expression originally given by V.E. Zakharov. For the most convenient formulation for H_{ww} (see Craig & Sulem [CS]) consider potential flow in a fluid region $R(\eta) = \{0 \le x \le 2\pi, \ -h \le y \le \eta(x)\}$, which means that the velocity potential satisfies

$$\Delta \phi(x,y) = 0 , \qquad \partial_y \phi(x,-h) = 0$$

in $R(\eta)$. Letting $\phi(x,\eta(x)) = \xi(x)$ be the Dirichlet data for ϕ on the top boundary, N the outward pointing unit normal to the free surface, and $G(\eta)\xi = \nabla\phi \cdot N\sqrt{1 + (\partial_x\eta)^2}$ the Dirichlet-Neumann mapping, then the Hamiltonian is given by the expression

$$(6) \qquad H_{ww}(\eta, \xi) = \int_0^{2\pi} \tfrac{1}{2}\xi G(\eta)\xi + \tfrac{g}{2}\eta^2 \ dx .$$

With our present knowledge, system (5),(6) is too difficult to fit into our analytic framework. It does remain one of our goals, as a mathematically challenging and physically relevant open problem.

Exercise 1: Show that equations (1) through (5) are in their claimed form as Hamiltonian systems, with the given Hamiltonian functions and symplectic structures. Do the linear stability analysis about the zero solution, and explain that equations (2) (3) and (5) are linearly stable, (1) is linearly well-posed, and that (4) is linearly ill-posed.

§3. Results

I plan to explain the majority of the results in the context of the nonlinear wave equation (1). The Hamiltonian can be written

$$H(u, p) = H_2(u, p) + H_3(u) + H_4(u) + \cdots$$

where H_ℓ is homogeneous of degree ℓ in its arguments. Truncate to retain only $H_2(u, p) = \int \tfrac{1}{2}p^2 + \tfrac{1}{2}(\partial_x u)^2 + \tfrac{1}{2}g_1(x)u^2 dx$. The resulting system is of course the linearization of (1) about the equilibrium solution $u = 0$,

$$(7) \qquad \partial_t^2 v = \partial_x^2 v - g_1(x)v .$$

Using the eigenfunctions $\{\psi_j(x)\}$ and the eigenvalues $\{\omega_j^2\}$ of the operator $L(g_1) = -(d/dx)^2 + g_1(x)$ with the appropriate boundary conditions (Dir.) or (Per.), a time periodic solution to (7) is given by

$$(8) \qquad\qquad v_j(x,t) = r\cos(\omega_j t + \xi)\psi_j(x)$$

with frequency ω_j and 'action' r^2. We assume for convenience that all eigenvalues satisfy $\omega_j^2 > 0$; in any case there will be only finitely many exceptions for each choice of g_1, and the rest of the analysis will work independently of them. The general solution for $L^2(0,\pi)$ initial data is

$$v(x,t) = \sum_{j=1}^{\infty} r_j \cos(\omega_j t + \xi_j)\psi_j(x) , \qquad \sum_{j=1}^{\infty} r_j^2 < +\infty ,$$

which is time quasiperiodic (QP) if the set of frequencies $\{\omega_{j_\ell};\ r_{j_\ell} \neq 0\}$ has a finite frequency basis over the rationals. In general such an expression is almost periodic (AP) in time. Our main question is whether any of these periodic, quasiperiodic or almost periodic solutions persist for the nonlinear problem (1). For most nonlinearities the answer is yes, at least for some of the above motions. Let us focus on periodic motions with frequency close to one of the linear frequencies ω_{j_0}.

THEOREM 1. *(Craig & Wayne [CW1] (1993)) There is a set $\mathcal{G}$ in the space of all nonlinearities such that if $g \in \mathcal{G}$ then there exists a smooth Cantor family of periodic solutions to (1). More specifically, there exists a tolerance r_0, C^∞ functions $(u(x,t;r), \Omega(r))$ for $r \in [0, r_0)$, and a Cantor set $\mathcal{C}$ such that for $r \in \mathcal{C}$, then $u(x,t;r)$ is a time periodic solution to (1) with frequency $\Omega(r)$. Moreover, u is analytic with respect to (x,t), and with a choice of phase ξ,*

$$|u(x,t;r) - r\cos(\Omega(r)t + \xi)\psi_{j_0}(x)| < C|r|^2$$
$$|\Omega(r) - \omega_{j_0}| < C|r|^2 .$$

The nonlinear solution is given as a graph over the unperturbed orbits of the linear solution of (7), forming a Cantor set foliated by invariant circles.

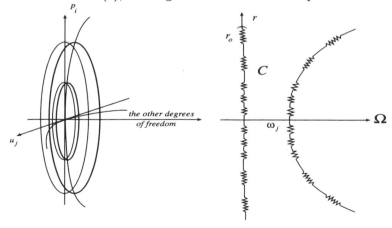

The set of acceptable nonlinearities $\mathcal{G}$ is large in the set of analytic nonlinearities, indeed it is open and dense. In fact the conditions defining $\mathcal{G}$ depend only upon the coefficients $g_1(x), g_2(x)$ and $g_3(x)$, the 3-jet of g. The condition on g_1 implies that the linear frequency sequences $\{\omega_j\}$ avoid exceptionally resonant cases, and those on (g_1, g_2, g_3) ask that the problem be genuinely nonlinear. For example, the potentials $g_1(x) = m^2$ a constant, $g_2 = 0$ and $g_3(x)u^3 = a_0 u^3$ are included in $\mathcal{G}$ for all $a_0 \neq 0$ and for an open and dense set of parameters m^2 of full Lebesgue measure.

This result has a direct analog when applied to the nonlinear Schrödinger equation (2), where the behavior of the frequency sequence $c|j|^2 \leq \omega_j \leq C|j|^2$ makes certain aspects of the problem somewhat easier, [CW4]. I will give more details of the proof of Theorem 1, which is by an approach that is nontraditional in KAM theory. First however I want to paraphrase several earlier results along similar lines.

THEOREM 2. *(Kuksin [K] (1987)(1993), Wayne [W] (1990)) There is a set $\mathcal{G}_1$ such that if $g \in \mathcal{G}_1$ then equation (1) with Dirichlet boundary conditions has Cantor families of quasiperiodic solutions.*

The version of E. Wayne gives $\mathcal{G}_1$ as a subset of the analytic nonlinearities which is not dense, although it is of large measure for an appropriate sense of measure. S. Kuksin's version of this theorem is through the study of equations of the form

$$(9) \qquad\qquad \partial_t u = i\omega(D, a)u + f , \qquad 0 \leq x \leq 2\pi$$

with $\omega(j, a) \sim |j|^p$, and nondegenerate in a certain sense with respect to a finite dimensional set of parameters a. The wave equation (1) with Dirichlet boundary conditions can be shown to fit into his framework, as can (2) and (3), however the frequency asymptotics exclude the resonances and near resonances of the periodic boundary conditions (Per.). In fact, a recent preprint of Kuksin & Pöschel [KP] shows that the nonlinear Klein-Gordon equation with Dirichlet boundary conditions fits into the framework of (9), using in part a global transformation to fourth order Birkhoff normal form.

The methods developed for the proof of these results also provide a lot of detail about these solutions to nonlinear wave equations. One result which builds upon Theorem 1 is that linear resonance produces instabilities.

DEFINITION 3. *A set $\{\omega_{j_\ell}\}$ of linear frequencies is in resonance if there is a relationship*

$$\omega_{j_1} = \frac{\omega_{j_2}}{k_2} = \cdots = \frac{\omega_{j_N}}{k_N} ,$$

for integer pairs (j_ℓ, k_ℓ).

An example of a two degree of freedom system in linear resonance $\omega_1 : \omega_2 = 1 : 1$ is given by the Hamiltonian

$$(10) \qquad H(x, y) = \tfrac{1}{2}(y_1^2 + y_2^2) + \tfrac{\omega^2}{2}(x_1^2 + x_2^2) + a_0(x_1^2 + x_2^2)^2 + a_1 x_1^4 + a_2 x_2^4 ,$$

for $(x_1, x_2, y_1, y_2) \in \mathbf{R}^4$.

Exercise 2: Find the families of time periodic solutions of the Hamiltonian system

$$\dot{x} = \partial_y H \,, \qquad \dot{y} = -\partial_x H \,,$$

with frequency close to the linear frequency ω. Calculate the linear stability of these periodic solutions.

Equation (10) is a model problem for the nonlinear Klein-Gordon equation

$$(11) \qquad \partial_t^2 u = \partial_x^2 u - m^2 u + (1 + a_0(x))u^3 \,, \qquad u(x,t) = u(x + 2\pi, t)$$

for which there is a $1 : 1$ resonance for every nonzero frequency $\omega_{2j-1} = \omega_{2j} = \sqrt{j^2 + m^2}$.

THEOREM 4. *(Craig, Kuksin & Wayne [CKW] (1994)) For the coefficient in equation (11) take $a_0(x) = a \cos(2jx)$. Then there exist smooth Cantor families of periodic solutions which have at least two normal hyperbolic directions.*

Here is a schematic picture of the situation:

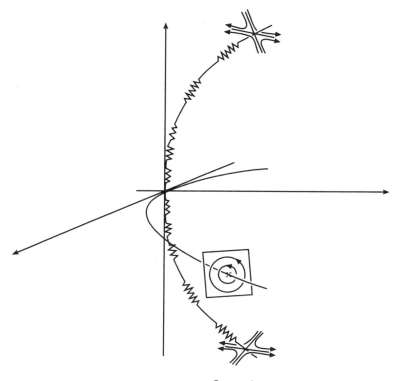

figure 2

It seems to be a general principle that in the infinite dimensional phase space for these PDE, the traveling wave solutions (or their perturbations) are stable, elliptic periodic orbits, while the standing wave solutions are unstable or unstable after perturbation, periodic orbits with at least two normal hyperbolic directions.

§4. Small divisors

We will now get to work constructing some of these solutions. In order to solve (1) we take the point of view of embedding a torus into our function space

$$(12) \qquad S(x,\xi) = \sum_{j=1}^{\infty} s_j(\xi)\psi_j(x) = \sum_{j=1}^{\infty} \sum_{k \in \mathbf{Z}^N} s(j,k)e^{ik\cdot\xi}\psi_j(x) \ .$$

We ask that $\xi \in \mathbf{T}^N$, so that $s_j(\xi + \gamma) = s_j(\xi)$ for $\gamma \in \mathbf{Z}^N$. For this to be a real torus, the sequence is to satisfy the reality condition $\overline{s(j,k)} = s(j,-k)$. If the embedding is to be invariant under the flow of equation (1), and if the flow is to pull back to the linear flow $\xi \to \xi + t\Omega$ on the torus $\mathbf{T}^N$, the function $S(x,\xi)$ must satisfy

$$(13) \qquad (\Omega \cdot \partial_\xi)^2 S - \partial_x^2 S + g(x,S) = 0$$

In terms of the eigenfunction expansion for S in (12) the sequence of Fourier coefficients $\{(j,k)\}$ must satisfy an equation on the lattice $\mathbf{Z}^{N+1}$;

$$(14) \qquad ((\Omega \cdot k)^2 - \omega_j^2)s(j,k) + W(s)(j,k) = 0$$

with a nonlinear (and nonlocal of course) function $W(s)$ defined through the nonlinearity g, which satisfies in some appropriate norm $\|W(s)\| \le O(\|s\|^2)$.

We will continue this discussion for the case of time periodic solutions, setting the torus dimension $N = 1$ and linearizing (14) about zero. We find

$$(15) \qquad ((\Omega \cdot k)^2 - \omega_j^2)s(j,k) = 0$$

with one solution $s_0(j,k) = \frac{1}{2}re^{i\xi k_0}\delta(j - j_0, k - k_0) + \frac{1}{2}re^{-i\xi k_0}\delta(j - j_0, k + k_0)$, and $\Omega = (\omega_{j_0}/k_0)$, corresponding to the solution (8) of the linear wave equation (7), $S(x,\xi) = r\cos(k_0\xi)\psi_{j_0}(x)$.

Exercise 3: Make this correspondence between the PDE and the lattice problem for the general quasiperiodic solution.

Assume that there is no exact linear resonance for ω_{j_0}, or that there is at most a finite resonance, so that the set $N = \{(j_\ell, \pm k_\ell); (\omega_{j_\ell}/k_\ell) = \omega_{j_0}\} \subseteq \mathbf{Z}^2$ is finite. Let Q be the orthogonal projection onto the subspace $\ell^2(N)$ in $\ell^2(\mathbf{Z}^2)$, the subspace of the linearly resonant modes, and let $P = 1 - Q$. Then the solution to equation (14) can be decomposed as

$$(16) \qquad s(j,k) = (Qs + Ps)(j,k) \equiv s_0(j,k) + u(j,k) \ .$$

This organizes the equation (14) as a Lyapunov-Schmidt decomposition into resonant and nonresonant components;

$$(17) \qquad \text{diag}\,((\Omega k)^2 - \omega_j^2)_N \ s_0 + QW(s_0 + u) = 0$$

$$(18) \qquad \text{diag}\,((\Omega k)^2 - \omega_j^2)_{\mathbf{Z}^2 \setminus N} \ u + PW(s_0 + u) = 0 \ .$$

The classical Lyapunov-Schmidt method would be to solve (18) for a sequence $u = u(j, k; s_0, \Omega)$, and then to find (s_0, Ω) to satisfy (17). The latter is the analog of the 'frequency map' of the classical aproach to the KAM theory.

The small divisor problem is already evident in the diagonal elements of the operator $V(\Omega) = \text{diag}\,((\Omega k)^2 - \omega_j^2)$. The standard procedure to solve (18) might be to use the inverse $V(\Omega)^{-1} = \text{diag}\,((\Omega k)^2 - \omega_j^2)^{-1}$, however for typical frequency sequences $\{\omega_j\}$ and most Ω, the eigenvalues of $V(\Omega)$ (simply its diagonal elements) are dense in $\mathbf{R}$, and in particular they accumulate at zero. To overcome the inherent losses in the problem, we will adopt the Nash-Moser modification of the Newton iteration method to solve (18).

Let $B_n = \{(j, k);\ |j| + |k| \leq L_0 2^n\} \backslash N \subseteq \mathbf{Z}^2$ be a sequence of lattice domains which eventually exhaust $\mathbf{Z}^2$. Then we propose to solve the approximate equations

$$(V(\Omega) + \partial_u PW(s_0 + u_n))_{B_n} v_n = -(V(\Omega) u_n + PW(s_0 + u_n))_{B_n}$$
(19) $$u_{n+1} = u_n + v_n$$

which is Newton's method $F'(u_n) v_n = -F(u_n)$, $u_{n+1} = u_n + v_n$, restricted to the sequence of approximate domains B_n. Now we may take advantage of the rapid convergence properties of the iteration. However the problems with the linearized operators are somehow worse even than $V(\Omega)^{-1}$, as we have to find the inverse of $(V(\Omega) + P \partial_u W)_{B_n}$, which has off-diagonal terms as well.

There is a strong analogy with known problems in statistical mechanics and localization theory. There the issue is the competition between the closeness to resonance of random potential wells, which enhances tunnelling, and the distance between them, which supresses it. To emphasize this in our small divisor problems, we will use the terminology that

$$H_{B_n} = H_{B_n}(\Omega, u) = (V(\Omega) + \partial_u PW(s_0 + u))_{B_n}$$

is the 'Hamiltonian' operator, and its inverse

$$G_{B_n} = (V(\Omega) + \partial_u PW(s_0 + u))_{B_n}^{-1}$$

is the 'Green's function'. Typically in our KAM problems, $V(\Omega)$ is diagonal and has small divisors recognizable by the fact that its spectrum is dense at zero, and $\partial_u PW$ gives rise to off-diagonal terms which decay in the off-diagonal distance. Also typical is the fact that for any subdomain A, H_A is self-adjoint.

§5. The Green's function

The last section of this lecture is a description of our analysis of the Green's function that is at the heart of the convergence proof for the iteration scheme (19).

DEFINITION 5. *A lattice site $x = (j, k) \in \mathbf{Z}^2$ is singular if*

$$|V(\Omega)| = |(\Omega^2 k^2 - \omega_j^2)| < d \, ,$$

and otherwise it is regular. A singular region is a group of singular sites, which we usually take to be connected.

Consider a selfadjoint operator W' on a lattice region $A \subseteq \mathbf{Z}^N$ (which in the course of our iteration will be the restriction to A of $\partial_u PW$).

LEMMA 6. *If $A \subseteq \mathbf{Z}^N$ is a region containing no singular sites, and if $\|W'\|_{op} \leq r < d$, then the Green's function $G_A = (V(\Omega) + W')^{-1}$ exists, and satisfies*

$$\|G_A\|_{op} \leq \frac{1}{d - r} \, .$$

PROOF. The Neumann series for G_A converges in operator norm,

$$G_A = (V(\Omega) + W')^{-1} = V^{-1} \sum_{j=0}^{\infty} (-1)^j (V^{-1} W')^j \, .$$

Our approximate solutions u_n correspond to analytic functions of (x, ξ), therefore we may consider sequences which decay exponentially, measured with the norm

$$\|u\|_\sigma^2 = \sum_{(j,k) = x \in \mathbf{Z}^2} |u(x)|^2 e^{2\sigma |x|} < +\infty \, .$$

To go along with this, we manufacture an operator norm

$$\|W'\|_\sigma = \sup_{x \in \mathbf{Z}^2} \sum_{y \in \mathbf{Z}^2} |W'(x, y)| e^{\sigma |x - y|} \, .$$

Operators with finite σ-norm have matrix elements which decay exponentially off-diagonal; that is, if $\|W'\|_\sigma \leq r_0$ then $|W'(x,y)| \leq r_0 e^{-\sigma |x-y|}$. Furthermore the norm has the algebraic property that $\|W_1 W_2\|_\sigma \leq \|W_1\|_\sigma \|W_2\|_\sigma$, and Lemma 6 works equally well with respect to this norm as with the regular operator norm.

The diagonal elements of $V(\Omega)$ are dense, so it is clear that there will be singlar sites in B_n for n large, and the hypotheses of Lemma 6 will not hold. What we must do is to insure that some control of the Green's function on B_n will remain. The goal is to show that for a Cantor set of parameters (r, Ω), there is an inductive choice $\sigma/2 < \sigma_n < \sigma$ such that $\|G_{B_n}\|_{\sigma_n} < 1/\delta_n$, and furthermore all of the other inductive constants stay under control.

To illustrate the methods to control the Green's function, I will work with a model problem, which will have the full character of the estimates at one iteration step. We make the following assumptions on the singular sites of a domain B and their geometry.

(1) The singular sites of B are contained in a set of singular regions S_j, so that

$$\cup_j S_j \qquad \text{contains all of the singular sites}$$
$$B \backslash \cup_j S_j \qquad \text{contains regular points only.}$$

(2) The singular regions are of bounded size,

$$\text{diam}\,(S_j) \leq C_0 \ .$$

(3) For singular regions, the Hamiltonian operator has spectrum bounded away from zero,

$$\text{dist}\,(\text{spec}\,(H_{S_j}), 0) > \delta \ ,$$

where $\delta \ll d$. In practice in the induction, the tolerance for small denominators $\delta = \delta_n \to 0$.

(4) Singular regions are separated from each other

$$\text{dist}\,(S_j, S_m) > 3\ell \ .$$

In the induction, $\ell = \ell_n$ will be a divergent series, but will grow less rapidly than the radius of the regions B_n.

(5) Finally, we consider disks $C(S_j)$ of radius ℓ about each singular region S_j. By requirement (4) these are pairwise disjoint. We also ask that $H_{C(S_j)}$ has spectrum bounded away from zero,

$$\text{dist}\,(\text{spec}\,(H_{C(S_j)}), 0) > \delta \ .$$

The singular regions of the lattice $\mathbf{Z}^2$ will look like this:

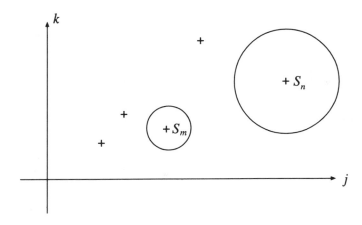

figure 3

Conditions (1) through (5) will not hold for all of the values of the parameters (r, Ω). At the nth iteration step, the parameters for which any of (1) to (5) fail must be excised from consideration and the approximate solution smoothly interpolated through the excision. This is the origin of the Cantor nature of the sets on which convergence of the procedure is achieved, and the C^∞ but not analytic nature of the bifurcation branches.

THEOREM 7. *If the above hypotheses are satisfied, $\|W'\|_\sigma \le r \ll d$ and δ, γ and ℓ are chosen appropriately, then the Green's function G_B exists, and satisfies*

$$(20) \qquad |G_B(x,y)| \le \frac{1}{\delta}\left(\frac{C}{\gamma^p}\right) e^{-(\sigma-2\gamma)|x-y|} .$$

Proof: We will use a resolvant expansion for G_B, based on the block diagonal decomposition of the Hamiltonian

$$(21) \qquad H_B = H_A \oplus_j H_{S_j} + \Gamma ,$$

where Γ consists of the off-diagonal coupling terms which involve lattice sites in A coupling to those in S_j, and couplings between different S_j's. In particular, if both $x, y \in A$ or else $x, y \in S_j$, then

$$\Gamma(x,y) = 0 .$$

Inverting (21), the generalized resolvant identity for this block decomposition is that

$$(22) \qquad G_B = G_A \oplus_j G_{S_j} + G_A \oplus_j G_{S_j} \Gamma G_B ,$$

which we may iterate to arrive at the formal expression,

$$(23) \qquad G_B = G_A \oplus_j G_{S_j} + \sum_{m=1}^{\infty} G_A \oplus_j G_{S_j} (\Gamma G_A \oplus_j G_{S_j})^m .$$

The object is to prove that under assumptions (1) through (5) this expression is convergent.

The first step will be to see that the term which involves no coupling of the blocks obeys the correct estimate (20).

$$|G_A \oplus_j G_{S_j}(x,y)| \quad \begin{aligned} &\le \frac{1}{d}e^{-\sigma|x-y|} &&\text{for } x, y \in A \\ &\le \frac{1}{\delta} &&\text{for } x, y \in S_j \\ &= 0 &&\text{if } x, y \text{ are not in the same block} \end{aligned}$$

This is in agreement with the bound (20).

The expression involving one coupling operation between blocks is

$$G_A \oplus_j G_{S_j} \Gamma G_A \oplus_j G_{S_j} ,$$

which we will consider by cases. Since there are potentially two occurrances of G_{S_j}, the object will be to avoid factors of $1/\delta^2$ in the estimate, which for $\delta \ll r$ would call into question the convergence of the series.

In case $x \in A$ and $y \in S_j$ (or vice versa),

$$|G_A \oplus_j G_{S_j} \Gamma G_A \oplus_j G_{S_j}(x,y)| = |\sum_{\substack{z \in A \\ w \in S_j}} G_A(x,z)\Gamma(z,w)G_{S_j}(w,y)|$$

$$\leq \sum_{\substack{z \in A \\ w \in S_j}} \frac{1}{d}e^{-\sigma|x-z|}re^{-\sigma|z-w|}\frac{1}{\delta} .$$

Diminishing the exponential rate of decay by a small amount γ,

$$\leq \frac{r}{d}\frac{1}{\delta}e^{-(\sigma-\gamma)|x-w|}\sum_{\substack{z \in A \\ w \in S_j}} e^{-\gamma|x-z|}e^{-\gamma|z-y|}$$

$$\leq \frac{1}{\delta}\left(\frac{r}{d}\right)e^{-(\sigma-\gamma)|x-y|}\left(\frac{C}{\gamma^{p/2}}\right) ,$$

where we used that $|w-y| < C_0$ for both $w,y \in S_j$.

In case that both $x,y \in A$, or $x,y \in S_j$ the same block, we have $G_A \oplus_j G_{S_j}\Gamma G_A \oplus_j G_{S_j}(x,y) = 0$, since the interaction operator Γ strictly enforces that the lattice sites (z,w) be in distinct blocks. The final case exhibits the need for assumption (4); we take $x \in S_j$ and $y \in S_k$, $k \neq j$. Then

$$|\sum_{\substack{z \in S_j \\ w \in S_k}} G_{S_j}(x,z)\Gamma(z,w)G_{S_k}(w,y)| \leq |\sum_{\substack{z \in S_j \\ w \in S_k}} \frac{1}{\delta}re^{-\sigma|z-w|}\frac{1}{\delta}|$$

(24) $$\leq \frac{1}{\delta}\left(\frac{r}{d}\right)e^{-(\sigma-\gamma)|x-y|}\left(\sum_{\substack{z \in S_j \\ w \in S_k}} \frac{d}{\delta}e^{-\gamma|z-w|}\right) .$$

Since $z \in S_j$ and $w \in S_k$ with $k \neq j$, there is a jump between lattice sites included in the exponential sum of at least distance ℓ, and if we demand that

$$\frac{d}{\delta}e^{-\gamma\ell/2} < 1 ,$$

the result is that

(25) $$\sum_{\substack{z \in S_j \\ w \in S_k}} \frac{d}{\delta}e^{-\gamma|z-w|} < C\frac{d}{\delta}e^{-\gamma\ell/2} \leq C ,$$

so that in this case too the bound is $(1/\delta)(r/d)e^{-(\sigma-\gamma)|x-y|}$.

The factors $(r/d)^m < 1$ will be the convergence factors in the sum (23). In order to continue the process of estimates of the terms of it, we are forced to study both of the local Hamiltonians H_{S_j} and $H_{C(S_j)}$. The following result illustrates the role played by the assumption (5).

LEMMA 8. *Under the above hypotheses, the local Hamiltonians in a neighborhood of a singular region S satisfy*

$$(26) \qquad |G_{C(S)}(x,y)| \leq \frac{1}{\delta}\frac{C}{\gamma^p}e^{-(\sigma-\gamma)|x-y|} \ .$$

Proof: If both $x,y \in S$, using assumption (5) and the fact that $H_{C(S)}$ is self-adjoint,

$$|G_{C(S)}(x,y)| < \frac{1}{\delta} \ ,$$

which is a sufficient for (26) as diam (S) is bounded. However if one of $x,y \notin S$ this simple bound will not give us an estimate of exponential decay. To do better we again proceed with the resolvent identities,

$$H_{C(S)} = H_D \oplus H_S + \Gamma \ ,$$

where $D = C(S)\backslash S$. The Green's function will satisfy

$$(27) \qquad \begin{aligned} G_{C(S)} &= G_D \oplus G_S + G_D \oplus G_S\Gamma G_{C(S)} \\ &= G_D \oplus G_S + G_D \oplus G_S\Gamma G_D \oplus G_S + G_D \oplus G_S\Gamma G_{C(S)}\Gamma G_D \oplus G_S \ . \end{aligned}$$

When $x \notin S$ but $y \in S$,

$$G_{C(S)} = G_D \oplus G_S + G_D \oplus G_S\Gamma G_{C(S)} \ ,$$

and the first term vanishes as x,y are not in the same block. The second term is controlled by an exponential sum, giving

$$\begin{aligned} |G_{C(S)}(x,y)| &\leq \sum_{\substack{z\in D \\ w\in S}} \frac{1}{d}e^{-\sigma|x-z|}re^{-\sigma|z-w|}\frac{1}{\delta} \\ &\leq \left(\frac{r}{d}\right)\frac{1}{\delta}\left(\frac{C}{\gamma^p/2}\right)e^{-(\sigma-\gamma)|x-y|} \ , \end{aligned}$$

which is sufficient for (26).

If both $x,y \in D$, then we need to use the second expression in (27). The first term does not involve S, and is bounded safely by $(1/d)e^{-\sigma|x-y|}$ using Lemma 6. The second term vanishes as the interaction Γ demands a change of block, and we are left with;

$$\begin{aligned} |G_{C(S)}(x,y)| &\leq \frac{1}{d}e^{-\sigma|x-y|} \\ &\quad + |\sum_{\substack{z\in D \ w\in S \\ p\in S \ q\in D}} G_D(x,z)\Gamma(z,w)G_{C(S)}(w,p)\Gamma(p,q)G_D(q,y)| \\ &\leq \frac{1}{d}e^{-\sigma|x-y|} + \left(\frac{r}{d}\right)^2\frac{1}{\delta}\sum_{z\in D}e^{-\sigma|x-z|}e^{-\sigma|z-w|}\sum_{q\in D}e^{-\sigma|p-q|}e^{-\sigma|q-y|} \\ &\leq \frac{1}{d}e^{-\sigma|x-y|} + \left(\frac{r}{d}\right)^2\frac{1}{\delta}e^{-(\sigma-\gamma)|x-w|}e^{-(\sigma-\gamma)|p-y|}\left(\frac{C}{\gamma^p}\right) \ . \end{aligned}$$

Since in this expression we have both $w, p \in S$, the distance between them is bounded, and again we satisfy the criterion of the lemma, which finishes the proof.

This almost completes the proof of Theorem 7, which needs only one more step. To make a good estimate of the terms of the resolvant expansion of degree $m \geq 2$, we use a slightly more sophisticated version of the resolvant expansion, in which we alternate the generalized resolvant identity with respect to the decompositions $B = A \cup_j S_j$ and $B = A' \cup_j C(S_j)$, where $A' = B \backslash \cup_j C(S_j)$. That is, we can write

$$H_B = H_{A'} \oplus_j H_{C(S_j)} + \Gamma'$$

and therefore alternating the use of the two decompositions, we write

$$
\begin{aligned}
G_B &= G_A \oplus_j G_{S_j} + G_A \oplus_j G_{S_j} \Gamma G_{A'} \oplus_j G_{C(S_j)} \\
&\quad + \sum_{m=1}^{\infty} (G_A \oplus_j G_{S_j} \Gamma G_{A'} \oplus_j G_{C(S_j)} \Gamma')^m \\
&\quad \times (G_A \oplus_j G_{S_j} + G_A \oplus_j G_{S_j} \Gamma G_{A'} \oplus_j G_{C(S_j)}) \, .
\end{aligned}
$$
(28)

The convergence of this series with uniform exponential bounds gives the proof of Theorem 7, and this in turn can be easily deduced from the following Lemma.

LEMMA 9. *Estimates on the principal term of the series (28).*

$$|G_A \oplus_j G_{S_j} \Gamma G_{A'} \oplus_j G_{C(S_j)} \Gamma'(x, y)| \leq \left(\frac{r}{d}\right)^2 \left(\frac{C}{\gamma^p}\right) e^{-(\sigma - 2\gamma)|x - y|} \, .$$

The proof of this is of similar character as the above arguments, in which we consistently avoid factors of $(1/\delta)$ which would ruin the convergence estimates. We do this by insuring that each summand associated with a lattice site in either S_j or $C(S_j)$ must enter the sum associated with a long path $(x; z; w; p; q; r; y)$ from x to y of length at least ℓ. The estimate is then similar to (24),(25). I will leave the precise details of the argument to the reader, as a somewhat combinatorial exercise.

REFERENCES

Principal Reading:

[CW1] W. Craig & C.E. Wayne, *Newton's method and periodic solutions of nonlinear wave equations*, Commun. Pure Applied Math. **46** (1993), 1409-1501.

[K1] S. Kuksin, *Nearly integrable infinite-dimensional Hamiltonian systems Springer Lecture Notes in Math. 1556*, Springer Verlag, Berlin, 1993.

Further reading:

[CS] W. Craig & C. Sulem, *Numerical simulation of gravity waves*, Journal Comp. Physics **108** (1993), 73-83.

[CW2] W. Craig & C.E. Wayne, *Nonlinear waves and the KAM theorem: nonlinear degeneracies*, 'Large Scale structures in nonlinear physics' Proceedings Villefrache, France J.-D. Fourier and P.-L. Sulem, ed's. Springer Lecture Notes in Physics 392, Springer Verlag, Berlin, 1991.

[CW3]   W. Craig & C.E. Wayne, *Nonlinear waves and the* 1 : 1 : 2 *resonance*, Singular
        limits of dispersive waves; Proceedings of the NATO conference at ENS-Lyon, 1991,
        N. Ercolani, D. Levermore and D. Serre ed's., Series B: Physics, vol. 320 , Plenum,
        1994.
[CW4]   W. Craig & C.E. Wayne, *Periodic solutions of nonlinear Schrödinger equations and
        the Nash Moser method*, ETH preprint (1993).
[CKW]   W. Craig, S. Kuksin & C.E. Wayne, *in preparation* (1994).
[FS]   J. Fröhlich & T. Spencer, *Absence of diffusion in the Anderson tight binding model
        for large disorder or low energy*, Commun. Math. Physics **88** (1983), 151-184.
[K2]   S. Kuksin, *Hamiltonian perturbations of infinite-dimensional linear systems with an
        imaginary spectrum*, Funts. Anal. Prilozh. **21:3** (1987), 22-37; English translation in:
        Functl. Anal. Applications **21** (1987), 192-205.
[KP]   S. Kuksin & J. Pöschel, *preprint* (1994).
[P]   J. Pöschel, *Small divisors with spatial structure*, Commun. Math. Physics **127** (1990),
        351-393.
[VB]   M. Vittot & J. Bellissard, *Invariant tori for an infinite lattice of coupled classical
        rotors*, Preprint CPT-Marseille (1985).
[W]   C.E. Wayne, *Periodic and quasiperiodic solutions of the nonlinear wave equation via
        KAM theory*, Commun. Math. Physics **127** (1990), 479-528.

MATHEMATICS DEPARTMENT, BROWN UNIVERSITY, PROVIDENCE, RHODE ISLAND 02912

E-mail address: craigw@math.brown.edu

Lectures in Applied Mathematics
Volume **31**, 1996

GLOBAL CENTER MANIFOLDS AND
SINGULARLY PERTURBED EQUATIONS:
A BRIEF (AND BIASED) GUIDE
TO (SOME OF) THE LITERATURE

Nancy Kopell

Center manifolds arise when a critical point of a differential equation has $k \geq 1$ eigenvalues with zero real part. The center manifold has dimension k and is tangent to the eigenspace of the k eigenvalues with zero real part [1].

In general, unlike stable and unstable manifolds, center manifolds are not unique, and can be constructed only locally. Indeed, there are examples which show that the more differentiable one requires the center manifold to be, the smaller may be the center manifold that can be constructed [1, p 29,30]. Some important examples for which a center manifold *is* global are so-called singularly perturbed equations. These are equations of the form

$$(1a) \qquad \epsilon \, dx/dt = f(x,y) \qquad x = (x_1, \ldots, x_n)$$

$$(1b) \qquad dy/dt = g(x,y) \qquad y = (y_1, \ldots, y_m)$$

where $\epsilon << 1$. Here the $\{x_i\}$ are the "fast variables" and the $\{y_i\}$ are the slow variables.

Equation (1) can be rewritten as

$$(2a) \qquad dx/d\tau = f(x,y)$$

$$(2b) \qquad dy/d\tau = \epsilon \, g(x,y)$$

where $\tau = t/\epsilon$. In this time scale, in the limit as $\epsilon \to 0$ there is a manifold of critical points defined by

$$(3) \qquad f(x,y) = 0.$$

Partially supported by grant NSF-DMS 8901913. AMS Subject classification: 34B15, 34C30, 34C35, 34C37, 34E15.

Suppose for some region of (x, y) the matrix $\partial f / \partial x$ has eigenvalues with real part bounded away from zero. In such a region, the manifold (3) of critical points is called "normally hyperbolic" and is a center manifold parameterized by y.

For $\epsilon > 0$, most of these critical points disappear generically. However, there is still a connection to center manifolds as follows. One appends the equation

$$(2c) \qquad\qquad d\epsilon/d\tau = 0$$

to (2). Then in any region in which the points satisfying (3) are normally hyperbolic, (2a-c) has an $m+1$ dimensional center manifold in x, y, ϵ space; this center manifold is localized around $f(x, y) = 0$, $\epsilon = 0$. It can be shown [2] that for ϵ sufficiently small, such manifolds can be chosen globally, parameterized by y. The center manifolds of (2) are often called "slow manifolds." (Usage varies; sometimes "slow manifold" refers to the $\epsilon \to 0$ limit of the center manifold, i.e. $f(x, y) = 0$.)

The importance of the slow manifolds is that they can be used to construct (allegedly) approximate solutions to (1) or (2). These so-called "singular solutions" are unions of solutions to a pair of simpler equations, the "slow equations" and the "fast equations." The slow equations are

$$(4) \qquad\qquad dy/dt = g(x, y)$$

where $x = x(y)$ satisfies $f(x, y) = 0$. The fast equations are

$$(5) \qquad\qquad dx/d\tau = f(x, y)$$

where y is a constant. There is a large applied math literature that discusses "matching conditions" under which solutions to (4) and (5) can be pieced together to get an object that satisfies some extra conditions, such as periodicity, boundary conditions at finite values of t, or limits as $t \to \pm\infty$. (See, e.g. [3-5]; [3] is a good introduction.) The singular solution is the zeroth order version of this pieced-together object. Better (formal) approximations can be calculated from (1) and (2) according to the theory of "matched asymptotic expansions" ([3-5]). Especially for the $\epsilon = 0$ version, the construction of the singular object is in general much easier than the construction of the actual solution to (1) or (2).

In much of the applied math literature, there is a blanket assumption that the singular object is indeed an approximation to some actual solution to (1) or (2), with the desired extra conditions. However, it is also well-known that this assumption is occasionally grievously incorrect. (See papers in [6] on "resonance" for a well-studied example and a geometric treatment.) This uncertainty has given rise to another large literature on methods for figuring out circumstances under which the existence of a singular solution implies that there is an actual one for which the singular solution is an approximation. The methods invented to do this come from dynamical systems, topology, functional analysis, and nonstandard analysis. See [7,8] for references to some of that literature. The methods I find most conceptual and most powerful are the geometric ones, involving both topological techniques and dynamical systems techniques. Here I shall deal only with the latter.

One problem that has been much analyzed from many points of view is the existence of traveling wave solutions to nerve-conduction equations, a class of parabolic PDE's. The desired solution is a homoclinic orbit to the associated ODE obtained by using the traveling wave ansatz. The singular solution associated with this problem has two slow pieces and two fast pieces. A description of the singular solution can be found in [7,9,10].

The first rigorous geometric treatment was given by C. Carpenter [9], working under the direction of C. Conley. Her proof was a topological one, based on fixed point theorems. In 1980, R. Langer [11] produced a proof for a simple subset of the class studied by Carpenter; Langer's proof was a dynamical systems version in which the homoclinic orbit is constructed as the intersection of the stable and unstable manifolds of a critical point. The advantage of Langer's version was that it proved local uniqueness, and gave further information that led to the proof by C. Jones of the stability of the homoclinic orbit as a solution to the original PDE [12]. The disadvantage was that it seemed very complicated and not easily generalizable.

The essential difficulty in Langer's proof arises in a very large class of problems involving the construction of solutions to singularly perturbed equations by dynamical systems methods. In many cases (e.g. boundary value problems, homoclinic orbits, heteroclinic orbits) the desired trajectory is to be constructed as the transverse intersection of a pair of invariant manifolds. To establish the transversality involves tracking the position and tangent planes of the relevant manifolds over global distances. Since the desired actual solution (if it exists) is near the singular solution, it must pass close to the slow manifolds described above. This means that the invariant manifolds whose intersection is the desired solution also must pass close to the slow manifolds. (The relevant manifolds in the case of a homoclinic orbit are the stable and unstable manifold of the critical point to which the orbit is homoclinic.) The actual trajectory in general takes time of the order $\tau = O(1/\epsilon)$ to pass by the slow manifold, so careful tracking is needed to keep control of estimates.

In [13] and [14], Jones et al. resolved this difficulty by proving the "exchange lemma," a tool for tracking invariant manifolds as they pass close to a slow manifold. This result gives estimates on the position of the invariant manifold after a sojourn of time $\geq O(1/\epsilon)$ near the slow manifold, providing that some transversality hypothesis is satisfied. This hypothesis is on submanifolds of the simpler fast system (5), and hence is generally easier to verify than hypotheses about the full system. A proof of the existence of a particular orbit for a particular set of equations then reduces to verifying a set of transversality conditions for that problem. (For a verification in the case of FitzHugh-Nagumo nerve conduction equations, see [10]). The exchange lemma machinery then constructs an orbit to the full equations.

The work in [13] and [14] requires that the manifold being tracked have one higher dimension than the number of positive eigenvalues of a critical point on the slow manifold (i.e. of the matrix $\partial f/\partial x$). This constraint is satisfied for nerve-conduction equations, but not for many other kinds of singularly perturbed equations. In [15], S.K. Tin generalized [13] and [14] by removing this constraint. With the more general result in [15], Tin et al. [8] were able to prove in a fairly simple way a general theorem about boundary value problems. (See also [16]). The tracking techniques also apply to some perturbed Hamiltonian systems, in which the relevant manifolds of the fast system are not transversal at $\epsilon = 0$, but for which Melnikov methods yield a weaker form of transversality [17]. Finally, the tracking techniques are beginning to be applied to equations other than singularly perturbed systems (e.g. weakly perturbed Hamiltonian systems) [15].

REFERENCES

1. J. Carr, *Applications of Centre Manifold Theory,* Applied Mathematical Sciences, vol. 35, Springer Verlag, New York, 1981.
2. N. Fenichel, *"Geometric singular perturbation theory for ordinary differential equations",* J. Diff. Equa. **31** (1979), 53-98.
3. C.C. Lin and L. Segel, *Mathematics Applied to Deterministic Problems in the Natural Sciences,* MacMillan, 1974.
4. W. Eckhaus, *Asymptotic Analysis of Singular Perturbations,* North Holland, 1970.
5. J.D, Cole, *Perturbation Methods in Applied Mathematics,* Blaisdell, 1968.
6. R.E. Meyer and S.V. Parter (eds.), *Singular Perturbations and Asymptotics,* Math. Res. Center Symposia and Advanced Sciences Series, Academic press, 1980.
7. N. Kopell, *"Dynamical systems and the geometry of singularly perturbed differential equations"* , From Topology to Computation: Proceedings of the Smalefest (1993), Springer- Verlag, New York, 545-556.
8. S.-K. Tin, N. Kopell and C.K.R.T. Jones, *"Invariant manifolds and singularly perturbed boundary value problems",* SIAM J. Numer. Anal. **31** (1994), 1558-1576.
9. G.A. Carpenter, *"A geometric approach to singular perturbation problems with applications to nerve impulse equations",* J. Diff. Equa. **23** (1977), 335-367.
10. C.K.R.T. Jones, N. Kopell and R. Langer, *"Construction of the FitzHugh-Nagumo Pulse using differential forms"* **37,** (1991), Springer-Verlag, New York, 101-116.
11. R. Langer, *"Existence and uniqueness of pulse solutions to the FitzHugh-Nagumo equations",* Ph.D. thesis, Northeastern Univ., 1980.
12. C.K.R.T. Jones, *"Stability of the travelling wave solution of the FitzHugh-Nagumo system",* Trans. Amer. Math. Soc. **286** (1984), 431-469.
13. C.K.R.T. Jones and N. Kopell, *"Tracking invariant manifolds with differential forms in singularly perturbed systems",* J. Diff. Equa. **108** (1994), 64-88.
14. C.K.R.T. Jones, T.J. Kaper and N. Kopell, *"Tracking invariant manifolds up to exponentially small errors",* To appear in SIAM J. Math. Anal.
15. S.-K. Tin, *"On the dynamics of tangent spaces near a normally hyperbolic invariant manifold",* Ph.D. thesis, Brown Univ., 1994.
16. X.-B. Lin, *"Heteroclinic bifurcation and singularly perturbed boundary value problems",* J. Diff. Equa. **84** (1990), 319-382.
17. T.J. Kaper and G. Kovacic, *"Multi-bump orbits homoclinic to resonance bands",* Los Alamos National Laboratory Technical Report # LAUR 93-2918 (1993).

Lectures in Applied Mathematics
Volume 31, 1996

Melnikov Analysis for Pde's

David W. McLaughlin *
Courant Institute of Mathematical Sciences
New York University
New York, New York 10012

Jalal Shatah †
Courant Institute of Mathematical Sciences
New York University
New York, New York 10012

Abstract

Melnikov analysis for pde's is described, primarily through the example of a perturbed nonlinear Schroedinger equation. This technique, when combined with geometric singular perturbation theory, normal form analysis, and integrable theory, enables one to prove the persistence of homoclinic orbits for pde's. The article focuses upon results that we have obtained, in collaboration with Y. Li and S. Wiggins, since the summer school at the Mathematical Sciences Research Institute. The proofs illustrate dynamical systems methods for pde. In the article, we outline the steps in the proofs, while always emphasizing the intuition behind each argument.

1991 *Mathematics Subject Classification.* Primary 35B25, 35Q55.
*Funded in part by AFOSR-90-0161 and by NSF DMS 8922717 A01.
†Funded in part by by NSF DMS 9401558.

1 Introduction

In this article we will describe the use of Melnikov Integrals, together with methods from invariant manifold theory and geometric singular perturbation theory, in the construction of global solutions to partial differential equations (pde's). These methods have been developed in dynamical systems theory of ordinary differential equations [23], [62], and have only recently been modified for and extended to the infinite dimensional setting of pde's. In fact, the use of Melnikov integrals together with geometric perturbation theory for the rigorous mathematical study of pde's is really very new, with most of the developments having occured since the summer school at the Mathematical Sciences Research Institute. Our purpose here is to describe, briefly, these very recent developments and to provide a short list of relevant literature.

As with most methods as they are being developed, this one begins from explicit examples. Over the past year and a half, several rich pde examples have been investigated, and the study of one pde has been completed. It is clear that these examples could be extended and developed into general methods for classes of pde's, a generalization which will certainly take place over the next few years. Here we will focus on that pde which is the most completely understood [the nonlinear Schroedinger (NLS) equation], and mention the other examples with references.

Melnikov integrals are defined for perturbations of Hamiltonian systems,

$$\vec{q}_t = J\nabla H + \epsilon \vec{G}(\vec{q}), \tag{1.1}$$

where $\vec{q}$ belongs to a linear inner product space $\mathcal{F}$ (the "phase space"), the Hamiltonian H is a real valued functional on $\mathcal{F}$, the perturbation $\vec{G} : \mathcal{F} \to \mathcal{F}$, J is the symplectic matrix

$$J = \begin{pmatrix} 0 & 1 \\ -1 & 0 \end{pmatrix}$$

and ϵ is a small positive parameter. Let $\vec{h}(t)$, $t \in (-\infty, +\infty)$, denote a periodic, quasiperiodic, or homoclinic orbit of the unperturbed ($\epsilon = 0$) Hamiltonian system. Furthermore, let I denote a real valued functional on $\mathcal{F}$,

$$I : \mathcal{F} \to \mathbf{R},$$

which is a constant of the motion for the unperturbed Hamiltonian system, i.e, which "Poisson commutes with the Hamilitonian H,"

$$\{H, I\}_J := (grad H, J grad I) = 0.$$

Here $(\cdot, \cdot)$ denotes the inner product on the phase space $\mathcal{F}$.

Definition 1.1 *The Melnikov Integral (based upon I) is defined as*

$$M_I \equiv \lim_{T \to \infty} \frac{1}{2T} \int_{-T}^{T} (grad I, \vec{G})\,|_{\vec{h}(t)}\,dt, \qquad (1.2)$$

where the unperturbed orbit $\vec{h}(t)$ is periodic or quasiperiodic. In the case that the orbit $\vec{h}(t)$ is homoclinic, the temporal integral is replaced by

$$\lim_{j \to \infty} \int_{T_j^-}^{T_j^+} \cdots dt,$$

for suitable sequences of real numbers $\{T_j^-\}$ and $\{T_j^+\}$, tending to $-\infty$ and $+\infty$ respectively.

Remark 1.1 *i*) The Melnikov integral for periodic orbits appears in the method of averaging [54] [23]. For quasiperiodic orbits, this definition of the Melnikov function may be found in [56].

ii) The sequences $\{T_j^-\}$ and $\{T_j^+\}$ are used [31] to define conditionally the Melnikov integral in those cases where the improper integral itself does not converge. In our own work, the constant of the motion I can be chosen to guarantee convergence, and the extra freedom given by the sequences $\{T_j^-\}$ and $\{T_j^+\}$ has not been needed.

Melnikov integrals are used to assess the fate of the orbit $\vec{h}$ under the perturbation, ($\epsilon > 0$, but small). "Does $\vec{h}$ persist (as a periodic, quasiperiodic, or homoclinic orbit) for the perturbed system, or does the perturbation destroy $\vec{h}$?" The Melnikov integral, *together with geometric analysis*, can provide answers to such persistence questions.

As is clear from its definition, the Melnikov integral M_I provides an estimate of the change in the value of the constant of motion I over the perturbed orbit. Without an additional geometric setting, this change provides very little information about persistence. Nevertheless, for specific applications it is wise to understand very well the behavior of M_I as a function of the parameters of the problem *before* developing the geometric setting. We have learned this point of view from experience, and have advocated it for some time [48], [51]. For example, if the Melnikov function never vanishes, one can often quickly infer (even with mathematical rigor [56]) that persistence is impossible. Once the parameter dependence of the Melnikov function M_I is thoroughly understood, the real mathematical challenge remains – to use M_I together with geometric analysis to establish positive results about persistence. This complete goal has been achieved for several specific pde's, as described below.

Melnikov analysis was developed [54], [31], [32] for ordinary differential equations where it has been a very successful tool as is well documented in the literature. See, for example, [23] [62]. It was first extended to pde's by Holmes and Marsden, who developed a general setting [30], which they then applied to a nonlinear beam equation and to a perturbed sine-Gordon equation [29]. This work initiated the use of Melnikov methods in the context of pdes, where it proved the existence of solutions with extremely irregular behavior in time. However, numerical studies [9] of perturbed sine-Gordon equations did not observe this particular type of irregular behavior. Rather a very different type of chaotic phenomena is seen numerically.

Remark 1.2 Frequently in the literature the expression "Melnikov analysis for pde's" is used to describe the following situation: The original pde (in one space x and one time t) is reduced to a finite dimensional ode in x through a traveling wave ansatz. This ode is then analysed with Melnikov methods for ode's. In this article, we do not use this terminology, and we restrict the expression "Melnikov analysis for pde's" to describe analysis of the original pde as an infinite dimensional dynamical system.

2   Background

Our own studies of chaotic behavior in pde's began with the numerical investigation of the damped and driven sine-Gordon equation [9],

$$u_{tt} - u_{xx} + \sin u = \epsilon[-\alpha u_t + \Gamma \cos \omega t], \qquad (2.1)$$

with even, periodic boundary conditions,

$$\begin{aligned} u(-x, t) &= u(x, t) \\ u(x + L, t) &= u(x, t). \end{aligned}$$

This equation provides a very natural starting point for the study of chaotic behavior in pdes, since it can be viewed as the continuum limit of coupled pendula, and since the damped and driven pendulum is a prototype [5] for chaotic behavior in ode's. These numerical experiments, as well as those for the closely related damped and driven nonlinear Schroedinger equation (NLS),

$$-2iq_\tau + q_{yy} + \left(\frac{1}{2}qq^* - 1\right)q = i\epsilon\left[\alpha q + \Gamma\right], \qquad (2.2)$$

$$\begin{aligned} q(-y, \tau) &= q(y, \tau) \\ q(y + l, \tau) &= q(y, \tau), \end{aligned}$$

are described in detail in the survey [52], a preprint of which was made available at the summer school. In addition to a description of the numerical experiments, that survey describes the unperturbed integrable theory, it's connection to the perturbed numerical experiments, and our initial steps toward a Melnikov analysis for natural finite dimensional discretizations of the perturbed pde. Since that detailed survey is now available, in this article we will restrict ourselves to a very brief summary of the numerical experiments and focus upon very recent analytical developments for the pde which were completed after that survey was written.

A broad overview of the numerical experiments is as follows: In both the perturbed sine-Gordon and NLS cases, for certain ranges of parameter values, the long time behavior is chaotic. For the sine-Gordon case, this chaotic behavior occurs at small amplitude $|u(x,t)| << 1$; hence, one expects, for the NLS equation, very similar chaotic oscillations, which indeed are also observed. (In fact, movies of the numerical experiments have been made which display such chaotic behavior [5], [36].) In both the NLS and sine Gordon cases, several key features include (See Figure 2.1 and Figure 2.2):

1. Temporal chaos is observed at small values of the perturbation parameter ϵ;

2. Spatially, the wave forms are very regular and consist of coherent localized solitary waves which oscillate chaotically in time t;

3. Both dynamical systems diagnostics (Poincare Sections, Lyapanov and information dimensions) and the spectral transform of the unperturbed integral theory indicate that the "dimension of the attractor" can be very small. In the simplest case, the wave consists in one solitary wave interacting with a long wavelength background;

4. Spectral transform measurements of the unperturbed integrable theory also establish that the fundamental instability which causes the chaotic behavior is the modulational instability of this long wavelength background (or mean), a classical instability well known in the theory of nonlinear waves [61];

5. In this simplest case, the lone solitary wave jumps, irregularly in time t, between a location at the center ($x = 0$) and location at the edge ($x = L/2$), the only two spatial locations allowed by even periodic boundary conditions.

This last point makes feasible the existence of a "symbol dynamics" on the two symbols, [C (=center) and E (=edge)]. Mathematically, one anticipates the possibility that an invariant set exists on which the dynamics is topologically equivalent to a Bernoulli shift on two symbols; that is, on which the dynamics is as random as a sequence of coin tosses.

In any event, the observed chaotic behavior consists in the interaction of coherent localized spatial structures with each other and with a long wavelength mean. Moreover, the modulational instability of this long wavelength

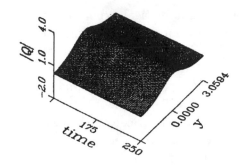

Figure 2.1: Unperterbed Solitary Wave

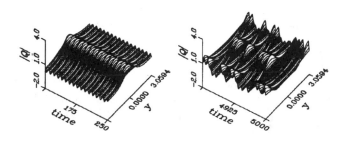

Figure 2.2: Instabilities induced by increasing Γ

mean is central to the chaotic behavior. Neither of these crucial characteristic features is present in the initial existence theory in [30]. Further analysis is required.

3　A Perturbed NLS Equation

We study a perturbed nonlinear Schroedinger equation (PNLS) of the form

$$iq_t = q_{xx} + 2[q\bar{q} - \omega^2]q + i\epsilon[\widehat{D}q - 1], \tag{3.1}$$

where the constant $\omega \in (\frac{1}{2}, 1)$, ϵ is a small positive constant, and $\widehat{D}$ is a *bounded* negative definite linear operator on the Sobolev space $H^1_{e,p}$ of even, 2π periodic functions which are square integrable with square integrable first derivative. Specific examples of the dissipation operator $\widehat{D}$ include the discrete Laplacian and a "smoothed Laplacian" given by

$$\widehat{D}q = -\alpha q - \beta\widehat{B}q, \tag{3.2}$$

where the operator $\hat{B}$ has symbol given by

$$b(k) = \begin{cases} k^2 & k < \kappa \\ 0 & k \geq \kappa. \end{cases}$$

This pde is well posed in $H^1_{e,p}$ [60]. In fact, the solution

$$q(t; \epsilon) \; = \; F^t_\epsilon(q_{in})$$

has several derivatives in q_{in} and in the parameters such as ϵ, with the exact number of derivatives increasing with decreasing ϵ.

Our analysis of this equation begins with two observations: *First,* when $\epsilon = 0$, the unperturbed NLS equation is a completely integrable soliton equation. *Second,* the "plane of constants" Π_c,

$$\Pi_c := \{q(x,t) \,|\, \partial_x q(x,t) \equiv 0\} \,,$$

is an invariant plane for PNLS. In each of these two cases, $[\epsilon = 0 \text{ or } q \in \Pi_c]$, the behavior of solutions $q(\cdot, t)$ can be described completely. In the first case, this description is accomplished through the spectral transform of completely integrable "soliton mathematics"; in the second case, it is accomplished through "phase plane analysis". In the jargon of the theory of dynamical systems, our methods will be a form of "local-global" analysis, where at times the term "local" will mean close to the plane Π_c, and at other times "local" will mean close to the integrable case. In any event, throughout our global arguments, control is achieved either because of proximity to (i) the plane Π_c or (ii) $\epsilon = 0$.

3.1   Motion on the Invariant Plane

On the invariant plane Π_c, the equation takes the form

$$iq_t \; = \; 2\left[q\bar{q} - \omega^2\right]q \; - \; i\epsilon\left[\alpha q + 1\right], \tag{3.3}$$

where it is assumed that the dissipation operator $\hat{D}$ acts invariantly on Π_c as

$$\hat{D}q = -\alpha q,$$

for α a positive constant. Equivalently, in terms of polar coordinates

$$q := \sqrt{I} \, \exp i\theta,$$

these equations take the form

$$\begin{aligned} I_t &= -2\epsilon\left[\alpha I + \sqrt{I} \cos\theta\right] \\ \theta_t &= -2(I - \omega^2) + \frac{\epsilon}{\sqrt{I}} \sin\theta. \end{aligned} \tag{3.4}$$

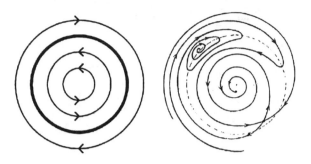

Figure 3.1: Phase Plane Diagram of the ODE

When $\epsilon = 0$, the unperturbed orbits on Π_c are nested circles, with S_ω a circle of fixed point given by $I = \omega^2$. For $\epsilon > 0$, the perturbed orbits on Π_c are very different (see Figure 3.1). First, only three fixed points exist: O, which is a deformation of the origin; Q, a saddle which deforms from the circle S_ω; and P, a spiral sink which also deforms from the circle S_ω. Formulas for the sink P and the saddle Q, together with their associated growth rates, are given by

$$
\begin{aligned}
I_p &= \omega^2 + \frac{\epsilon}{2\omega}\sqrt{(1 - \alpha^2\omega^2)} + O(\epsilon^2) \\
\theta_p &= -\tan^{-1}\frac{\sqrt{1 - \alpha^2\omega^2}}{\alpha\omega} - \pi + O(\epsilon)
\end{aligned}
\tag{3.5}
$$

$$
\begin{aligned}
I_q &= \omega^2 - \frac{\epsilon}{2\omega}\sqrt{(1 - \alpha^2\omega^2)} + O(\epsilon^2) \\
\theta_q &= \tan^{-1}\frac{\sqrt{1 - \alpha^2\omega^2}}{\alpha\omega} - \pi + O(\epsilon)
\end{aligned}
\tag{3.6}
$$

$$
\begin{aligned}
\sigma_p &= \pm 2i\sqrt{\epsilon\omega}\,[1 - \alpha^2\omega^2]^{\frac{1}{4}} - \epsilon\alpha + O(\epsilon^{\frac{3}{2}}) \\
\sigma_q &= \pm 2\sqrt{\epsilon\omega}\,[1 - \alpha^2\omega^2]^{\frac{1}{4}} - \epsilon\alpha + O(\epsilon^{\frac{3}{2}}).
\end{aligned}
\tag{3.7}
$$

While the circle of fixed points S_ω for the unperturbed ($\epsilon = 0$) problem does not persist as a circle of fixed points, motion near S_ω remains slow for small positive ϵ. Introducing the variable J

$$
J = I - \omega^2
$$

equation (3.4) is written as

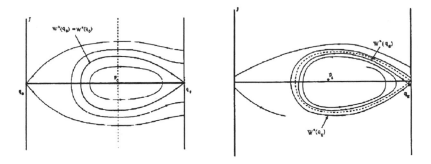

Figure 3.2: Phase Plane Diagram of the ODE in the j θ coordinates

$$
\begin{aligned}
J_t &= -2\epsilon\left[\alpha(J+\omega^2)+\sqrt{J+\omega^2}\,\cos\theta\right] \\
\theta_t &= -2J+\frac{\epsilon}{\sqrt{J+\omega^2}}\sin\theta.
\end{aligned}
\tag{3.8}
$$

In order to describe the slow flow close to this circle S_ω, we rescale the coordinates

$$
\begin{aligned}
\tau &= \nu t \\
J &= \nu j.,
\end{aligned}
\tag{3.9}
$$

where $\nu = \sqrt{\epsilon}$. This rescaling is suggested by the expressions (3.7) for the growth rates σ_p and σ_q. In these scaled coordinates, the equations (3.4) on the plane Π_c take the form of an $O(\nu)$ perturbation of the conservative system

$$
\begin{aligned}
j_\tau &= -2(\alpha\omega^2+\omega\cos\theta) \\
\theta_\tau &= -2j\,.
\end{aligned}
$$

Thus, we see that near the circle S_ω, the slow motion is approximated as a driven pendulum, (see Figure 3.2) with energy

$$
E(j,\theta):=\frac{1}{2}j^2-\omega\left(\sin\theta+\alpha\omega\theta\right).
\tag{3.10}
$$

This completes the information about motion on the invariant plane that we will use in this general overview. More information is needed for the detailed mathematical arguments. This additional information may be found in [37], [40],[52],[53],[44].

3.2   Two Results from the Integrable Theory of NLS

The unperturbed ($\epsilon=0$) NLS equation is a Hamiltonian system on the function space $H^1_{e,p}$,

$$
-iq_t=\frac{\delta}{\delta\bar{q}}\,H,
\tag{3.11}
$$

with the Hamiltonian H given by

$$H = \int_0^{2\pi} \left[q_x \bar{q}_x - (q\bar{q})^2 + 2\omega^2 q\bar{q} \right] dx.$$

It is well known that this is a completely integrable Hamiltonian system, a fact whose verification begins from the *Lax pair*

$$\varphi_x = U^{(\lambda)} \varphi \qquad (3.12)$$

$$\varphi_t = V^{(\lambda)} \varphi, \qquad (3.13)$$

where

$$U^{(\lambda)} := i\lambda\sigma_3 + i \begin{pmatrix} 0 & q \\ \bar{q} & 0 \end{pmatrix};$$

$$V^{(\lambda)} := i\left[2\lambda^2 - (q\bar{q} - \omega^2) \right] \sigma_3 + \begin{pmatrix} 0 & 2i\lambda q + q_x \\ +2i\lambda\bar{q} - \bar{q}_x & 0 \end{pmatrix},$$

and where σ_3 denotes the third Pauli matrix $\sigma_3 := diag(1, -1)$. This over determined system is compatible $(\partial_t \varphi_x = \partial_x \varphi_t)$ if and only if the coefficient q satisfies the NLS equation. Consequently, one can use this linear system to develop representations of solutions $q(x, t)$ of NLS.

Focusing attention upon the "spatial flow" (3.12), the integration of the NLS equation is accomplished through the spectral theory of the differential operator $\hat{L} = \hat{L}(q)$,

$$\hat{L} := -i\sigma_3 \frac{d}{dx} - \begin{pmatrix} 0 & q \\ -\bar{q} & 0 \end{pmatrix},$$

which is viewed as an operator on $L^2(\mathbb{R})$, with dense domain H^1. "Soliton mathematics" provides a complete description of the solutions of this unperturbed integrable system; however, from that complete theory, we will only need two results:

1. Expressions for homoclinic orbits;

2. A natural constant of motion $\mathbf{F}$, together with an expression for the Melnikov function $M_{\mathbf{F}}$.

Remark 3.1 It is natural to inquire the degree to which integrable results are required to carry out "local-global" analysis for pde's. In fact, no integrable information is required, provided one has (i) an expression for the unperturbed homoclinic orbit and (ii) a constant of the motion I which is related to the geometry of the pde in a manner that allows both the definition and geometric use of a Melnikov function M_I. However, integrable theory provides not one, but a rich class of homoclinic orbits. Moreover, as will be shown below, the constant of motion $\mathbf{F}$ from integrable theory is so natural from a geometric perspective that, when available, it should definitely be used to define the Melnikov function. We take the viewpoint that integrable theory sets the standard toward which other more general theories of unperturbed systems can strive.

3.2.1   Integrable Homoclinic Orbits

Fix a solution of the NLS equation, $q(x,t) \in H^1_{e,p}$, for which ν is a periodic (or antiperiodic) eigenvalue of the operator $\hat{L}(q)$, which is purely imaginary and has geometric multiplicity 2. As described in [42], this eigenvalue could be associated with a linear instability of $q(x,t)$, and we assume this is indeed the case. As described in detail in the surveys [52], [49], [50] and in the original articles [16] and [42], Bäcklund (Darboux) transformations at (q, ν) can be used to construct orbits homoclinic to (a torus translate) of $q(x,t)$: Denote two linearly independent solutions of the Lax pair at $(q, \lambda = \nu)$ by $(\vec{\phi}^+, \vec{\phi}^-)$, and the general solution of the linear system at (q, ν) by

$$\vec{\phi}(x,t;\nu;c_+,c_-) = c_+\vec{\phi}^+ + c_-\vec{\phi}^-. \tag{3.14}$$

We use $\vec{\phi}$ to define a transformation matrix [58] G by

$$G = G(\lambda; \nu; \vec{\phi}) := N \begin{pmatrix} \lambda - \nu & 0 \\ 0 & \lambda - \bar{\nu} \end{pmatrix} N^{-1}, \tag{3.15}$$

where

$$N := \begin{bmatrix} \phi_1 & -\bar{\phi}_2 \\ \phi_2 & \bar{\phi}_1 \end{bmatrix} \tag{3.16}$$

Then we define Q and $\vec{\Psi}$ by

$$Q(x,t) := q(x,t) + 2(\nu - \bar{\nu})\,\frac{\phi_1\bar{\phi}_2}{\phi_1\bar{\phi}_1 + \phi_2\bar{\phi}_2} \tag{3.17}$$

and

$$\vec{\Psi}(x,t;\lambda) := G(\lambda;\nu;\vec{\phi})\,\vec{\psi}(x,t;\lambda), \tag{3.18}$$

where $\vec{\psi}$ solves the Lax pair at (q, λ). Formulas (3.17) and (3.18) are the Bäcklund transformations for the potential and eigenfunctions, respectively.

We have the following

Theorem 3.1 *(i) $Q(x,t)$ is an solution of NLS, with spatial period 2π, which is homoclinic to $q(x,t)$ in the sense that $Q(x,t) \longrightarrow q_{\theta_\pm}(x,t)$, exponentially as $\exp(-\sigma_\nu|t|)$ as $t \longrightarrow \pm\infty$. Here $q_{\theta_\pm}$ is a "torus translate" of q, σ_ν is the nonvanishing growth rate associated to the complex double point ν, and explicit formulas can be developed for this growth rate and for the translation parameters $\theta_\pm$.*

(ii) $\vec{\Psi}(x,t;\lambda)$ solves the Lax pair at (Q,λ).

Remark 3.2 This theorem is quite general, constructing homoclinic solutions from a wide class of starting solutions $q(x,t)$. The orbit $Q(x,t)$ is homoclinic to a torus which itself possesses rather complicated spatial and temporal structure, and is not just a fixed point; nevertheless, the homoclinic orbit typically has still more complicated spatial structure than its "target torus". These Bäcklund formulas provide coordinates for the stable and unstable manifolds of the critical tori; thus, they provide explicit representations of the critical level sets which consist in "whiskered tori"[4].

For this work a very important **special example** is the case $q(x,t)$ constant, independent of x :

$$q(x,t) = c \exp\{-i[2(c^2 - \omega^2)t - \gamma]\}.$$

A standard stability calculation shows that $q(x,t)$ linearly unstable, with positive growth rate σ for the linearized "cos x mode" given by

$$\sigma = \sqrt{4c^2 - 1}.$$

In this case, two linearly independent solutions of the Lax pair are given by

$$\begin{pmatrix} \psi_1^{(\pm)} \\ \psi_2^{(\pm)} \end{pmatrix} = \exp\{\pm i[\kappa(\lambda)(x + 2\lambda t)]\} \times$$

$$\begin{pmatrix} c \exp\{-i[2(c^2 - \omega^2)t - \gamma]/2\} \\ (\pm\kappa(\lambda) - \lambda) \exp\{i[2(c^2 - \omega^2)t - \gamma]/2\} \end{pmatrix}, \quad (3.19)$$

where $\kappa(\lambda)$ is given by

$$\kappa(\lambda) = \sqrt{\lambda^2 + c^2}.$$

From these eigenfunctions, one sees that

$$\nu = \frac{i}{2}\sqrt{4c^2 - 1}$$

is a purely imaginary double (antiperiodic) eigenvalue, which is indeed associated to an instability. Hence, the Darboux transformation will produce a homoclinic orbit:

$$q_h^{\pm} = \left\{\frac{\cos 2p - i\sin 2p \tanh\tau \pm \sin p \operatorname{sech}\tau\cos x}{1 \mp \sin p \operatorname{sech}\tau\cos x}\right\} q, \quad (3.20)$$

where

$$\tau = \sigma(t + t_0)$$
$$e^{ip} = \frac{1 + i\sigma}{2c}.$$

Here $\pm$ labels a symmetric pair of homoclinic orbits. (Notice that $-\cos x = \cos(x + \pi)$, which shows that one sign $(+)$ represents an excitation centered at $x = 0$, while the other sign $(-)$ represents an excitation centered at $x = \pi$.)

If we specialize to $c = \omega$, q lies on the circle of fixed points S_ω, and the orbit Q is homoclinic to this circle. Thus, from one point of view, (3.20) provides an explicit representation of a "whiskered circle"; while from another viewpoint, it provides an explicit representation of the unstable manifold

$$W^u(S_\omega) = W^s(S_\omega) = \bigcup_{\gamma, t_0, \pm} q_h^\pm (t; \gamma, t_0, c).$$

When examining the "whiskers" for this example, one notes that while Q is homoclinic to the circle S_ω, it is actually heteroclinic to pairs of fixed points on S_ω, with the individual fixed points separated by a "phase shift" of $-4p$,

$$e^{-4ip} = \left[\frac{1 - i \sqrt{4c^2 - 1}}{2c} \right]^4. \tag{3.21}$$

(Here $c = \omega$, although this formula for the phase shift is valid more generally and actually applies for an open set of values for c.) This phase shift will play an important role in the singular behavior of the persistent homoclinic orbits, as well as in the construction of these persistent orbits as singular deformations of integrable orbits (3.20).

3.2.2   The Constant F

In order to introduce a constant of the motion which geometrically is the most natural for constructing a Melnikov integral, we must describe some spectral theory of the operator $\hat{L}$. This material is described in detail in [42], [52].

Since the coefficient q is a periodic function of x, Floquet theory can be used to characterize this spectrum. Floquet theory begins from the *fundamental matrix* $M = M(x; \lambda; q)$, which is defined as the 2×2 matrix valued solution of the linear problem

$$\hat{L}\psi = \lambda\psi,$$

whose initial value at $x = 0$ is the identity matrix. Next, one introduces the transfer matrix T

$$T(\lambda; q) = M(2\pi; \lambda; q).$$

Then the spectrum $\sigma(\widehat{L})$ can be characterized as the set of all λ for which the 2×2 matrix T has eigenvalues on the unit circle. Since the $\det T = 1$, this is in turn determined by a single scalar function called the *Floquet discriminant*:

$$\Delta \; : \; \mathbb{C} \times H^1_{e,p} \to \mathbb{C} \quad by \quad \Delta(\lambda; q) := tr[T(\lambda; q)].$$

In terms of Δ, the spectrum is then given in terms of the Floquet discriminant by

$$\sigma(\widehat{L}(q)) = \{\lambda \in \mathbb{C} \mid \Delta(\lambda, q) \text{ is real and } -2 \le \Delta \le +2\}.$$

The Floquet discriminant $\Delta(\lambda; q, \bar{q})$,

$$\Delta \; : \; \mathbb{C} \times H^1_{e,p} \times H^1_{e,p} \to \mathbb{C},$$

is entire in λ, q, and $\bar{q}$. Its first variation admits the following representation:

$$\delta\Delta(\lambda; q, \bar{q}) = \int_0^{2\pi} \left[\frac{\delta\Delta}{\delta q(x)} \delta q(x) + \frac{\delta\Delta}{\delta \bar{q}(x)} \delta\bar{q}(x) \right] dx,$$

where

$$\frac{\delta}{\delta q(x)} \Delta(\lambda; q, \bar{q}) = -\frac{i}{2} tr\left[M^{-1}(x) \begin{pmatrix} 0 & 1 \\ 0 & 0 \end{pmatrix} M(x + 2\pi) \right]$$

$$\frac{\delta}{\delta \bar{q}(x)} \Delta(\lambda; q, \bar{q}) = -\frac{i}{2} tr\left[M^{-1}(x) \begin{pmatrix} 0 & 0 \\ 1 & 0 \end{pmatrix} M(x + 2\pi) \right]. \quad (3.22)$$

Here $M(x) = M(x; \lambda^c, q, \bar{q})$ denotes the fundamental matrix. To verify this representation, one [52] proceeds as follows:

$$(\widehat{L} - \lambda) M = 0, \quad M(0) = I$$

$$(\widehat{L} - \lambda) \delta M = \begin{pmatrix} 0 & \delta q \\ -\delta\bar{q} & 0 \end{pmatrix} M, \quad \delta M(0) = 0.$$

One then solves for $\delta M(x)$ by "variation of parameters", which together with the definition,

$$\delta\Delta = tr\, \delta M(2\pi),$$

produces the representation.

The Floquet discriminant generates an infinite family of constants of the motion for the NLS system, as the following proposition states:

Proposition 3.1 *(i) Floquet discriminants Poisson commute:*

$$\{\Delta(\lambda; q, \bar{q}), \Delta(\lambda'; q, \bar{q})\} = 0 \quad \forall \lambda, \lambda',$$

where the Poisson bracket is defined as

$$\{F, G\} = \int_0^{2\pi} i \left(\frac{\delta F}{\delta q} \frac{\delta G}{\delta \bar{q}} - \frac{\delta F}{\delta \bar{q}} \frac{\delta G}{\delta q} \right) dx.$$

(ii) $\Delta(\lambda; q, \bar{q})$ *is a constant of the motion for the NLS equation since its Poisson bracket with the Hamiltonian H vanishes:*

$$\{\Delta(\lambda), H\} = 0 \quad \forall \lambda.$$

Thus, $\Delta(\lambda; \bar{q})$ *generates an infinite family of NLS constants of motion, one for each* λ.

The proof of this proposition, may be found in the survey [52].

We will define the invariant **F** in terms of the Floquet discriminant $\Delta(\lambda; \bar{q})$: Fix a potential $q_o \in H^1_{e,p}$ which has a *purely imaginary* critical point λ^c,

$$\frac{\partial}{\partial \lambda} \Delta(\lambda; q_o)\Big|_{\lambda^c} = 0.$$

Let $N_b = N_b(q_o)$ denote a small neighborhood of q_o, and consider the critical point as a functional on this neighborhood, $\lambda^c = \lambda^c(q)$:

$$\frac{\partial}{\partial \lambda} \Delta(\lambda; q)\Big|_{\lambda^c(q)} = 0; \quad \lambda^c(q_o) = \lambda^c.$$

Definition of the Invariant F: In terms of this purely imaginary critical point, we introduce an important invariant:

$$\boldsymbol{F} : N_b \to \mathbb{R} \text{ given by } \boldsymbol{F} := \Delta(\lambda^c(q); q). \tag{3.23}$$

The functional $\boldsymbol{F}$ is smooth, provided $\frac{d^2}{d\lambda^2} \Delta(\lambda, q) \neq 0$ for all $q \in N_b$. To verify this smoothness, one calculates

$$\frac{\delta \boldsymbol{F}}{\delta q} = \frac{\delta}{\delta q} \Delta(\lambda^c(q); q) = \Delta'(\lambda^c(q); q)\frac{\delta \lambda^c}{\delta q} + \frac{\delta \Delta}{\delta q}$$

$$= \frac{\delta \Delta}{\delta q}(\lambda; q)\Big|_{\lambda = \lambda^c(q)},$$

provided $\lambda^c(q)$ is differentiable. But $\lambda^c(q)$ is smooth, as the following calculation shows:

$$\Delta'(\lambda^c(q); q) = 0$$
$$\Delta''(\lambda^c(q); q)\frac{\delta \lambda^c}{\delta q} + \frac{\delta \Delta'}{\delta q} = 0 .$$

If $\Delta''(\lambda^c(q); q) \neq 0$, one can continue to differentiate to any order.

Critical Points of F: The critical points of the functional $\boldsymbol{F}$ will play important roles. To understand these, it will be useful to develop a formula for the gradient $\boldsymbol{F}'(q)$ in terms of *Bloch functions*.

Remark 3.3 The specific eigenfunctions (3.19) used in the example $q(x) = c$, independent of x, were Bloch functions.

Let $\vec{\psi}^{\pm}(x, \lambda)$ denote Bloch functions, that is eigenfunctions of $\hat{L}$ at $[q, \lambda]$ which are defined (up to normalization) by the transfer condition across one period:

$$\vec{\psi}(x + 2\pi, \lambda) = \rho(\lambda)\vec{\psi}(x, \lambda). \tag{3.24}$$

Here $\rho(\lambda)$ denotes the Floquet multiplier, which is related to the Floquet discriminant by

$$\rho(\lambda) = \frac{1}{2}\left[\Delta(\lambda) + \sqrt{\Delta^2(\lambda) - 4}\right]. \tag{3.25}$$

The functions ρ and $\vec{\psi}$ are well defined on the Riemann surface for $(\lambda, \sqrt{\Delta^2(\lambda) - 4})$, and $\vec{\psi}^{\pm}(x, \lambda)$ denote the values of $\vec{\psi}$ on the two sheets over λ. At branch points (*simple* periodic or antiperiodic points), the two sheets touch and $\vec{\psi}^{\pm}$ become linearly dependent. (This is compatible with the fact that at a simple eigenvalue, the eigenspace is one dimensional.) At *real* multiple points, $\vec{\psi}^{\pm}$ remain linearly independent, while at *complex* multiple points they *may, but need not, become dependent.* These two possibilities at the complex multiple points are a key to this nonselfadjoint spectral problem.

In any case, for fixed λ, these Bloch eigenfunctions can be represented explicitly in terms of the columns of the fundamental matrix $M(x; \lambda) = column\{Y^{(1)}(x; \lambda), Y^{(2)}(x; \lambda)\}$:

$$\vec{\psi}^{\pm}(x; \lambda) = \alpha^{\pm}\{M_{21}(1; \lambda)Y^{(1)}(x; \lambda) + [M_{22}(1; \lambda) - \rho^{\pm}(\lambda)]Y^{(2)}(x; \lambda)\}, \tag{3.26}$$

where $\alpha_{\pm}$ denotes normalization constants.

The gradient of $\mathbf{F}$ admits a beautiful representation in terms of these Bloch functions, which follows from equation (3.22):

Proposition 3.2

$$\text{grad } \mathbf{F}(q, \bar{q}) = i\frac{\sqrt{\Delta(\lambda^c)^2 - 4}}{W[\psi^+, \psi^-]}\left[\begin{array}{c} \psi_2^+(x; \lambda^c)\psi_2^-(x; \lambda^c) \\ \\ -\psi_1^+(x; \lambda^c)\psi_1^-(x; \lambda^c) \end{array}\right]. \tag{3.27}$$

From this representation, one obtains the following

Theorem 3.2 *The potential q is a critical point of the functional $\mathbf{F}$ if and only if $\lambda^c(q)$ is a multiple point with geometric multiplicity 2.*

Remark 3.4 We note that if λ^c were a real multiple point, it's geometric multiplicity is always 2. Thus, a potential q for which $\lambda^c(q)$ is a real double point is a critical point of $\mathbf{F}$. However, since $\lambda^c(q)$ is a complex double point, the geometric multiplicity may be either 2 or 1, and q may or may not be a critical point of $\mathbf{F}$.

Critical Tori: To understand the critical points of F, we first note that, generically, the NLS level sets are tori $T^\infty = S \times S \times S \cdots$ of infinite dimension, with the radius of the jth circle measured by $r_j = |\Delta(\lambda_j^c) \mp 2|$. For q_* for which $\lambda_j^c(q_*)$ is a multiple point $(\Delta(\lambda_j^c) = \pm 2)$, $r_j = 0$ and the jth circle is "pinched off". In this case, the function q_* resides on a torus which is singular in the sense that it has one dimension less than maximal because $r_j = 0$. Consider the singular torus associated with $r = \Delta(\lambda^c) \mp 2 = 0$. Any q_* on this singular torus is a critical point of F.

Whiskered Tori and Unstable Manifolds: The existence of these critical tori is guaranteed by inverse spectral theory which can be used to construct some of them in terms of finite genus theta functions. In reference [42] we have further studied these critical tori, by calculating an explicit representation of the Hessian of F at q_*. If the multiple point $\lambda^c(q_*)$ were real, the Hessian shows that F is either a maximum or a minimum at q_*. However, since $\lambda^c(q_*)$ is a purely imaginary double point, F has a saddle structure at q_*. In the latter case, if F were the NLS Hamiltonian, q_* would be an unstable fixed point; however, F is only in involution with the NLS Hamiltonian H. As such, q_* lies on a singular torus which could be unstable (hyperbolic) under NLS dynamics. If the complex double point $\lambda^c(q_*) = \nu$ is indeed associated to an NLS instability, the Lax pair for the NLS flow can be used in a Bäcklund transformation $q_* \to q_h$, to provide a representation of the whiskers for the critical tori on which q_* resides. In this manner one learns the very important fact that

$$W^{cs}(q_*) = W^{cu}(q_*) = \{q \mid F(q) = \pm 2\},$$

with the $\pm$ sign depending upon whether $\lambda^c(q_*)$ is a periodic or antiperiodic eigenvalue.

Remark 3.5 To understand these singular tori more completely, one [42] must study the entire sequence of invariants

$$F_j(q) = \Delta(\lambda_j(q); q) \qquad \forall j = \cdots, -2, -1, 0, 1, 2, \cdots.$$

However, for our purposes here, we don't need this degree of generality.

4 The Melnikov Function $M_{\mathbf{F}}$

In this section we construct the Melnikov function. First, we obtain a useful formula for the gradient $\mathbf{F}'$, evaluated on a whisker q_h.

4.1 $\mathbf{F}'(q_h)$

Let q_* lie on a critical unstable torus with the instability associated with the purely imaginary double point ν, and with whisker q_h represented by the

Bäcklund formulas (3.17). An explicit representation of the gradient $F'(q_h)$ can also be constructed with Bäcklund formulas (3.17, 3.18).

We begin from equation (3.27) for the grad F,

$$\frac{\delta F}{\delta \vec{q}} = \lim_{\lambda \to \nu} \ i \frac{\sqrt{\Delta^2 - 4}}{W[\Psi^{(+)}, \Psi^{(-)}]} \begin{pmatrix} \Psi_2^{(+)} \Psi_2^{(-)} \\ -\Psi_1^{(+)} \Psi_1^{(-)} \end{pmatrix} \qquad (4.1)$$

where $\vec{\Psi}^{\pm}(x, \lambda)$ are a Floquet basis at $(\vec{Q}_H, \nu)$. In [42] we compute this limit using the Bäcklund formulas. The result is

$$\frac{\delta F}{\delta \vec{q}} = C_\nu \ \frac{c_+ c_- W[\psi^{(+)}, \psi^{(-)}]}{|\vec{\phi}|^4} \begin{pmatrix} \bar{\phi}_1^2 \\ -\bar{\phi}_2^2 \end{pmatrix}, \qquad (4.2)$$

where the constant C_ν is given by

$$C_\nu := i(\nu - \bar{\nu}) \ \sqrt{\Delta(\nu) \Delta''(\nu)}$$

Remark 4.1 Since $\vec{\phi} = c_+ \vec{\psi}^{(+)} + c_- \vec{\psi}^{(-)}$, one sees explicitly from this formula that $\frac{\delta F}{\delta \vec{q}}|_{(q_h, \nu)} \to 0$ as $c_+/c_- \to 0, \infty$. Also, since the eigenfunctions $\vec{\psi}^{(+)}$ and $\vec{\psi}^{(-)}$ at the complex double point ν grow or decay exponentially,

$$\vec{\psi}^{(\pm)} \approx \exp(\pm \sigma_\nu t), \quad t \to \infty,$$

the formula also shows explicitly that grad $F|_{(q_h, \nu)} \to 0$ as $t \to \infty$. (The vector field grad F must vanish because, in these limits, the point $\vec{q}_h$ on the whisker tends to a critical function of F.)

In the case of $q_* = c \ \exp\{-i [2(c^2 - \omega^2)t - \gamma]\} = c \, e^{i\theta}$, explicit formulas can be used to produce the following representation of $F'(q_h)$, which is valid when acting upon even functions of x:

$$\frac{\delta F}{\delta q} = 2\pi \ \sin^2 p \ \mathrm{sech}^2 \tau \ \frac{[(\mp \sin p \cosh \tau \pm i \cos p \sinh \tau) \cos x + 1]}{[1 \mp \sin p \ \mathrm{sech} \ \tau \cos x]^2} \ c \, e^{i\theta},$$

$$\frac{\delta F}{\delta \bar{q}} = \overline{\frac{\delta F}{\delta q}}, \qquad (4.3)$$

where $\tau = \sigma(t - t_0)$, $\tan p = \sigma$, and $\sigma = \sqrt{4c^2 - 1}$. From this representation, we see explicitly that

$$F'(q_h(t)) \neq 0,$$
$$\lim_{t \to \pm\infty} F'(q_h(t)) \to 0,$$

at an exponential $[\exp(-\sigma|t|)]$ rate of approach. Thus, q_* is indeed critical, while the whisker $q_h(t)$ is not.

4.2 The Melnikov Integral

In this subsection we use these Bäcklund formulas for gradF to develop an explicit representation of the Melnikov integral. First, we write the perturbed NLS equation in the form

$$q_t = iH'(q) + \epsilon G(q) \tag{4.4}$$

where $H'(q) = -q_{xx} - 2[q\bar{q} - \omega^2]q$ and where $G(q)$ denotes the perturbation. Usually we will consider a particular example of this perturbation:

$$G(q) = -\alpha q - \beta \hat{B} q - 1, \tag{4.5}$$

where the operator $\hat{B}$ has a symbol given by:

$$b(k) = \begin{cases} k^2 & k < \kappa \\ 0 & k \geq \kappa \end{cases}$$

Definition 4.1 In this pde setting, the Melnikov function is defined by

$$M_F = \int_{-\infty}^{\infty} \langle F'[q_h(t)], G[q_h(t)] \rangle \, dt. \tag{4.6}$$

Next we specialize the orbit $q_h(t)$ to one homoclinic to a circle of fixed points: Recall that, on the plane of constants Π_c, the unperturbed orbits take the form

$$q = r_b \, e^{-i[2(r_b^2 - \omega^2)t - \theta_b]}$$

thus, $r_b = \omega$ represents the circle of fixed points S_ω. Orbits homoclinic to S_ω are denoted $q_h = q_\omega$ and are given by

$$q_\omega(t) = \left[\frac{\cos 2p - i \sin 2p \tanh \tau + \sin p \operatorname{sech} \tau \cos x}{1 - \sin p \operatorname{sech} \tau \cos x} \right] \times$$
$$\omega \exp i[\theta_b - 2p] \tag{4.7}$$

where

$$\tan p = \sqrt{4\omega^2 - 1}$$
$$\tau = (\tan p)(t + t_o).$$

While the orbit q_ω approaches the circle S_ω as $t \to \pm\infty$, it approaches the fixed point $\omega \exp i\theta_b$ as $t \to -\infty$ and, as $t \to +\infty$,

$$q_\omega(t) \longrightarrow \omega \, e^{i(\theta_b - 4p)}.$$

Thus, the (heteroclinic) orbit experiences a *phase shift* of $-4p$

$$e^{-4ip} = \left[\frac{1 - i\sqrt{4\omega^2 - 1}}{2\omega} \right]^4.$$

With these ingredients, we assemble the final expression for the Melnikov integral:

Proposition 4.1 *For the specific perturbation (4.5) and homoclinic orbit q_ω, equation (4.7), the Melnikov integral takes the form*

$$M_F = M_F\left(\alpha,\ \beta,\ \theta_b\right) \ = \ \int_{-\infty}^{\infty} \langle \mathbf{F}'(q_\omega(t)),\ G(q_\omega(t)) \rangle\, dt \qquad (4.8)$$

$$= \ [\alpha M_\alpha + \beta M_\beta + M(\theta_b)],$$

where

$$M_\alpha \ = \ \int_{-\infty}^{\infty} \langle \mathbf{F}'(q_\omega(t)),\ q_\omega(t) \rangle\, dt$$

$$M_\beta \ = \ \int_{-\infty}^{\infty} \langle \mathbf{F}'(q_\omega(t)),\ \widehat{B} q_\omega(t) \rangle\, dt$$

$$M(\theta_b) \ = \ \int_{-\infty}^{\infty} \langle \mathbf{F}'(q_\omega(t)),\ 1 \rangle\, dt\ .$$

More explicitly,

$$M_\alpha \ = \ \int_{-\infty}^{\infty} d\tau \int_0^{2\pi} dx\ \frac{4\pi\,\omega^2 \sin^2 p_o\ sech\ \tau}{\sigma\, A^3}\ \times$$
$$[sech\ \tau + \sin p_o \tanh^2 \tau \cos x - \sin^2 p_o\ sech\ \tau(2 + \cos^2 x)$$
$$+ 2\sin^3 p_o\ sech\ ^3\tau \cos x]$$

$$M_\beta \ = \ \int_{-\infty}^{\infty} d\tau \int_0^{2\pi} dx\ \frac{4\pi\omega^2 \sin^2 p_o\ sech\ \tau}{\sigma A^5}\ \times$$
$$\left[\sin p_o\ sech\ \tau \cos x - \sin^2 p_o\ sech\ ^2\tau(1 + \sin^2 x)\right] \times$$
$$[2\ sech\ \tau - \sin p_o\ sech\ ^2\tau \cos x - 2\sin^2 p_o\ sech\ \tau$$
$$+ 2\sin^3 p_o\ sech\ ^2\tau \cos x] + O(\sin^{\kappa - 2}\ p_o)$$

$$M(\theta_b) \ = \ \cos(\theta_b - 2p_o) \int_{-\infty}^{\infty} d\tau \int_0^{2\pi} dx\ \frac{4\pi\omega \sin^2 p_o\ sech\ \tau}{\sigma A^2}\ \times$$
$$[\ sech\ \tau - \sin p_o \cos x]$$

where $p_o = \tan^{-1} \sqrt{\omega^2 - 1}$, $A = [1 - \sin p_o\ sech\ \tau \cos x]$ and where the $O(\sin^{\kappa-2}\ p_o)$ term in the M_β equation is due to the fact that we used $-\partial_x^2$ instead of $\widehat{B}$ in our computation. Thus, the final expression for the Melnikov integral is of the form

$$M(\alpha,\ \beta,\ \theta_b) = \alpha M_\alpha + \beta M_\beta + M_1 \cos(\theta_b - 2p_o), \qquad (4.9)$$

where M_α, M_β, and M_1 are functions of ω only.

Clearly, for small α and β, this Melnikov function has simple zeros as a function of θ_b. At issue, of course, is the geometric meaning of these zeros.

5   Persistent Homoclinic Orbits for PNLS

Simple zeros of the Melnikov function (4.9) enable us to prove the following persistence theorem:

Theorem 5.1 *The perturbed NLS equation (3.1) possesses a symmetric pair of orbits which are homoclinic to the saddle fixed point Q, provided the parameters lie on a codimension 1 set in parameter space which is approximately described by*

$$\alpha \simeq E(\omega)\beta.$$

Remark 5.1 *Calculations in the proof of this theorem provide formulas which describe, approximately for small ϵ, characteristic properties of these homoclinic orbits such as the constant $E(\omega)$ and an important "take-off" point.*

These two homoclinic orbits differ by the location of a transient spatial structure – a solitary wave which is located either at the center $(x = 0)$ or the edge $(x = \pi)$ of the periodic box. As such, this theorem provides a key necessary step toward a symbol dynamics for the pde.

The proof of this theorem is organized with "local-global" analysis. It involves normal forms for the perturbed NLS equation [47], invariant manifold theory for NLS and geometric singular perturbation theory, combined with integrable theory and Melnikov analysis. It is described in mathematical detail in [44]. Here we outline the steps in the proof, emphasizing the intuition behind each step.

The steps in the proof are organized in the subsections which follow:

1. **Preliminary set up** including (i) motion on Π_c, (ii) coordinates near Π_c, (iii) linear stability and time scales, and (iv) a normal form.

2. **Local arguments** including (i) persistent invariant manifolds, (ii) fiber representations, and (iii) the height of $W^s(Q)$.

3. **Global arguments** including (i) integrable theory, (ii) the first Melnikov measurement, and (iii) the second Melnikov measurement.

When working through this overview of the proof, keep in mind that, throughout the proof, control is obtained in one of two ways - - either the orbits are (i) close to the invariant plane Π_c, or they are (ii) close to the integrable case. Also keep in mind that the arguments will be a form of "shooting", where the goal will be to force an orbit to "hit" target manifolds of high dimension, but in an infinite dimensional space. To make these manifolds easy targets, we make them very large in the sense that they will be codimension 1. In any case, the overview of the proof follows:

5.1   Preliminary Set Up

There are four preliminary steps before the proof really begins. These are
(i) a study of motion on the invariant plane Π_c: (ii) the definition of local
coordinates near Π_c; (iii) the use of these local coordinates for a linear stability analysis which identifies two time scales which result from a gap in the
linearized spectrum; and (iv) the development of a near identity transformation which places the local coordinates in a "normal form". Each of these
preliminary "set-up steps", while tedious, are essential to the proof.

5.1.1   Motion on Π_c

Motion on the plane has already been described in Section 3.1, and is summarized by the figures 3.1 and 3.2. In particular, we remind the reader of
the two fixed points near the unit circle – the saddle Q and the sink P, with
their growth rates on the plane Π_c given by (3.7). As described in Section
3.1, when restricted to a "resonance band" near the circle S_ω, the motion will
be slow. In particular, in the resonance band, control is achieved by the energy of the driven pendulum, equation (3.10). In particular, with that energy
we can obtain and control global representations of the stable and unstable
manifolds, restricted to the plane Π_c, of the saddle point Q – $W^s(Q)$ and
$W^u(Q)$.

5.1.2   Coordinates Near Π_c

The next preliminary step is to introduce coordinates which will be useful
locally near the plane Π_c, or more specifically, near the circle S_ω. This entails
introducing coordinates (J, θ, f), where θ is the angle on S_ω, J is a measure
of distance from S_ω on the plane Π_c, and f is in the orthogonal compliment
of Π_c. These coordinate changes,

$$q \to (\rho, \theta, f) \to (J, \theta, f),$$

are arrived at in the following manner:

First, we introduce coordinates (ρ, θ, f) given by

$$q := [\rho(t) + f(x, t)] \exp i\theta(t), \tag{5.1}$$

where ρ and θ are polar coordinates on the plane Π_c, and $f \in \Pi_c^\perp$, i.e. f has
spatial mean zero.

The L^2 norm is a constant of motion for the unperturbed ($\epsilon = 0$) flow;
therefore the perturbed equations can be somewhat simplified by replacing
the variable ρ with the variable I :

$$I := \frac{1}{2\pi} \int_0^{2\pi} q\bar{q}dx \; = \; \frac{1}{2\pi} \int_0^{2\pi} [\rho^2 + f\bar{f}] \, dx \; = \rho^2 + \langle f\bar{f} \rangle. \tag{5.2}$$

Finally since we are working in a neighborhood of the circle of fixed points S_ω which corresponds to $I = \omega^2$, it will be convenient to introduce the variable J defined by

$$J = I - \omega^2 \tag{5.3}$$

In terms of these variables, the perturbed NLS equation takes the form

$$
\begin{aligned}
J_t &= -2\epsilon \left[\alpha(J + \omega^2) + \sqrt{J + \omega^2} \cos \theta\right] + \mathcal{E}_1(J, \theta, f; \epsilon) \\
\theta_t &= -2J + \epsilon(J + \omega^2)^{-1/2} \sin \theta + \mathcal{E}_2(J, \theta, f; \epsilon) \\
if_t &= L_\epsilon f + V_\epsilon f + \omega Q_3(f) + \mathcal{E}_3(J, \theta, f; \epsilon),
\end{aligned}
\tag{5.4}
$$

where

$$
\begin{aligned}
L_\epsilon f &= f_{xx} + i\epsilon \hat{D} f + 2\omega^2 (f + \bar{f}) \\
V_\epsilon f &= 2J (f + \bar{f}) + \epsilon \frac{\sin \theta}{\sqrt{J + \omega^2}} f ,
\end{aligned}
$$

and where $\mathcal{E}_k$ are 2π-periodic functions in θ, of order

$$
\begin{aligned}
\mathcal{E}_1(J, \theta, f; \epsilon) &= O(\epsilon f^2) \\
\mathcal{E}_2(J, \theta, f; \epsilon) &= O(f^2) \\
\mathcal{E}_3(J, \theta, f; \epsilon) &= O(J f^2 + f^3)
\end{aligned}
$$

for small J and f.

5.1.3 Linear Stability and Time Scales

The linearization of equations (5.4) is a perturbation of the linear system

$$
\begin{aligned}
J_t &= 0 \\
\theta_t &= -2J \\
if_t &= L_\epsilon f.
\end{aligned}
\tag{5.5}
$$

This is a constant coefficient linear system whose growth rates are easy to compute. First, it is convenient to introduce real coordinates:

$$f = u_R + iu_I = (u_R, u_I)^T := u,$$

in terms of which the linear system takes the form

$$
\begin{aligned}
J_t &= 0 \\
\theta_t &= -2J \\
u_t &= L_\epsilon u .
\end{aligned}
\tag{5.6}
$$

Thus, we must consider the eigenvalue problem

$$L_\epsilon e = \lambda e$$

for the eigen-pairs $\{e(x), \lambda\}$. Using Fourier expansions, one finds for $j = 1$:

$$
\begin{aligned}
e_{s,u} &= \frac{1}{2\sqrt{\pi}\,\omega}\,(1, \mp\sigma)^T \cos x, \\
|\sigma^\epsilon_{s,u}| &= |\pm\sigma - O(\epsilon)|,
\end{aligned}
\tag{5.7}
$$

where

$$\sigma = \sqrt{4\omega^2 - 1}. \tag{5.8}$$

For $j \geq 2$, the eigenvalues come in complex conjugate pairs, with negative real part:

$$\lambda_j = i\Omega_j - O(\epsilon),$$

where

$$\Omega_j = j\sqrt{j^2 - 4\omega^2} > 0. \tag{5.9}$$

Thus, one identifies a very important *gap in the spectrum*. In the "$\cos x$" direction in function space, the growth and decay rates are $O(\epsilon^0)$, while all other growth and contraction is slow. Specifically, on the plane Π_c, the expansion and contraction are $O(\sqrt{\epsilon})$. All other modes $\{\cos jx, j \geq 2\}$ decay at the slow $O(\epsilon)$ rate. This gap in the growth rates will play a central role throughout the analysis.

In terms of the eigen-basis, the mean zero function u may be written as

$$u(x) = v_u e_u(x) + v_s e_s(x) + v_o(x) \tag{5.10}$$

where v_u and v_s are real scalars, and where $v_o(x) \in [\,span\,\{\Pi_c, e_u, e_s\}]^\perp$. In terms of these variables the linear equations (5.6) split into

$$
\begin{aligned}
J_t &= 0 \\
\theta_t &= -2J \\
v_{u,t} &= \sigma^\epsilon_u v_u \\
v_{s,t} &= -\sigma^\epsilon_s v_s \\
v_{o,t} &= L_\epsilon v_o
\end{aligned}
\tag{5.11}
$$

Thus, we explicitly see that, for $\epsilon = 0$ the linear equations have one unstable $(\cos x)$ direction e_u, one stable $(\cos x)$ direction e_s, and an infinite number of center directions (J, θ, v_o). Combining these center variables as $v_c = (J, \theta, v_o)^T$, the linear equations can be written as

$$
\begin{aligned}
v_{u,t} &= \sigma^\epsilon_u v_u \\
v_{s,t} &= -\sigma^\epsilon_s v_s \\
v_{c,t} &= A v_c
\end{aligned}
\tag{5.12}
$$

where A is defined from equations (5.11).

In a δ neighborhood of the circle of fixed points S_ω, the nonlinear equation can be viewed as a perturbation of the linear equation (5.12). Under the flow of this linear equation, and for $\epsilon = 0$, S_ω has one dimensional stable and unstable manifolds, together with a codimension 2 center manifold. We focus our attention on the center manifold $E^c(S_\omega)$, together with the center-stable $E^{cs}(S_\omega)$ and center-unstable $E^{cu}(S_\omega)$ manifolds:

$$
\begin{aligned}
E^{cs}(S_\omega) &= span\{e_u\}^\perp \\
E^{cu}(S_\omega) &= span\{e_s\}^\perp \\
E^c(S_\omega) &= span\{e_u, e_s\}^\perp.
\end{aligned}
$$

We will be interested in the persistence of these manifolds, under both the nonlinearity and the perturbations.

5.1.4 A Normal Form of the Equations

The stable manifold of the saddle Q, $W^s(Q)$, is a codimension 2 manifold for which we will need global control. On the plane Π_c, this control is achieved with the pendulum energy (3.10). Off the plane, we need to estimate its "height" in order to complete the last step in the argument call the "Second Measurement".

In order to illustrate the issue, consider the following model problem:

$$
f_t = -if - \epsilon f + |f|f.
$$

The origin is a stable fixed point for this equation. How large is it's stable manifold, $W^s(0)$? A quick calculation shows that

$$
f = \epsilon \exp(-it)
$$

is a nearby periodic solution, and that $W^s(0)$ cannot be any larger than $O(\epsilon)$.

Next, consider the similar model problem

$$
f_t = -if - \epsilon f + f^2,
$$

which is also quadratically nonlinear. The former example, together with the following scaling calculation, both indicate that if $|f| \geq O(\epsilon)$, the nonlinear term *might* effect the decay:

$$
\begin{aligned}
f &= \epsilon \tilde{f} \\
\tilde{f}_t &= -i\tilde{f} - \epsilon(\tilde{f} + \tilde{f}^2).
\end{aligned}
$$

On the other hand, if the nonlinearity was of higher power, the effect would be diminished:

$$\begin{aligned} g_t &= -ig - \epsilon f + f^3, \\ g &= \epsilon^\mu \tilde{g}, \\ \tilde{g}_t &= -i\tilde{g} - \epsilon(\tilde{g} + \epsilon^{2\mu-1}\tilde{f}^3). \end{aligned}$$

One sees that, for cubic nonlinearities, the decay will not be effected for $\mu > 1/2$. That is, g can be $O(\epsilon^{\frac{1}{2}+\delta})$ without effecting the decay.

For a quadratic nonlinearity, one is simply unsure of its effect without more detailed information about its specific nature. If it really doesn't effect the decay, we should be able to transform it away. Such a transformation approach is the philosophy behind "normal forms". In the second example, we introduce the "near identity transformation"

$$f = g + cf^2.$$

If we choose $c = -(i + \epsilon)^{-1}$, the equation gets mapped to

$$g_t = -ig - \epsilon g + O(g^3),$$

and we see that $|g| = O(\epsilon^{\frac{1}{2}+\delta})$ without effecting the decay.

In this example, the transformation is allowed because, in the definition of the constant c,

$$c = -\frac{1}{(i + \epsilon)},$$

the denominator never vanishes for small ϵ. In more general examples, vanishing or small denominators can provide obstacles to the existence of the transformation.

In our application to the perturbed NLS equation, we return to equation (5.4) and focus upon the "f equation",

$$if_t = L_\epsilon f + V_\epsilon f + \omega Q_3(f) + \mathcal{E}_3(J, \theta, f; \epsilon).$$

Here we consider the possibility of transforming away the quadratic nonlinearity $Q_3(f)$ with a near identity transformation of the form

$$\begin{aligned} g &= f + K(f, f) \qquad\qquad\qquad (5.13) \\ K(f, h) &:= K_{11}(f, h) + K_{1\bar{1}}(f, \bar{h}) + K_{\bar{1}1}(\bar{f}, h) + K_{\bar{1}\bar{1}}(\bar{f}, \bar{h}), \end{aligned}$$

where K are bounded bilinear maps $K : \Pi_c^\perp \times \Pi_c^\perp \longrightarrow \Pi_c^\perp$

$$\begin{aligned} K_{11}(f, h) &= \int\!\!\int K_{11}(x - y_1, x - y_2)\, f(y_1)\, h(y_2)\, dy_1 dy_2 \\ K_{1\bar{1}}(f, \bar{h}) &= \int\!\!\int K_{1\bar{1}}(x - y_1, x - y_2)\, f(y_1)\, \bar{h}(y_2)\, dy_1 dy_2, \end{aligned}$$

with similar expressions for $K_{\bar{1}1}$ and $K_{\bar{1}\bar{1}}$. In terms of Fourier expansions these bilinear maps can be written as

$$K_{11}(f,h) \;=\; \sum_{k+\ell \neq 0} \hat{K}_{11}(k,\ell)\,\hat{f}(k)\,\hat{h}(\ell)\,e^{i(k+\ell)x}$$

$$K_{1\bar{1}}(f,\bar{h}) \;=\; \sum_{k+\ell \neq 0} \hat{K}_{1\bar{1}}(k,\ell)\,\hat{f}(k)\,\bar{\hat{h}}(-\ell)\,e^{i(k+\ell)x}.$$

In this pde setting, vanishing denominators which would be obstructions to the transformation take the form of "resonances" for the linear equation

$$i\partial_t f = f_{xx} + 2\omega^2\,(f + \bar{f});$$

that is, simultaneous zeros of the pair of equations

$$\begin{aligned} k_1 + k_2 &= k_3 \\ \Omega_1 \pm \Omega_2 &= \pm\Omega_3\,, \end{aligned}$$

where the dispersion relation $\Omega_j = \Omega(k_j)$ is given by equation (5.9). However, some algebra shows that, in the space $\Pi_c^\perp$ (where $k \neq 0$), these equations have no solutions; that is, there are "no quadratic resonances".

Thus, a near identity transformation exists which transforms away the quadratic nonlinearity. In fact, this transformation can be found explicitly:

$$\begin{aligned} \hat{K}_{11}(k,\ell) &= -\frac{\omega}{k\ell} \\ \hat{K}_{1\bar{1}}(k,\ell) &= -\frac{\omega}{\ell(k+\ell)} \\ \hat{K}_{\bar{1}1}(k,\ell) &= -\frac{\omega}{k(k+\ell)} \\ \hat{K}_{\bar{1}\bar{1}}(k,\ell) &= 0\,. \end{aligned}$$

Note that $k \neq 0$, $\ell \neq 0$, and $\ell + k \neq 0$ since we are in the space $\Pi_c^\perp$. Moreover since

$$\sum |\hat{K}_{ab}(\ell,k)|^2 < \infty,$$

we have $K \in L^2(S^1 \times S^1)$, and this implies that K is a bounded bilinear map on $\Pi_c^\perp$

$$\|K(f,f)\|_{H^1} \leq C\|f\|_{H^1}^2$$

for all $f \in \Pi_c^\perp$. Finally we can invert the equation

$$g = f + K(f,f)$$

for f in a neighborhood of the zero to obtain

$$f = g + \mathcal{K}(g)$$

where $\mathcal{K}$ is of order $O(g^2)$.

Applying this near identity map places the full equations in their final form. The transformation

$$g = f + K(f, f) \, .$$

will eliminate the Q_3 term, but it will introduce new quadratic terms in the equation for g that have ϵ coefficients such as

$$\epsilon \hat{D} K(f, f),$$

due to the presence of terms like $\epsilon \hat{D} f$ in the f equation. Therefore, using (J, θ, g) as coordinates, the equations near S_ω are written as

$$
\begin{aligned}
J_t &= -2\epsilon \left[\alpha(J + \omega^2) + \sqrt{J + \omega^2} \cos \theta \right] + N_1(J, \theta, g; \epsilon) \\
\theta_t &= -2J + \epsilon(J + \omega^2)^{-1/2} \sin \theta + N_2(J, \theta, g; \epsilon) \qquad (5.14) \\
ig_t &= L_\epsilon g + W_\epsilon g + N_3(J, \theta, g; \epsilon) \, .
\end{aligned}
$$

where, for g in a neighborhood of zero, we have

$$
\begin{aligned}
N_1(J, \theta, g; \epsilon) &= O(\epsilon g^2) \\
N_2(J, \theta, g; \epsilon) &= O(g^2) \\
N_3(J, \theta, g; \epsilon) &= O(Jg^2 + \epsilon g^2 + g^3)
\end{aligned}
$$

Finally, since we will be working with invariant *real* manifolds in a neighborhood of the circle of fixed points S_ω, it will be convenient to introduce a *real* coordinate system:

$$u = (Re(g), Im(g))^T.$$

In terms of these variables the above equation takes its final form:

$$
\begin{aligned}
J_t &= -2\epsilon \left[\alpha(J + \omega^2) + \sqrt{J + \omega^2} \cos \theta \right] + N_1(J, \theta, u; \epsilon) \\
\theta_t &= -2J + \epsilon(J + \omega^2)^{-1/2} \sin \theta + N_2(J, \theta, u; \epsilon) \qquad (5.15) \\
u_t &= L_\epsilon u + V_\epsilon u + N_3(J, \theta, u; \epsilon).
\end{aligned}
$$

Here N_3 is interpreted as a two vector,

$$
\begin{aligned}
L_\epsilon &= \mathcal{J} \partial_x^2 - 4\omega^2 \mathcal{S} + \epsilon \hat{D} \\
V_\epsilon &= -4J\mathcal{S} + \epsilon \frac{\sin \theta}{\sqrt{J + \omega^2}} \mathcal{J},
\end{aligned}
$$

where

$$
\mathcal{J} = \begin{pmatrix} 0 & 1 \\ -1 & 0 \end{pmatrix} \qquad \mathcal{S} = \begin{pmatrix} 0 & 0 \\ 1 & 0 \end{pmatrix} .
$$

5.2   Local Arguments (Near Π_c)

In this section we describe the local parts of the argument. We will be interested in the fate of the manifolds

$$
\begin{aligned}
E^{cs}(S_\omega) &= span\{e_u\}^\perp \\
E^{cu}(S_\omega) &= span\{e_s\}^\perp \\
E^{c}(S_\omega) &= span\{e_u, e_s\}^\perp
\end{aligned}
$$

under nonlinearity and perturbation. It will turn out that these manifolds persist and become invariant manifolds for a "cut-off flow":

$$
\begin{aligned}
E^{cs}(S_\omega) &\to W_\epsilon^{cs} \\
E^{cu}(S_\omega) &\to W_\epsilon^{cu} \\
E^{c}(S_\omega) &\to W_\epsilon^{c} := \mathcal{M}.
\end{aligned}
$$

From the linear spaces, it is clear that these persistant manifolds will be cod 1, cod 1, and cod 2, respectively.

In order to establish the existence of these local invariant manifolds, as well as to coordinatize them in a manner useful for singular perturbation calculations, we begin by introducing a "cut-off flow": We start by fixing $\delta > 0$, and introducing a localization function ψ_δ,

$$
\psi_\delta : \mathbb{R} \to \mathbb{R}, \quad \psi_\delta(s) = \psi(s/\delta),
$$

where ψ is C^∞ and satisfies

$$
\psi(s) = \left\{ \begin{array}{ll} 1, & |s| \leq 1 \\ 0, & |s| \geq 2 \end{array} \right\}.
$$

Then we localize the full equations

$$
\begin{aligned}
J_t &= -2\epsilon[\alpha(J + \omega^2) + \sqrt{J + \omega^2}\cos\theta]\psi_\delta(J) + N_1(J, \theta, u; \epsilon)\psi_\delta(J, u) \\
\theta_t &= -2J + \epsilon[(J + \omega^2)^{-1/2}\sin\theta]\psi_\delta(J) + N_2(J, \theta, u; \epsilon)\psi_\delta(J, u) \\
u_t &= L_\epsilon u + [V_\epsilon u + N_3(J, \theta, u; \epsilon)]\psi_\delta(J, u),
\end{aligned} \tag{5.16}
$$

where

$$
\psi_\delta(J, u) = \psi_\delta(J)\,\psi_\delta(\|u\|_{H_1}).
$$

Note that we do not cut-off the variable θ; thus, the function $\psi_\delta(J, f)$ has the effect of cutting off the right hand sides whenever the phase point lies outside a neighborhood U_δ of the circle S_ω. Because of this localization, the right-hand side of equation (5.16) has a global Lipshitz constant of order $O(\epsilon + \delta)$.

This localization has the effect of keeping the flow unchanged in a δ-neighborhood of S_ω, while changing the nonlinear equations (5.15) to become globally a δ-perturbation of a linear constant coefficient system. Using $v := (v_u, v_s, v_c)^T$ as variables and the operator A defined in equation (5.12), we can write equations (5.16) as

$$
\begin{aligned}
v_{u,t} &= \sigma_u^\epsilon \, v_u + R_u^\delta \, (v \, ; \, \epsilon) \\
v_{s,t} &= -\sigma_s^\epsilon \, v_s + R_s^\delta \, (v \, ; \, \epsilon) \\
v_{c,t} &= A v_c + R_c^\delta (v \, ; \, \epsilon),
\end{aligned}
\tag{5.17}
$$

where $R^\delta (v \, ; \, \epsilon)$ and its first derivatives are of order $O(\delta + \epsilon)$.

5.2.1   Persistent Invariant Manifolds

We will show that the localized equation (5.17) has C^k $(k > 3)$ invariant manifolds that deform smoothly from E^{cs}, E^{cu}, and E^c. In turn, for the original equations, these manifolds will be locally invariant in a δ neighborhood of S_ω.

First we define *local invariance*:

Definition 5.1 *Given an open set O, a manifold $\mathcal{M}$ is called "locally invariant (in O)" under a flow F^t if, for every open interval I such that $F^I(q) \subset O$, $F^{t_*}(q) \in \mathcal{M}$ for one $t_* \in I$ $\implies$ $F^t(q) \in \mathcal{M} \, \forall t \in I$.*

Then we state the *persistence theorem*:

Theorem 5.2 *There exist a δ neighborhood U_δ of S_ω, an $\epsilon_0(\delta) > 0$, and an integer k such that $\forall \epsilon \in [0, \epsilon_0)$, equation (5.15) has a locally invariant (in U_δ) manifold of codimension 1,*

$$
W_\epsilon^{cs} = \left\{ v \in H^1 \, \middle| \, v_u = h_u(v_s, v_c; \, \epsilon) \right\},
\tag{5.18}
$$

where the function h_u is C^k in all of its arguments, and 2π-periodic in θ. Moreover for $\epsilon = 0$, W_0^{cs} intersects E^{cs} tangentially along S_ω.

Similarly, we have a locally invariant manifold given by

$$
W_\epsilon^{cu} = \left\{ v \in H^1 \, \middle| \, v_s = h_s(v_u, v_c; \, \epsilon) \right\}
\tag{5.19}
$$

where the function h_s is C^k in all of its arguments, and 2π-periodic in θ. Moreover for $\epsilon = 0$, W_0^{cu} intersects E^{cu} tangentially along S_ω.

The existence of a codimension 2 "slow manifold" $\mathcal{M}_\epsilon$ is then given by the following:

Corollary 5.1 *Let $\mathcal{M}_\epsilon$ denote the intersection*

$$\mathcal{M}_\epsilon = W_\epsilon^{cs} \bigcap W_\epsilon^{cu}.$$

Then $\mathcal{M}_\epsilon$ is a locally invariant (in U_δ) manifold of codimension 2,

$$\mathcal{M}_\epsilon = \left\{ v \in H^1 \,\middle|\, v_u = h_u^c(v_c; \epsilon), \; v_s = h_s^c(v_c; \epsilon) \right\} \tag{5.20}$$

where the functions $h_{u,s}^c$ are C^k in their arguments, and 2π-periodic in θ. Moreover for $\epsilon = 0$, $\mathcal{M}$ intersects E^c tangentially along S_ω.

There are essentially two methods to prove such a persistence theorem – the "graph transform method" of Hadamard [24] and the "integral equation approach" of Perron [57]. The graph transform approach for the NLS pde is developed in [45]. (This approach is developed in generality for dissipative pde's in [7]; the integral equation approach is developed for pde's in [6].) Here, since our NLS equation may be viewed as a perturbation of a constant coefficient linear equation, we prefer the integral equation approach. While it is less geometric, it is more explicit and more concrete than the graph transform method.

Remark 5.2 In this near conservative setting, the manifolds $\mathcal{M}$ and W_ϵ are naturally infinite dimensional. (In addition, their small codimension is important for our "shooting" arguments.) These manifolds are not compact; hence, dissipative methods such as those in [7] do not apply.

First we rewrite equation (5.17) in integral form:

$$
\begin{aligned}
v_u(t) &= \exp\left[\sigma_u^\epsilon(t - t_u)\right] v_u(t_u) + \int_{t_u}^t \exp\left[\sigma_u^\epsilon(t - s)\right] R_u^\delta(v(s); \epsilon)\, ds \\
v_s(t) &= \exp\left[-\sigma_s^\epsilon(t - t_s)\right] v_s(t_s) + \int_{t_s}^t \exp\left[-\sigma_s^\epsilon(t - s)\right] R_s^\delta(v(s); \epsilon)\, ds \\
v_c(t) &= \exp\left[At\right] v_c(0) + \int_0^t \exp\left[A(t - s)\right] R_c^\delta(v(s); \epsilon)\, ds. \tag{5.21}
\end{aligned}
$$

Most solutions of these equations will grow as $\exp \sigma_u^\epsilon t$ as $t \to +\infty$. However, solutions in the "center stable manifold" W_ϵ^{cs} cannot grow this rapidly. In fact, the gap in the growth rates allow one to characterize the invariant manifold W_ϵ^{cs} by

$$W_\epsilon^{cs} = \left\{ v \in H^1 \,\middle|\, \sup_{t \geq 0} \left(\exp\left[\frac{-\sigma}{n_0} t\right] \|F^t(v; \epsilon)\|_{H_1} \right) < \infty \right\}, \tag{5.22}$$

where $F^t(v; \epsilon)$ is the cut-off flow of equations (5.17) and n_0 is a fixed positive integer. Thus, for $v \in W_\epsilon^{cs}$, we have

$$\exp[-\sigma_u t_u] |v_u(t_u)| \longrightarrow 0 \qquad \text{as} \quad t_u \to +\infty.$$

Therefore for solutions on W_ϵ^{cs} the integral equation can be written as

$$v_u(t) = \int_{+\infty}^t \exp\left[\sigma_u^\epsilon(t-s)\right] R_u^\delta(v(s);\epsilon)\,ds$$

$$v_s(t) = \exp\left[-\sigma_s^\epsilon t\right]v_s + \int_0^t \exp\left[-\sigma_s^\epsilon(t-s)\right] R_s^\delta(v(s);\epsilon)\,ds$$

$$v_c(t) = \exp\left[At\right]v_c + \int_0^t \exp\left[A(t-s)\right] R_c^\delta(v(s);\epsilon)\,ds, \qquad (5.23)$$

The persistence theorem is then established with Newton iterations, using the space-time norm

$$\|v\|_\lambda = \sup_{t \ge 0} \exp\left\{\left[-\sigma\,t/\lambda\right]\|v(t)\|_{H^1}\right\}.$$

Once convergence has been established, the function h_u which represents the manifold W_ϵ^{cs} as a graph,

$$v_u = h_u(v_s, v_c; \epsilon),$$

is given by

$$h_u(v_s, v_c; \epsilon) = v_u(t=0) = \int_{+\infty}^0 \exp\left[\sigma_u^\epsilon(t-s)\right] R_u^\delta(v(s);\epsilon)\,ds.$$

Remark 5.3 The smoothness part of the argument is slightly more delicate, for an interesting reason. While solutions in W_ϵ^{cs} cannot grow as $\exp \sigma_u^\epsilon t$ as $t \to +\infty$, they can grow slowly (compared with $\exp \sigma_u^\epsilon t$). This slow growth, together with a quadratic nonlinearity, causes the estimates of the iterates associated to derivatives to degenerate. Fortunately, the gap in the spectrum still allows these terms to be controlled. In fact, the size of the gap determines how many derivatives the manifolds possess. (See [44] for details.)

5.2.2   Fiber Representations of the Normally Hyperbolic Invariant Manifolds

This gap in the spectrum also indicates the presence of two distinct time scales in the problem – motion on the plane Π_c experiences slow expansion and contraction on the νt time scale, where $\nu = \sqrt{\epsilon}$, while motion off the plane expands and contracts on the fast scale t. These two distinct scales make the long time behavior singular, and necessitate representations of the invariant manifolds appropriate for singular perturbation calculations. Such representations were introduced by Fenichel [17], [18], [19], [20] for finite dimensional dynamical systems. In this finite dimensional setting, recent uses of these "fiber representations" include [37], [40], [38], [27], [39], [34], [33]. Surveys of the finite dimensional situation are [35] and [63]. There is

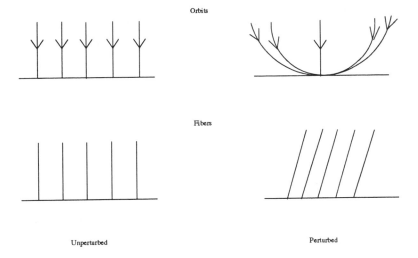

Figure 5.1: Orbits and Fibers

not much literature concerning these representations for pde's. Some earlier work includes [12], [41], [44], [7], [45].

For intuition about these Fenichel Fibers, consider the following trivial example with two time scales:

$$
\begin{aligned}
y_t &= -y \\
x_t &= -\nu(x + y),
\end{aligned}
$$

where $0 \leq \nu << 1$. Clearly, the orbits behave very differently when $\nu = 0$ than when $\nu > 0$. (See Figure 5.1.) For example, when $\nu = 0$, the x axis is a line of fixed points, while for $\nu > 0$, the origin is the only fixed point. However, in this latter case ($\nu > 0$) the x axis remains an invariant manifold on which the motion experiences slow contraction. Clearly, the orbits behave singularly as $\nu \to 0$. Nevertheless, the center stable manifold behaves smoothly with ν, but its representation as a union over orbits would not display this smoothness. We seek an alternative representation in terms of fibers.

The "stable fiber $\mathcal{F}_m^\nu$ through the base point m on the slow manifold $\mathcal{M}$" is defined as that equivalence class of points z (initial conditions) with the property that the trajectories initialized at z *rapidly* approach trajectories initialized at m. For our example, this means that, for $m \in \mathcal{M}$,

$$
\mathcal{F}_m^\nu := \left\{ z \,\middle|\, \| F^t(z; \nu) - F^t(m; \nu) \| = O[\exp(-t)] \text{ as } t \to +\infty \right\}. \quad (5.24)
$$

In this trivial example, one can calculate explicitly: Set $m = (x_m, 0)$ and

$z = (x, y)$ and obtain

$$F^t(z; \nu) = \left(\exp(-\nu t) \left[x - \frac{\nu}{1-\nu} y \right] + O[\exp(-t)], \ \exp(-t) y \right)$$
$$F^t(m; \nu) = (\exp(-\nu t) x_m, 0);$$

from which one sees that the fiber

$$\mathcal{F}_m^\nu = \left\{ (x, y) \,\middle|\, x - x_m = \frac{\nu}{1-\nu} y \right\}.$$

Clearly, in contrast to orbits, these fibers are smooth with ν.

This example can also be used to describe an alternative view of the fibers. One seeks a coordinate transformation which diagonalizes the slow flow in the sense that the equations for the slow variables do not depend upon upon the fast variables. In the context of this simple example, one changes variables

$$(x, y) \to (w, y)$$

so that, in terms of the new variables, the equations take the form

$$y_t = -y$$
$$w_t = -\nu w.$$

The transformation which accomplishes this is

$$x = f(w, y), \quad f(w, 0) = w,$$

where

$$f(w, y) = w + \frac{\nu}{1-\nu} y.$$

In the new coordinates (w, y), called "*fiber coordinates*", the fibers have been "straightened out",

$$\mathcal{F}_m = \{(w, y) \,|\, w = 0\}.$$

In the original coordinates (x, y) the transformation function provides a representation of the fiber. Indeed, fix a base point $m = (w, 0)$. Then the fiber through m is given by

$$\mathcal{F}_m = \left\{ (x, y) \,\middle|\, x = f(w, y) = w + \frac{\nu}{1-\nu} y \right\}.$$

Once a flow is *partially* diagonalized, it is clear that "points on a fiber suffer the same fast fate as that fiber's base point on the slow manifold"; that is (for this example),

$$F^t(\mathcal{F}_m) = F^t(m) + O[\exp(-t)], \quad t \to +\infty$$

In the literature, the prevalent view of fibers is the geometric one of an equivalence class of points, primarily because of the pioneering work of Fenichel. Here the perturbed NLS equation can be viewed as a perturbation of a constant coefficient linear problem. Because of this simplifying feature, we prefer the alternative view as a partial diagonalization of the flow which decouples the slow dynamics. In this NLS setting [44], this view allows us to construct the fibers with an integral equation framework which is very similar to the Perron method that we used for persistence. (The original "equivalence class" point of view is implemented for perturbed NLS equations in [41], [45].)

With integral equation methods, one [44] can establish the following fibration theorem:

Theorem 5.3 *For all $\epsilon \in [0, \epsilon_0]$ the C^k manifold W_ϵ^{cu} admits , through a C^{k-1} transformation, coordinates*

$$\eta_c \in E^c$$
$$\eta_u \in [-\eta_0, \eta_0]$$

such that the submanifold $\mathcal{M}_\epsilon$ corresponds to $\eta = 0$, and the flow on W_ϵ^{cu} decouples in the following manner:

$$\dot{\eta}_u = [\sigma_u^\epsilon + \Gamma^\delta(\eta_u, \eta_c; \epsilon)]\, \eta$$
$$\dot{\eta}_c = A\,\eta_c + \bar{S}_c^\delta(\eta_c; \epsilon)$$

where η, Γ^δ, $\bar{S}_c^\delta$ and their first derivatives are of order $O(\epsilon + \delta)$. The transformation itself is also C^{k-1} with respect to parameters such as ϵ. A similar statement holds for W_ϵ^{cu}.

The proof of this theorem, which proceeds by integral equation methods, may be found in [44]. It involves comparing solution trajectories through an arbitrary point with trajectories through a base point on the slow manifold. This comparison requires enough preliminary notation that the argument, while straight foreward, is somewhat tedious. As in the argument for persistence, the smoothness part is more delicate than existence.

Remark 5.4 One reason for the loss of derivative is due to the simple iteration procedure that we use to show the persistence and the fibration of the manifold. We use a simple Newton's iteration procedure applied to functions defined on noncompact domain.

These fibers allow general motion to be tracked by following only motion through the base point on the slow manifold. In addition, they are nicely behaved with respect to the parameter ϵ, as well as with respect to the base point. As such, they are very useful in singular perturbation calculations.

5.2.3   Height of $W^s(Q)$

The final local information concerns the "height of the stable manifold $W^s(Q)$".
On the plane Π_c, this manifold is controlled, globally, with the energy of the
driven pendulum (3.10). Here we discuss an estimate of its height over the
plane.

The near identity or normal form transformation discussed earlier in Section 5.1.4 enables us [44] to use integral equation methods to establish the
following

Theorem 5.4 *The point Q has a local stable manifold in $\mathcal{M}_\epsilon$ which can be
parametrized by* (θ, v_o);

$$\mathcal{W} = \{(J, \theta, v_o) \mid J = f(\theta, v_o)\}$$

for all $\|v_o\|_{H^1} \in [0, \epsilon^{3/4}]$.

Remark 5.5 Note that, in this theorem, one is restricted to lie within the
slow manifold $\mathcal{M}_\epsilon$,
$$\mathcal{W} = W^s(Q) \cap \mathcal{M}_\epsilon.$$

The theorem states that the "wall" $\mathcal{W}$ is tall enough, $O(\epsilon^{3/4})$, to permit a
successful "shooting argument" in the second measurement.

5.3   Global Arguments

With this material about the local behavior near the plane Π_c in hand, we can
turn to the global issues from integrable theory, and to the first and second
measurements.

5.3.1   Integrable Theory

The global material from integrable theory has already been discussed in
Section 3.2. It consists in representations of homoclinic orbits, a natural
constant of the motion $\mathbf{F}$, and a Melnikov function M_F which is based upon
$\mathbf{F}$. Here we restrict ourselves to a discussion of the reason the invariant $\mathbf{F}$ is
so natural an invariant upon which to base the Melnikov function.

Remark 5.6 1. For integrable cases, the explicit representations of the homoclinic orbits provide beautiful and explicit representations of the Fenichel
fibers [41], [43], [45].

2. As discussed earlier, $\mathbf{F}$ is critical at unstable tori; hence, $\mathbf{F}'$ will vanish
exponentially as the whisker approaches its unstable torus. This provides
exponential convergence of the Melnikov integrals and is one reason to base
the Melnikov function on $\mathbf{F}$. However, this convergence, which itself is due to
geometric reasons, is not the main structural reason that $\mathbf{F}$ is natural.

Recall that, for the integrable NLS equation, the center-stable manifold of the circle of fixed points S_ω admits the following global characterization:

$$W_0^{cs}(S_\omega) = \left\{ q \in H_{e,p}^1 \,|\, \mathbf{F} = -2 \right\}.$$

Thus, $grad\mathbf{F}$ is normal to $W_0^{cs}(S_\omega)$. As this manifold is cod 1, the distance of a general point from it can be assessed with one measurement in the $grad\mathbf{F}$ direction. Under perturbation, the manifold $W_0^{cs}(S_\omega)$ persists to the manifold W_ϵ^{cs}. Since this deformation is smooth (at least C^k), $grad\mathbf{F}$ remains transversal to the perturbed manifold W_ϵ^{cs}. As this perturbed manifold is also cod 1, one can use the $grad\mathbf{F}$ direction to develop a "signed distance" of a general point from W_ϵ^{cs}. Indeed, this is the Melnikov function of the first measurement!

5.3.2   First (Melnikov) Measurement

Melnikov analysis was first used for pde's by Holmes and Marsden [30], [29] to establish the transversal intersection of the stable and unstable manifolds when one of them is one dimensional. However, when the perturbation introduces new slow saddle directions (i.e., when the problem is singular), their methods are not sufficient to establish the transversal intersection of these manifolds. For a truncation of NLS to a singular problem in four dimensions, Kovacic and Wiggins [37], [40] were the first to combine Fenichel fibers with Melnikov analysis. We [41], [44] then developed this combination for pde's.

Consider the perturbed NLS equation in the form

$$q_t = iH'(q) + \epsilon\, G(q)$$

where $H'(q) = -q_{xx} - 2\,[\,q\bar{q} - \omega^2\,]\,q$ and $G(q) = -\alpha\,q - \beta\,\widehat{B}\,q - 1$. In this "first measurement" we [44] will construct a Melnikov function Δ whose zeros correspond to orbits that do not lie in the invariant plane Π_c, are asymptotic to the saddle point Q in backward time, and asymptotic to the locally invariant manifold $\mathcal{M}_\epsilon$ in forward time. This orbit is of the type depicted schematically in Figure 5.2. Notice that as the orbit leaves the saddle Q, the phase point moves very slowly and remains near the plane Π_c for a long time. Then it suddenly "takes-off" and rapidly flies away, only to return near the slow manifold $\mathcal{M}_\epsilon$, which it slowly approaches in foreward time. The very slow motion, on both the "take-off" and "landing" sides, will be described with Fenichel Fibers.

The argument begins by considering the unstable manifold $W^u(Q)$ of the saddle Q. This manifold is two dimensional, with a slow direction in the plane $\Pi_c \subset \mathcal{M}_\epsilon$ and the other fast direction off of the plane, tangent to $\cos x$. On the plane Π_c, this unstable manifold is a curve $\mathcal{C}^u$ which is controlled by the energy (3.10) of the driven pendulum. This curve $\mathcal{C}^u$ lies near the circle

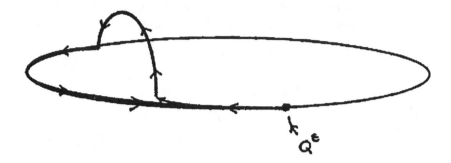

Figure 5.2: Schematic Diagram of the Homoclinic Orbit

of fixed points S_ω, and can be viewed as a graph over this circle. That is, the curve C^u can be parameterized by an angle θ_b.

Since $W^u(Q)$ is a submanifold of W^{cu}_ϵ, it inherits the Fenichel Fibration over the curve C^u:

$$W^u(Q) = \bigcup_{m \in C^u} \{ \mathcal{F}^u_m(\cdot\,;\epsilon) \}.$$

In particular, we will fix a base point θ_b on the curve C^u, and consider two distinct fibers each with this same base point – the perturbed fiber $\mathcal{F}^u_m(\cdot\,;\epsilon)$ and its unperturbed neighbor $\mathcal{F}^u_m(\cdot\,;0)$. (See 5.3 Fixing a short positive distance $\delta > 0$ along these fibers enables us to define the unperturbed and perturbed "take-off" points, $\mathcal{T}^0$ and $\mathcal{T}^\epsilon$, each of which is a function (through m) of the base angle θ_b:

$$\begin{aligned} \mathcal{T}^0 &= \mathcal{F}^u_m(\delta; 0) \\ \mathcal{T}^\epsilon &= \mathcal{F}^u_m(\delta; \epsilon). \end{aligned} \tag{5.25}$$

In this manner, we have used fibers to describe the slow motion and have made a precise identification two take-off points, one on the unstable manifold $W^u(Q)$ and the other on the unperturbed manifold $W^u_0(\Pi_c) = W^s_0(\Pi_c)$.

Next, we consider two orbits initialized at these take off points:

$$\begin{aligned} q_h(t) &= F^t(\mathcal{T}^0; 0) \\ q_\epsilon(t) &= F^t(\mathcal{T}^\epsilon; \epsilon). \end{aligned} \tag{5.26}$$

The orbit $q_h(t)$ is an unperturbed integrable orbit, which certainly asymptotes to the plane Π_c in forward time. In fact, analytical expressions precisely and concretely determine its fate on the plane. On the other hand, the orbit $q_\epsilon(t)$ is a solution of the perturbed NLS equation. Our goal is to determine if a take-off angle θ_b can be chosen to guarantee that this orbit will asymptote to the slow manifold $\mathcal{M}_\epsilon$ as $t \to +\infty$.

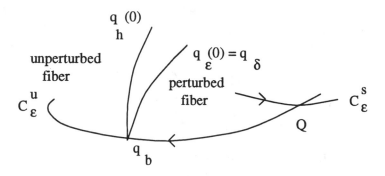

Figure 5.3: The Initial Points on the W_ϵ^u Fibers

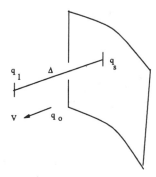

Figure 5.4: Schematic Diagram of the First Measurement

From the explicit expression for the integrable homoclinic orbit $q_h(t)$, one sees that there is a time $T_*(\delta)$ such that $dist(q_h(t), \Pi_c) \leq \delta$ for all $t \geq T_*$. At T_* we define two "landing points"

$$\ell_0 = q_h(T_*) = F^{T_*}(\mathcal{T}^0; 0)$$
$$\ell_\epsilon = q_\epsilon(T_*) = F^{T_*}(\mathcal{T}^\epsilon; \epsilon).$$

By finite time perturbation theory, these two landing points are $O(\epsilon)$ apart. The unperturbed landing point $\ell_0 \in W_0^s(\Pi_c)$. We hope to choose the take-off angle θ_b so that the perturbed landing point $\ell_\epsilon \in W_\epsilon^{cs}$.

The manifold W_ϵ^{cs} is codimension 1 and is ϵ close to the integrable W_0^{cs}; moreover, the latter is characterized by $\{q \in H^1 | \boldsymbol{F}(q) + 2 = 0\}$. Therefore, we can define

$$\Delta := \langle \boldsymbol{F}'(\ell_0), \ell_\epsilon - q_s \rangle$$

as a measure of the signed distance between ℓ_ϵ and q_s, where q_s denotes the intersection of the line through ℓ_ϵ and the manifold W_ϵ^{cs}. (See Figure 5.4).

To actually calculate Δ, we define, for $t \leq 0$, the orbits

$$q_*(t) = q_h(t + T_*)$$
$$q_u(t) = q_\epsilon(t + T_*).$$

Similarly, for $t \geq 0$, we define $q_s(t)$ to be the solution of the perturbed problem with initial data q_s, and we extend the definition of $q_*(t)$ to all t. These orbits allow us to introduce the definitions

$$\begin{aligned}
\Delta^-(t) &= \langle F'(q_*(t), q_u(t) - q_*(t) \rangle \qquad t \leq 0 \\
\Delta^+(t) &= \langle F'(q_*(t), q_s(t) - q_*(t) \rangle \qquad t \geq 0 \\
\Delta &= \Delta^-(0) - \Delta^+(0)
\end{aligned}$$

Finally, as in the finite dimensional case [62], differential equations for $\Delta^-(t)$ and $\Delta^+(t)$ enable us to estimate the signed distance Δ in terms of the Melnikov integral:

Proposition 5.1 *The distance Δ is given by*

$$\Delta = \epsilon M_F(\theta_b) + O(\epsilon^2),$$

where

$$M_F = \int_{-\infty}^{\infty} \langle F'(q_*(t)), G(q_*(t)) \rangle \, dt.$$

The detailed proof of this estimate may be found in [44].

This proposition establishes the geometric significance of the Melnikov function which was calculated in Section 4.2. For the readers convenience, we summarize the results of those calculations. Let $q_\omega(t)$ denote an orbit of the integrable pde which is homoclinic to the circle of fixed points S_ω and $\sqrt{\epsilon}$ close to the integrable homoclinic orbit $q_*(t)$.

Corollary 5.2 *The distance Δ has an expansion in ϵ given by*

$$\Delta = \epsilon M_F(\alpha, \beta, \theta_b) + O(\epsilon^{3/2})$$

$$\begin{aligned}
M_F(\alpha, \beta, \theta_b) &= \int_{-\infty}^{\infty} \langle F'(q_\omega(t)), G(q_\omega(t)) \rangle \, dt \\
&= -[\alpha M_\alpha + \beta M_\beta + M_0 \cos(\theta_b - 2p_o)] \qquad (5.27)
\end{aligned}$$

where

$$\begin{aligned}
M_\alpha &= \int_{-\infty}^{\infty} \langle F'(q_\omega(t)), q_\omega(t) \rangle \, dt \\
M_\beta &= \int_{-\infty}^{\infty} \langle F'(q_\omega(t)), \widehat{B} q_\omega(t) \rangle \, dt \\
M_0 \cos(\theta_b - 2p_o) &= \int_{-\infty}^{\infty} \langle F'(q_\omega(t)), 1 \rangle \, dt \, .
\end{aligned}$$

More explicitly,

$$M_\alpha = \int_{-\infty}^{\infty} d\tau \int_0^{2\pi} dx \frac{4\pi\,\omega^2 \sin^2 p_o \, \text{sech}\, \tau}{\sigma A^3} \times$$

$$[\text{sech}\,\tau + \sin p_o \tanh^2 \tau \cos x - \sin^2 p_o \, \text{sech}\,\tau(2 + \cos^2 x)$$
$$+ 2\sin^3 p_o \, \text{sech}^3\tau \cos x]$$

$$M_\beta = \int_{-\infty}^{\infty} d\tau \int_0^{2\pi} dx \frac{4\pi\omega^2 \sin^2 p_o \, \text{sech}\, \tau}{\sigma A^5} \times$$

$$\left[\sin p_o \, \text{sech}\,\tau \cos x - \sin^2 p_o \, \text{sech}^2\tau(1 + \sin^2 x)\right] \times$$

$$[2\,\text{sech}\,\tau - \sin p_o \, \text{sech}^2\tau \cos x - 2\sin^2 p_o \, \text{sech}\,\tau$$
$$+ 2\sin^3 p_o \, \text{sech}^2\tau \cos x] + O(\sin^{\kappa-2} p_o)$$

$$M_0 = \int_{-\infty}^{\infty} d\tau \int_0^{2\pi} dx \frac{4\pi\omega \sin^2 p_o \, \text{sech}\, \tau}{\sigma A^2} \times$$

$$[\,\text{sech}\,\tau - \sin p_o \cos x]$$

where $p_o = \tan^{-1}\sqrt{\omega^2 - 1}$, $A = [1 - \sin p_o \, \text{sech}\,\tau \cos x]$ and where the $O(\sin^{\kappa-2} p_o)$ term in the M_β equation is due to the fact that we used $-\partial_x^2$ instead of $\hat{B}$ in our computation.

Assuming that M_α of M_β are nonzero, then the function

$$M(\alpha,\, \beta,\, \theta_b) = \alpha M_\alpha + \beta M_\beta + M_0 \cos(\theta_b - 2p_o) \qquad (5.28)$$

has nondegenerate zero. This implies that we can choose our parameters so that $\Delta = 0$, i.e.,

$$\ell_\epsilon = q_s \in W_\epsilon^{cs}\,,$$

and the first measurement is complete.

5.3.3   Second Measurement

As a consequence of the first measurement, the "take-off angle" θ_b can be chosen so that the landing point ℓ_ϵ lives in W_ϵ^{cs}, and thus the orbit asymptotes to $\mathcal{M}_\epsilon$. However, we do not know if it asymptotes to the saddle point Q. In order to ascertain that it does return to Q, we perform a second measurement.

The necessity of a second measurement was first understood in finite dimensional discretizations of this NLS system – first for a four dimensional system [38],[53], and then for the more delicate $2N + 2$ dimensional system [41], [43].

This second measurement makes substantial use of Fenichel Fibers. In order to determine the foreward fate of the orbit, one does not follow the orbit through the landing point ℓ_ϵ. Rather, one follows the orbit through the base point of that stable fiber on which ℓ_ϵ resides.

This second measurement is also a shooting argument. While the stable manifold $W^s(Q)$ is cod 2 as a submanifold of the entire space, it is cod 1 as a submanifold of W^{cs}_ϵ. Moreover, as a consequence of the first measurement, the landing point ℓ_ϵ resides in W^{cs}_ϵ. Thus, a one dimensional shooting argument should suffice for the second measurement.

Denote the base points of the landing fibers in the perturbed and unperturbed cases by b_ϵ and b_0, respectively:

$$\ell_\epsilon \in \mathcal{F}^s_{b_\epsilon}(\cdot\,; \epsilon)$$
$$\ell_0 \in \mathcal{F}^s_{b_0}(\cdot\,; 0).$$

The unperturbed base point is known to lie on the plane $b_0 \in \Pi_c$. On the other hand, the perturbed base point is only restricted to reside in the cod 2 slow manifold $b_\epsilon \in \mathcal{M}_\epsilon$. (In the four dimensional model problem [53], [38], both base points lie on the plane Π_c which makes the second measurement easier.) However, because of the smoothness of the fibers with respect to ϵ, we do know that the two base points b_ϵ and b_0 are ϵ close.

The issue is whether the base point $b_\epsilon \in \mathcal{M}_\epsilon$ actually lies on the stable manifold $\mathcal{W} = W^s(Q) \cap \mathcal{M}_\epsilon$. In [44], we establish

Proposition 5.2 *The distance from the landing base point b_ϵ to the stable manifold $\mathcal{W}$ can be measured as a function of the take-off angle θ_b by*

$$d = \omega\left[2\sin 2p_o \cos(\theta_b - 2p_o) + 4\alpha\omega p_o\right] + O\left(\sqrt{\epsilon}\right)$$

The proof of this second measurement involves

1. the smoothness of the fibers to estimate that the perturbed and unperturbed base points are close;

2. the energy (3.10) of the driven pendulum to measure distances on the plane Π_c;

3. the height of the wall $\mathcal{W}$ as estimated by the normal form calculation to ensure that projections on the plane accurately capture the landing base point as it oscillates across the wall (as a function of the "shooting parameters");

4. the explicit formula for the integrable phase shift which expresses the landing angle as a function of the take-off angle θ_b.

Details of this second measurement are described in [44].

In summary, the first and second measurement produce a pair of equations which must be satisfied for a homoclinic orbit to persist:

$$\cos(\theta_b - 2p_o) = -\frac{2\alpha\omega p_o}{\sin 2p_o} \tag{5.29}$$

$$\beta = -\alpha \left[M_\alpha + \frac{2\omega p_o}{\sin 2p_o} \, M_0 \right] \Big/ M_\beta \tag{5.30}$$

By the implicit function theorem, for ϵ small we can solve $d = \Delta = 0$ in a small neighborhood of the point given by equations (5.29) and (5.30). Finally, β can be shown to be positive (so that the perturbation is indeed dissipative) either numerically for wide range of parameter values, or with a completely analytic argument if we assume that $p_o = \tan^{-1}\sqrt{4\omega^2 - 1}$ is small enough i.e., ω close to $1/\sqrt{2}$.

The simultaneous zeros of d and Δ determine (i) the take- off angle θ_b as a function of the external parameters and (ii) one relationship amongst the external parameters. Thus, the homoclinic orbit is shown to persist (as a singular deformation of the integrable orbit) provided the external parameters are restricted in a cod 1 manner.

Remark 5.7 Notice that both measurements are used *together* to remove the "cut-off". The first measurement establishes a heteroclinic connection between the saddle Q and the slow manifold $\mathcal{M}_\epsilon$, a manifold which has only been shown to persist for the cut-off flow. However, the second measurement, which shows that the orbit returns to the saddle Q, can be used, together with control provided by the unperturbed homoclinic orbit, to show that the orbit homoclinic to Q, once it returns near $\mathcal{M}_\epsilon$, never leaves the region where the cut-off function is the identity. In this manner, the theorem establishes the existence of a orbit homoclinic to the saddle Q for the original perturbed NLS equation, *without any cut-off*.

Remark 5.8 Once the persistence of one homoclinic orbit is established, the symmetry $x \to x + \pi$ shows the persistence of a second homoclinic orbit. One member of this pair represents a coherent spatial excitation located at $x = 0$, and the other represents an excitation at $x = \pi$. In addition, equations (5.29) and (5.30) can possess several distinct solutions in some parameter ranges, each with its symmetric partner. Thus, the analysis shows that many homoclinic orbits can persist for the perturbed NLS equation.

6   Remarks on Symbol Dynamics

The simplest chaotic behavior which was observed in the numerical experiments for the perturbed NLS equation consisted of a single solitary spatial excitation which jumps, irregularly in time, between the two distinct spatial locations at $x = 0$ and $x = \pi$. These numerical experiments, together with the persistence of a *symmetric pair* of homoclinic orbits, suggests a "symbol dynamics" explanation of this phenomena.

More precisely, the term "symbol dynamics" refers to the existence of an invariant set in the phase space which is topologically equivalent to a set of all symbol valued sequences. In our setting, these sequences would take the values of C (center) or E (edge), and the dynamics is represented as a shift on this sequence space. As such, the dynamics, when restricted to the invariant set, is as random as a sequence of "coin tosses".

In finite (usually very small) dimensional, the existence of such an invariant set is established by constructing a "Smale horseshoe", [59], [55], [23], [62]. Such constructions are carried out for orbits homoclinic to the saddle Q for the four dimensional truncation [53] and for the $2N + 2$ dimensional truncation in [41], [46]. A similar construction is in progress for the pde.

Symbol dynamics is very appealing because it demonstrates the existence of chaotic motions which last for all time. However, it has some drawbacks. First, it occurs on a very small set in phase space, which is not shown to be (and is likely not) a stable set. As such, this type of chaos my not be observable. Moreover, the behavior depends on parameters in a bifurcation fashion. Often the parameter values required to show the existence of the horseshoe are very far from the values of the parameters at which chaotic behavior is observed in numerical experiments. (For example, in our analytical results [44], an addition dissipation $\beta > 0$ is required which satisfies a cod 1 constraint. However, in the numerical experiments [52], chaotic behavior is observed for $\beta = 0$, over a range of α values.) Finally, the construction of the horseshoe is almost always done for generic abstract models, rather than for a fixed specific dynamical system. To us, this generic situation seems to be a severe limitation of the practice of the method. (Sometimes numerical constructions are used to verify the generic assumptions for specific examples, and thus to overcome this limitation. However, in high dimensional singular situations, such numerical verifications are extremely difficult – if not impossible.)

Recently, Haller has been developing an alternative perspective which he has applied to finite dimensional discretizations of the perturbed NLS equation [27], [25], and which he is currently extending to the pde [26]. In his work, using very similar geometric perturbation methods to those summarized here, he constructs a large class of heteroclinic orbits from the saddle Q to (for example) the sink P. He fixes a finite, but arbitrary, positive integer N, as well as any sequence of C's and E's of length N, and then establishes the existence of a heteroclinic orbit which recovers this pattern. While only transient behavior, the length of the transients is arbitrarily long. In any case, this set of heteroclinic orbits certainly demonstrates very complicated dynamics which depends sensitively upon initial conditions. Moreover, as the second measurement is not required to force the orbit to return to the saddle

Q, these heteroclinic orbits exist for a full open set of external parameter values, without any cod 1 restriction.

7 Other Equations and Related Work

The importance of this work is not in the construction of one particular homoclinic orbit for one specific equation. Rather, it is the development Melnikov methods, together with geometric singular perturbation theory, for near conservative pde's. We believe that these will become general methods which will be applicable to a wide class of problems. For example, the methods should be useful for the study (i) perturbations of other integrable pde's such as the sine-Gordon [22], [8], and Davey-Stewartson equations; (ii) discretizations of integrable equations [2], [3], [51], [10]; (iii) systems without even symmetry [1]; (iv) the Ginzburg- Landau equation [56], and many others. Connections between Melnikov analysis and the methods of inertial manifolds [13], [14], [15], should be developed. Melnikov methods should also be useful in the study of conservative perturbations of conservative pde's, which possess interesting transport and diffusion properties [21]. Finally, these methods should also play a role in the description of the spatial, as well as temporal, chaos [28], [11], [52].

Clearly this work for pde's, which combines Melnikov methods with geometric singular perturbation theory, is just beginning. The study of perturbations of the NLS equation shows that the methods are useful tools for pde's. Their generality, and general usefulness, is yet to be determined.

Acknowledgement: *Both of us, and especially DWMc, wish to acknowledge many years of close and fruitful collaboration with Yanguang Li and Steve Wiggins on all aspects of this project. We also have benefited from many discussions with George Haller and, more recently, with Peter Bates.*

References

[1] M. J. Ablowitz, B. Herbst, and C. Schober. The NLS Equation Asymptotic Perturbations and Chaotic Structures. *Preprint, Univ of Colorado*, 1995.

[2] M. J. Ablowitz and B. M. Herbst. Numerically Induced Chaos in the Nonlinear Schrödinger Equation. *Phys Rev Lett*, 62:2065–2068, 1989.

[3] M. J. Ablowitz and B. M. Herbst. On Homoclinic Structure and Numerically Induced Chaos for the Nonlinear Schrödinger Equation. *SIAM Journal on Applied Mathematics*, 50:339–351, 1990.

[4] V. I. Arnold. Instabilities Of Systems With Several Degrees Of Freedom. *Sov Math Dokl*, 5:581–585, 1964.

[5] G. Baker and J. Gollub. *Chaotic Dynamics an Introduction*. Cambridge University Press (Cambridge), 1990.

[6] J. Ball. Saddle Point Analysis for an Ordinary Differential Equation in a Banach Space. *Nonlinear Elasticity (ed. by R. Dickey, Academic Press)*, pages 93 – 160, 1973.

[7] P. Bates and K. Lu. *Private Communication*, 1995.

[8] B. Birnir and R. Grauer. An Explicit Description of the Global Attractor of the Damped and Driven Sine-Gordon Equation. *Communications Mathematical Physics*, 162:539–590, 1994.

[9] A. R. Bishop, M. G. Forest, D. W. McLaughlin, and E. A. Overman II. A Quasi-Periodic Route to Chaos in a Near-Integrable pde. *Physica D*, 23:293–328, 1986.

[10] A. Calini, N. Ercolani, D. W. McLaughlin, and C. M. Schober. Melnikov Analysis of Numerically Induced Chaos in the NLS Equation. *Physica D*, to appear, 1995.

[11] C. Chow. *Spatiotemporal Chaos in the Nonlinear Three Wave Interaction*. PhD thesis, M I T, 1992.

[12] S. Chow, X. Lin, and K. Lu. Smooth Invariant Foliations in Infinite Dimensional Spaces. *Journal of Differential Equations*, 94:266–291, 1991.

[13] P. Constantin. A Construction of Inertial Manifolds. *Contemporary Mathematics*, 99, 1989.

[14] P. Constantin, C. Foias, B. Nicolaenko, and R. Temam. *Integral Manifolds and Inertial Manifolds for Dissipative Partial Differential Equations*. Springer-Verlag (New York), 1989.

[15] C. Doering, J. D. Gibbon, D. D. Holm, and B. Nicolaenko. Inertial Manifolds for the Ginzburg-Landau Equation. *Nonlinearlity 1*, 1:279–309, 1989.

[16] N. M. Ercolani, M. G. Forest, and D. W. McLaughlin. Geometry of the Modulational Instability, Part III: Homoclinic Orbits for the Periodic sine- Gordon Equation. *Physica D*, 43:349–84, 1990.

[17] N. Fenichel. Persistence and Smoothness of Invariant Manifolds for Flows. *Ind University Math J*, 21:193–225, 1971.

[18] N. Fenichel. Asymptotic Stability with Rate Conditions. *Ind Univ Math J*, 23:1109–1137, 1974.

[19] N. Fenichel. Asymptotic Stability with Rate Conditions II. *Ind Univ Math J*, 26:81–93, 1977.

[20] N. Fenichel. Geometric Singular Perturbation Theory for Ordinary Differential Equations. *J Diff Eqns*, 31:53–98, 1979.

[21] M. Forest, C. Goedde, and A. Sinha. Chaotic Transport and Integrable Instabilities in a Nearly Integrable Hamiltonian Discrete Sine Gordon Lattice. *Physica D*, 67:347–386, 1993.

[22] R. Grauer and B. Birnir. The Center Manifold and Bifurcations of the Sine-Gordon Breather. *Physica D*, 56, 1994.

[23] J. Guckenheimer and P. Holmes. Nonlinear Oscillations, Dynamical Systems, and Bifurcations of Vector Fields. *Appl Math Sci*, 42, 1983.

[24] J. Hadamard. Sur Literation et les Solutions Asymptotiques des Equations Differentielles. *Bull. Soc. Math. France*, 29:224–228, 1901.

[25] G. Haller. Multipulse Homoclinic Orbits for Multidimensional Systems. in preparation 1995.

[26] G. Haller. Orbits Homoclinic to Resonances: The Hamiltonian Pde Case. in preparation 1995.

[27] G. Haller and S. Wiggins. Orbits Homoclinic to Resonances: The Hamiltonian Case. *Physica D*, 66:298–346, 1993.

[28] P. C. Hohenberg and B. I. Shraiman. Chaotic Behavior of an Extended System. *Physica D*, 37:109–115, 1989.

[29] P. Holmes. Space and Time Periodic Perturbations of the Sine Gordon Equation. *Proc of Warrick Conference on Turbulence and Dynamical Systems in Springer Lecture Notes in Mathematics*, 898:164–191, 81.

[30] P. Holmes and J. Marsden. A Partial Differential Equation with Infinitely Many Periodic Orbits and Chaotic Oscillations of a Forced Beam. *Archive for Rational Mechanics and Analysis*, 76:135–165, 1981.

[31] P. Holmes and J. Marsden. Horeshoes in Perturbations of Hamiltonian Systems with Two Degrees of Freedom. *Comm Math Phys*, 82:523–544, 1982.

[32] P. Holmes and J. Marsden. Melnikov's Method and Arnold Diffusion for Perturbations of Integrable Hamiltonian Systems. *Journal Mathematical Physics*, 23:669–675, 1982.

[33] C. Jones, T. Kaper, and N. Kopell. Tracking Invariant Manifolds up to Exponentially Small Errors. *SIAM J. Math. Anal.*, 1994.

[34] C. Jones and N. Kopell. Tracking Invariant Manifolds with Differential Forms. *J. Diff. Eq*, 1994.

[35] C.K.R.T Jones. Geometric Singular Perturbation Theory. *Preprint Brown University*, 1994.

[36] Y. Kevrekidis, D. W. McLaughlin, and M. Winograd. Movie. 1993.

[37] G. Kovacic. *Orbits Homoclinic to Resonances: Chaos in a Model of the Forced and Damped Sine-Gordon Equation*. PhD thesis, California Institue of Technology, 1989.

[38] G. Kovacic. Dissipative Dynamics of Orbits Homoclinic to a Resonance Band. *Phys Lett A*, 167:143–150, 1992.

[39] G. Kovacic. Singular Perturbation Theory for Homoclinic Orbits in a Class of Near Integrable Hamiltonian Systems. *J. Dyn. Diff. Eq.*, 5:559–597, 1993.

[40] G. Kovacic and S. Wiggins. Orbits Homoclinic to Resonances, with an Application to Chaos in a Model of the Forced and Damped Sine-Gordon Equation. *Physica D*, 57:185–225, 1992.

[41] Y. Li. *Chaotic Behavior in PDE's*. PhD thesis, Princeton University, 1993.

[42] Y. Li and D. W. McLaughlin. Morse and Melnikov Functions for NLS Pdes. *Commun. Math Phys.*, 162:175–214, 1994.

[43] Y. Li and D. W. McLaughlin. Homoclinic Orbits and Chaos in Discretized Perturbed NLS Systems Part I Homoclinic Orbits. *Journal Nonlinear Science*, submitted, 1995.

[44] Y. Li, D. W. McLaughlin, J. Shatah, and S. Wiggins. Persistent Homoclinic Orbits for Perturbed NLS Equations. *Communications Pure and Applied Mathematics*, submitted, 1995.

[45] Y. Li, D. W. McLaughlin, and S. Wiggins. *Invariant Manifolds and Their Fibrations for Perturbed NLS Pdes Graph Transform Approach*. Springer-Verlag (New York), in prep, 1995.

[46] Y. Li and S. Wiggins. Homoclinic Orbits and Chaos in Discretized Perturbed NLS Systems Part I I Symbol Dynamics. *Journal Nonlinear Science*, submitted, 1995.

[47] H. P. McKean and J. Shatah. The Nonlinear Schrödinger Equation and the Nonlinear Heat Equation Reduction to Linear Form. *Communications on Pure and Applied Mathematics*, 44:1067–1080, 1991.

[48] D. W. McLaughlin. Notes on the Periodic NLS Equation. *unpublished*, 1988.

[49] D.W. McLaughlin. Whiskered Tori for the NLS Equation. *Important Developments in Soliton Theory in Springer Series Nonlinear Dynamics*, pages 537–558, 1993.

[50] D.W. McLaughlin. Whiskered Tori and Chaotic Behavior for Noonlinear Waves. *Proc International Congress of Mathematicians, Zurich*, to appear, 1995.

[51] D.W. McLaughlin and C.Schober. Chaotic and Homoclinic Behavior for Numerical Discretizations of the NLS Equation. *Physica D*, 44:1067–1080, 1991.

[52] D.W. McLaughlin and E. A. Overman. Whiskered Tori for Integrable Pdes and Chaotic Behavior in Near Integrable Pdes. *Surveys in Appl Math 1*, 1995.

[53] D.W. McLaughlin, E. A. Overman, S. Wiggins, and C. Xiong. Homoclinic Orbits in a Four Dimensional Model of a Perturbed NLS Equation. *Dynamics Reports*, to appear, 1995.

[54] V. Melnikov. On the Stability of the Center for Time Periodic Oscillations. *Trans. Moscow Math.*, 12:1–57, 1963.

[55] J. Moser. *Stable and Random Motion in Dynamical Systems*. Princeton University Press, 1987.

[56] G. Cruz Pacheco, D. Levermore, and B. Luce. Melnikov Methods for Pde's Applications to Perturbed Nonlinear Schroedinger Equations. *Physica D*, submitted, 1995.

[57] O. Perron. Die Stabilitatsfrage bei Differentialgleichungssysteme. *Math. Zeit.*, 32:703–728, 1930.

[58] D. H. Sattinger and V. D. Zurkowski. Gauge Theory of Backlund Transformations. *Physica D*, 26:225–250, 1987.

[59] S. Smale. Differential Dynamical Systems. *Bull Amer Math Soc*, 73, 1967.

[60] W. Strauss. Nonlinear invariant wave equations. In G. Velo and A. S. Wightman, editors, *Invariant Wave Equations*, pages 197–249. Springer-Verlag, Berlin, 1978.

[61] G. Whitham. *Linear and Nonlinear Waves*. John Wiley and Sons, 1974.

[62] S. Wiggins. *Global Bifurcations and Chaos: Analytical Methods.* Springer-Verlag (New York), 1988.

[63] S. Wiggins. *Normally Hyperbolic Invariant Manifolds in Dynamical Systems*. Springer-Verlag (New York), 1994.

Section II

Exactly Integrable Systems

Lectures in Applied Mathematics
Volume **31**, 1996

Integrable Hamiltonian systems

Percy Deift, Courant Institute

These four lectures concern integrable Hamiltonian systems. The goal in the lectures is to give a rapid but elementary introduction to the subject, with a view to describing some recent results in the field: the reader is of course encouraged to consult the References for more details. The lectures are presented in informal "blackboard" style in order to preserve the tutorial atmosphere of the Summer School, and are divided up as follows.

Lecture 1. What is a Hamiltonian system? What is an integrable system?

Lecture 2. Examples of finite-dimensional integrable systems.

Lecture 3. Examples of infinite-dimensional integrable systems.

Lecture 4. Long-time behavior of integrable systems.

Lecture 1. (References: [G] — a physics text, [A] — a mathematical physics text, [AM] — a mathematical view of the foundations of mechanics.) The space $M = \mathbb{R}^{2n}$ is an example of a *symplectic manifold* i.e., it is an

- even dimensional manifold, with a

- non-degenerate 2-form w, i.e., $w(u, v) = 0$ for all $v \in T_m M$, implies $u = 0$,

- which is closed, i.e. $dw = 0$.

Exercise: Show that a symplectic manifold must be even dimensional.

For $\mathbb{R}^{2n}$, the standard 2-form is $w = \sum_{i=1}^{n} dx_i \wedge dy_i$. Clearly $dw = 0$. If $v = \sum_{i=1}^{n} \left(a_i \frac{\partial}{\partial x_i} + b_i \frac{\partial}{\partial y_i} \right) \equiv \binom{a}{b}$, $v' = \binom{a'}{b'}$, then a simple calculation shows that $w(v, v') = \left(\binom{a}{b}, J\binom{a'}{b'} \right)$, where $J = \begin{pmatrix} 0 & I \\ -I & 0 \end{pmatrix}$, I is the $n \times n$ identity, and $(\cdot, \cdot)$ denotes the standard inner product on $\mathbb{R}^{2n}$. From this it is clear that

$$w(v, v') = 0 \text{ for all } v' \Rightarrow J^T \binom{a}{b} = 0 \Rightarrow \binom{a}{b} = 0, \text{ i.e. } v = 0,$$

so w is non-degenerate.

Functions (i.e. 'Hamiltonians') $H \colon M \to \mathbb{R}$ generate vector fields through w in the following way. At any point $m \in M$, dH_m is a 1-form: hence $v \mapsto dH_m(v)$ is a linear map from $T_m M$ to $\mathbb{R}$. Thus, as w is non-degenerate, there exists a unique vector $v_H(m) \in T_m M$ such that

$$(1) \qquad dH(v) = w(v_H, v), \quad \text{for all} \quad v \in T_m M.$$

In the case of $\left(\mathbb{R}^{2n}, w = \sum_{i=1}^{n} dx_i \wedge dy_i \right)$, equation (1) becomes

$$dH(v) = \sum_{i=1}^{n} \left(\frac{\partial H}{\partial x_i} a_i + \frac{\partial H}{\partial y_i} b_i \right) = \left(\binom{H_x}{H_y}, \binom{a}{b} \right) = \left(v_H, J\binom{a}{b} \right).$$

1991 *Mathematics Subject Classification.* Primary 58F07; Secondary 35Q20, 15A18, 65F15.

Hence $\left(\begin{smallmatrix} H_x \\ H_y \end{smallmatrix}\right) = J^T v_H$, which implies $v_H = \left(\begin{smallmatrix} H_y \\ -H_x \end{smallmatrix}\right)$. Thus H gives rise to the standard Hamiltonian vector field

$$\frac{d}{dt}\begin{pmatrix} x \\ y \end{pmatrix} = v_H = \begin{pmatrix} H_y \\ -H_x \end{pmatrix}.$$

The fact that w is non-degenerate plays an obvious role, as above. The role that the closedness of w plays, is more subtle. If H, K are two Hamiltonians on M, we define their *Poisson bracket* $\{H, K\}$ through

$$\{H, K\} = w(v_H, v_K).$$

Note that $\{H, K\} = -\{K, H\}$. Clearly $\{H, K\} = dH(v_K) = v_K(H) = \frac{dH}{dt}$, where $\frac{dH}{dt}$ is the derivative of H in the direction of v_K. Thus

$$\frac{dH}{dt} = \{H, K\},$$

and in particular $\frac{dK}{dt} = \{K, K\} = 0$, as $\{\cdot, \cdot\}$ in skew i.e. K is a conserved quantity for the flow that it generates.

Exercise: Verify Leibniz's rule, i.e. $\{HK, L\} = H\{K, L\} + \{H, L\}K$ for all Hamiltonians H, K, L.

Observe that for $\left(\mathbb{R}^{2n}, \sum\limits_{i=1}^{n} dx_i \wedge dy_i \right)$,

$$\{H, K\} = \left(J\begin{pmatrix} H_x \\ H_y \end{pmatrix}, J\left(J\begin{pmatrix} K_x \\ K_y \end{pmatrix}\right) \right) = \left(\begin{pmatrix} H_x \\ H_y \end{pmatrix}, J\begin{pmatrix} K_x \\ K_y \end{pmatrix} \right)$$

$$= \sum_{i=1}^{n} \left(\frac{\partial H}{\partial x_i}\frac{\partial K}{\partial y_i} - \frac{\partial H}{\partial y_i}\frac{\partial K}{\partial x_i} \right),$$

which is the standard Poisson bracket on $\mathbb{R}^{2n}$.

Now it is an (**Exercise**) in the exterior calculus to show that up to a constant C, dw is a 3-form satisfying

$$dw(v_H, v_K, v_L) = C(\{\{H, K\}, L\} + \{\{K, L\}, H\} + \{\{L, H\}, K\})$$

for all Hamiltonians H, K, L. Thus

$$w \text{ is closed } \Leftrightarrow \{\cdot, \cdot\} \text{ satisfies the Jacobi identity, i.e.}$$

$$\{\{H, K\}, L\} + \{\{K, L\}, H\} + \{\{L, H\}, K\} = 0.$$

This leads to the following critical calculation for the commutator $[v_H, v_K]$ of Hamiltonian vector fields:

$$\begin{aligned}
[v_H, v_K](L) &= v_H(v_K(L)) - v_K(v_H(L)) \\
&= v_H(\{L, K\}) - v_K(\{L, H\}) \\
&= \{\{L, K\}, H\} - \{\{L, H\}, K\} \\
&= \{\{L, K\}, H\} + \{\{H, L\}, K\} \\
&= -\{\{K, H\}, L\}, \quad \text{by the Jacobi identity,} \\
&= v_{\{K, H\}}(L)
\end{aligned}$$

i.e.

(2) $$[v_H, v_K] = v_{\{K,H\}}.$$

Thus the commutator of two Hamiltonian vector fields is again a Hamiltonian vector field. Moreover the map, $H \mapsto v_H$, is an anti-isomorphism from the *Poisson algebra* of functions on M with product given by the Poisson bracket, into the algebra of vector fields with product given by the commutator of vector fields. *Most importantly*, we notice that two Hamiltonian vector fields commute $\Leftrightarrow$ their Hamiltonians Poisson commute. This is perhaps the most useful consequence of the fact that w is closed.

Exercise: Two vector fields v, v' commute $\Longleftrightarrow$ the flows ϕ_v^t, $\phi_{v'}^{t'}$ that they generate commute i.e., if $\frac{d}{dt} \phi_v^t = v(\phi_v^t)$, $\frac{d}{dt'} \phi_{v'}^{t'} = v'(\phi_{v'}^{t'})$, then $\phi_v^t \circ \phi_{v'}^{t'} = \phi_{v'}^{t'} \circ \phi_v^t$.

To summarize, we see that *symplectic manifolds* (M, w) are equivalent to non-degenerate *Poisson manifolds* $(M, \{\cdot, \cdot\})$ where $\{\cdot, \cdot\}$

- is bilinear from $C^\infty(\mathcal{M}) \times C^\infty(\mathcal{M}) \to C^\infty(\mathcal{M})$,

- satisfies Leibniz's rule $\{HK, L\} = H\{K, L\} + \{H, L\}K$,

- is non-degenerate, i.e. $\{H, K\} = 0$ for all functions K implies $H = \text{const.}$,

- satisfies the Jacobi identity.

Exercise: Prove the equivalence.

Question: How do symplectic manifolds arise? There are three main sources for symplectic manifolds:

(i) T^*X, i.e. cotangent bundles of manifolds

(ii) co-adjoint orbits of groups on their dual Lie algebras (Kostant-Kirillov 2-forms)

(iii) constrained systems.

(i) Let X be a manifold, let $x \in X$, and let $\alpha_x \in T_x^*X$. Let v be a vector field on T^*X. Let π denote the natural projection on T^*X to the base point x: thus $\pi(\alpha_x) = x$ and $\pi_* v \in T_x X$. Hence, as $\alpha_x \in T_x^*X$,

$$\theta(\alpha_x)(v) \equiv -\alpha_x(\pi_* v)$$

defines a natural 1-form θ on T^*X. Then $w \equiv d\theta$ defines a 2-form on T^*X, which is clearly closed, and which can easily (**Exercise**) be shown to be non-degenerate.

Exercise: On $\mathbb{R}^{2n} = T^*\mathbb{R}^n$, show that $\theta = -\sum_{i=1}^{n} y_i dx_i$ and hence

$$w = d\theta = \sum_{i=1}^{n} dx_i \wedge dy_i.$$

The above construction is very basic in mathematics and leads to the appearance of symplectic manifolds in many different mathematical situations.

(ii) Let $\mathcal{G}$ be a Lie algebra with Lie bracket $[\cdot, \cdot]$. Thus $[\cdot, \cdot]$ is

- bilinear, and

- satisfies the Jacobi identity, i.e. $[[x, y]z] + [[y, z], x] + [[z, x], y] = 0$, for all $x, y, z \in \mathcal{G}$.

Let G be the associated (connected) group and let $\mathcal{G}^*$ be the dual Lie algebra. Then G acts on $\mathcal{G}$ by the Ad-action,

$$\begin{cases} Ad: \mathcal{G} \to \mathcal{G} \\ Ad_g x = \left. \frac{d}{dt} \right|_{t=0} g e^{tx} g^{-1}, \quad g \in G, x \in \mathcal{G}. \end{cases}$$

Also G acts on $\mathcal{G}^*$ by adjointness

$$\begin{cases} Ad^*: \mathcal{G}^* \to \mathcal{G}^* \\ \langle Ad_g^* \alpha, x \rangle \equiv \langle \alpha, Ad_g x \rangle = \alpha(Ad_g x), \end{cases}$$

where $\langle \cdot, \cdot \rangle$ denotes the pairing of $\mathcal{G}^*$ and $\mathcal{G}$.

Exercise: Show that $Ad_{g_2 g_1}^* = Ad_{g_1}^* Ad_{g_2}^*$.

The *co-adjoint orbit* O_α through a point $\alpha \in \mathcal{G}^*$ is given by

$$O_\alpha = \{Ad_g^* \alpha : g \in G\}.$$

The remarkable fact is that O_α carries a non-degenerate 2-form, and hence is naturally a symplectic manifold (and hence is always an even dimensional manifold). To see what this 2-form is one can proceed functorially. Vector fields on O_α through the point $\beta = Ad_g^* \alpha \in O_\alpha$, are given by $\left. \frac{d}{dt} \right|_{t=0} Ad_{e^{tx}}^* \beta$ for arbitrary $x \in \mathcal{G}$. Indeed,

$$\left. \frac{d}{dt} \right|_{t=0} Ad_{e^{tx}}^* \beta = \left. \frac{d}{dt} \right|_{t=0} Ad_{e^{tx}}^* Ad_g^* \alpha = \left. \frac{d}{dt} \right|_{t=0} Ad_{g e^{tx}}^* \alpha \in T_\alpha O_\alpha.$$

Question: How should we define

$$w_\beta \left(\left. \frac{d}{dt} \right|_{t=0} Ad_{e^{tx}}^* \beta, \quad \left. \frac{d}{dt} \right|_{t=0} Ad_{e^{ty}}^* \beta \right)?$$

The only natural way is

$$\equiv \beta([x, y]) = \langle \beta, [x, y] \rangle,$$

and indeed with this definition (**Exercise**), O_α is indeed a symplectic manifold.

Example: Consider

$$\begin{aligned} G &= Gl^+(n, \mathbb{R}) = Gl(n, \mathbb{R}) \cap \{M : \det M > 0\}, \\ \mathcal{G} &= gl(n, \mathbb{R}) = M(n, \mathbb{R}). \end{aligned}$$

We can identify $\mathcal{G}^*$ as $M(n, \mathbb{R})$ through the non-degenerate pairing

$$\langle A, B \rangle \equiv \operatorname{tr} AB$$

i.e. the matrix A induces a linear map on $\mathcal{G}$ to $\mathbb{R}$ through

$$B \mapsto \operatorname{tr} AB,$$

and moreover every linear map $\ell(B)$ on $\mathcal{G}$ to $\mathbb{R}$ is of this form for some (unique) $A = A(\ell)$.

Now for $g \in G, x \in \mathcal{G}$, $Ad_g x = \frac{d}{dt}\big|_{t=0} ge^{tx}g = gxg^{-1}$, where the RHS can now be taken to be ordinary matrix multiplication. Also

$$\begin{aligned} \langle Ad_g^* A, x \rangle &= \langle A, Ad_g x \rangle = \langle A, gxg^{-1} \rangle = \text{tr } Agxg^{-1} = \text{tr } g^{-1}Ag\, x \\ &= \langle g^{-1}Ag, x \rangle. \end{aligned}$$

Thus $Ad^* gA = g^{-1}Ag$ and hence

$$\begin{aligned} O_A &= \{g^{-1}Ag: g \in G\} \\ &= \{\text{set of all (real) matrices that are } Gl^+(n,\mathbb{R})\text{-conjugate to } A\}. \end{aligned}$$

Also $\dim O_A$ can be any even number $\leq n^2 - n$.

Exercise: Compute all possible co-adjoint orbits O_A in the case of 2×2 matrices.

Now observe that each $x \in \mathcal{G}$ induces a function H_x on $\mathcal{G}^*$ in a natural way, as follows,

$$H_x(A) = \langle A, x \rangle = \text{tr } Ax.$$

For such functions H_x we have

$$\begin{aligned} dH_x\left(\frac{d}{dt}\Big|_{t=0} Ad_{e^{ty}}^* A\right) &= \frac{d}{dt}\Big|_{t=0} H_x(Ad_{e^{ty}}^* A) = \frac{d}{dt}\Big|_{t=0} H_x(e^{-ty}Ae^{ty}) \\ &= \frac{d}{dt}\Big|_{t=0} \text{tr}(e^{-ty}Ae^{ty}x) = \frac{d}{dt}\Big|_{t=0} \text{tr}(Ae^{ty}xe^{-ty}) \\ &= \text{tr}(A[y,x]) \\ &= w_A\left(\frac{d}{dt}\Big|_{t=0} Ad_{e^{ty}}^* A, \quad \frac{d}{dt}\Big|_{t=0} Ad_{e^{tx}}^* A\right). \end{aligned}$$

Thus

$$V_{H_x}(A) = -\frac{d}{dt}\Big|_{t=0} Ad_{e^{tx}}^* A = -\frac{d}{dt}\Big|_{t=0} e^{-tx}Ae^{tx} = [x, A].$$

But then this means that

$$\{H_x, H_y\} = w_A(V_{H_x}, V_{H_y}) = \text{tr } A[x, y].$$

The differential $dH_x(A)$ is by definition the (unique) functional on $\mathcal{G}^*$, and hence the unique element in $(\mathcal{G}^*)^* = \mathcal{G}$, such that

$$dH_x(A)(B) = \frac{d}{dt}\Big|_{t=0} H_x(A + tB) = \frac{d}{dt}\Big|_{t=0} \text{tr } x(A + tB) = \text{tr } xB,$$

which implies $dH_x(A) = x$, and we see that

$$\{H_x, H_y\}(A) = \text{tr } A[dH_x(A), dH_y(A)].$$

By linearity and Leibniz's rule we conclude that

$$\{H, K\}(A) = \text{tr } A[dH(A), dK(A)]$$

for arbitrary smooth functions on $\mathcal{G}^*$.

Exercise: Show that arbitrary smooth functions on $\mathcal{G}^*$ can be approximated by finite linear combinations of powers of functions of type H_x.

On more general dual Lie algebras the above formula becomes

$$\{H, K\}(\alpha) = \langle \alpha, [dH(\alpha)dK(\alpha)\rangle.$$

In the case of $Gl^+(n, \mathbb{R})$, for $H\colon M(n, \mathbb{R}) \to \mathbb{R}$, we have

$$\frac{d}{dt}\Big|_{t=0} H(A + tB) = \sum_{i,j} \frac{\partial H}{\partial A_{ij}} B_{ij} = \operatorname{tr} \nabla H^T(A)B$$

where $\nabla H(A)$ is the matrix with entries $\left(\frac{\partial H}{\partial A_{ij}}\right)$. Thus

$$\{H, K\}(A) = \operatorname{tr}\left(A[\nabla H^T(A), \nabla K^T(A)]\right).$$

Notice that what we have *really* constructed is a *Poisson manifold* i.e. $\mathcal{G}^*$ is a manifold with a bracket $\{\cdot, \cdot\}$ which satisfies all the conditions for a Poisson bracket, except that it is in general degenerate. However $\mathcal{G}^*$ is foliated by *symplectic leaves* i.e. submanifolds of $\mathcal{G}^*$ which are symplectic i.e. on which the 2-form becomes non-degenerate. These symplectic leaves are precisely the co-adjoint orbits. We will say much more about this construction in the next lecture.

Exercise: In the case that the base manifold X is a group, say G, the manifolds in (i) and (ii) are related. Basically the symplectic structure in (ii) is a pull-back of the structure in (i) to the identity in G. Make this explicit.

(iii) *Constrained systems*

Suppose we have n one-dimensional harmonic oscillators, $\ddot{x}_i + \lambda_i x_i = 0$, $1 \le i \le n$. This flow is generated by the Hamiltonian $H = \frac{1}{2}\sum_{i=1}^{n} y_i^2 + \lambda_i x_i^2$ on $\left(\mathbb{R}^{2n}, \sum_{i=1}^{n} dx_i \wedge dy_i\right)$. Suppose we now constrain the oscillators to lie on the sphere

$$\phi_1(x) = \sum_{i=1}^{n} x_i^2 - 1 = 0.$$

How would we describe the motion? Well, we would recall how we solved the problem in our first physics course of a particle moving along a wire, and proceed accordingly. The Hamiltonian version of this procedure is the following.

Let $\phi_2 = \sum_{i=1}^{n} x_i y_i \left(= \sum_{i=1}^{n} x_i \dot{x}_i\right)$. Let $X = \{(x, y) \in \mathbb{R}^{2n}\colon \phi_1 = \phi_2 = 0\}$. Clearly the constrained motion should lie on $X \subset \mathbb{R}^{2n}$. Moreover X is even dimensional and it carries a natural 2-form i^*w, the pull-back of w under the immersion $i\colon X \to \mathbb{R}^{2n}$. Alternatively, $w|X$ is just the restriction of w to $TX \subset T\mathbb{R}^{2n}$. As the operator d commutes with the restriction (or the pull-back) operation it is clear that $d(w|X)$ is closed. The only question is whether it is non-degenerate.

Exercise: Show that $\{\phi_1, \phi_2\} \neq 0$, and hence $w|X$ is non-degenerate.

The constrained flow is generated by the Hamiltonian $H = \frac{1}{2} \sum\limits_{i=1}^{n} (y_i^2 + w_i^2 x_i^2)$ restricted to X with 2-form $w|X$.

Exercise: Compute the equations of motion for the above constrained flow. This constrained system is called the Neumann system. As a reference for constrained motion, see for example [DLTr]. We will say more about such systems in the next lecture.

Question: How can we integrate a dynamical system

$$(3) \qquad \qquad \dot{x} = V(x)$$

in $\mathbb{R}^m$, say? Suppose we are really lucky and have $m - 1$ (independent) conserved quantities $\phi_1, \ldots, \phi_{m-1}$, so that $\frac{d}{dt}\phi_j(x(t)) = 0$, $j = 1, \ldots, m-1$, for solutions $x(t)$ of (3). Then we could solve for $m - 1$ of the variables in favor of the first one, say, and then we would be left with the equation $\frac{dx_1}{dt} = V(x_1, x_2(x_1), \ldots, x_m(x_1))$, which can then be integrated by quadrature.

Now in the theory of Hamiltonian, as opposed to general, systems a remarkable reduction occurs. One can "solve" systems of dimension $m = 2n$ which have only n (independent) integrals $\phi_1, \ldots, \phi_n$, provided that these integrals have the additional property,

$$\{\phi_i, \phi_j\} = 0, \qquad 1 \leq i, j \leq n,$$

i.e. the integrals *Poisson commute*. As the Hamiltonian H for the system is conserved, we may always take one of the integrals, say ϕ_1, equal to H.

The main theorem in the subject is the so-called Liouville-Arnold-Jost Theorem, which says the following:

A Hamiltonian vector field V_H is called *integrable* (or *completely integrable*) on a domain $D \subset M^{2n}$ if it possesses n integrals $\phi_1(= H)$, $\phi_2, \ldots, \phi_n$ which are linearly independent on D (i.e. $d\phi_i, \ldots, d\phi_n$ are linearly independent at all points of D) and which Poisson commute.

We require that D is invariant under the flow generated by H for *all* t. Any Hamiltonian system is locally integrable i.e. given any H and a point $m \in M^{2n}$, there exists a neighborhood B of m such that H has n Poisson commuting integrals in this neighborhood (**Exercise:** prove this). But in general the flow generated by H escapes from B, and so these local integrals teach us nothing about the global behavior of the flow. The invariance of D is essential.

Theorem 1 *(Liouville-Arnold-Jost) Suppose that $H = \phi_1$ is integrable on a domain $D \subset M$ with integrals $\phi = \{\phi_1, \ldots, \phi_n\}$ and suppose that $N_0 = \phi^{-1}(0) \subset M$ is compact and connected. Then*

(a) N_0 is an imbedded n-dimensional torus T^n

(b) there exists an open neighborhood $U(N_0) \subset M$ which can be coordinatized as follows: if $x = \{x_1, \ldots, x_n\}$ are variables on the torus $\mathbb{T}^n = \mathbb{R}^n/\mathbb{Z}^n$ and $y = (y_1, \ldots, y_n) \in D_1$, where $D_1 \subset \mathbb{R}^n$ is a domain containing the origin 0, there exists a diffeomorphism

$$\psi \colon \ \mathbb{T}^n \times D_1 \to U(N_0).$$

Moreover ψ is symplectic $\left(i.e.\ \psi^* w = \sum\limits_{i=1}^{n} dx_i \wedge dy_i \right)$ *and*

$$H \circ \psi = h(y_1, \ldots, y_n)$$

for some function h.

In particular, near a compact, connected level set $\phi^{-1}(0)$, the flow is extremely simple. Indeed, as ψ is symplectic, the flow generated by H is again Hamiltonian in the variables (x, y) of $\mathbb{T}^n \times D_1$ (**Exercise**: show this), and we find

$$\dot{x}_j = \frac{\partial}{\partial y_j}(H \circ \psi) = \frac{\partial}{\partial y_i}h, \quad \dot{y}_j = -\frac{\partial}{\partial x_j}(H \circ \psi) = -\frac{\partial}{\partial x_j}h = 0,$$

so that

$$x_j(t) = x_j(0) + t\frac{\partial h}{\partial y_j}(y_1, \ldots, y_n), \qquad y_j(t) = y_j(0).$$

Thus the system can be integrated explicitly and the solution is given by straight line motion on a torus. The variables $\{y_j\}$ are called the *actions* for the system and the $\{x_j\}$ are called the *angles*.

The Theorem not only tells us how to integrate the system (in terms of the variables (x, y), which may or may not be hard to construct), but perhaps more importantly, we can understand the qualitative behavior of the system.

The quantities $w = (w_1, \ldots, w_n) = \left(\frac{\partial h}{\partial y_1}, \ldots, \frac{\partial h}{\partial y_n} \right)$ are called the *frequencies* of the system. The neighborhood $U(N_0)$ in the Theorem is foliated by tori which we parameterized by the values of $y \in D_1$. As $w = w(y)$, and as y is conserved by the flow, the frequencies are constant on the torus. However, in general, they vary from torus to torus. We can distinguish tori according to whether the w_i's are *rationally independent* or *rationally dependent*.

Rationally independent: Here $\sum\limits_{i=1}^{n} j_i w_i = 0$ for $j_i \in \mathbb{Z} \Rightarrow j_i = 0$. In this case, by a well-known theorem of Kronecker, $\{x_0 + tw : t \in \mathbb{R}\}$ is dense in $\mathbb{T}^n$. Thus the orbit of the flow is dense on such tori and in fact the flow is quasi-periodic in time with n frequencies.

Rationally dependent: Here there exist $j_i \in \mathbb{Z}$, not all zero, such that $\sum\limits_{i=1}^{n} j_i w_i = 0$. Then the flow is restricted to a sub-torus of $\mathbb{T}^n$. For example if $n = 2$ and $w_1 - 2w_2 = 0$, then the flow is restricted to $\{(x_1, x_2) \in \mathbb{T}^2 : x_1 - 2x_2 = \text{const}(\bmod\ \mathbb{Z})\}$. Again the flow is almost periodic, but with fewer frequencies.

Thus the essential problem of describing the long-time behavior of integrable Hamiltonian systems is solved, in principle, provided that the invariant set $\phi^{-1}(0)$ is compact and connected. If $\phi^{-1}(0)$ is compact but not connected, we can just restrict our attention to each connected component. Also if $\phi^{-1}(0)$ is not compact, then the Theorem goes through provided it is known a priori that each ϕ_i generates a global flow, at least for data near $\phi^{-1}(0)$. In this case one learns that $\phi^{-1}(0)$ has a neighborhood $U(N_0)$ which is a thickening by an n-dimensional disk D_1 of a product of lines and circles,

$$U(N_0) = \mathbb{T}^k \times \mathbb{R}^{n-k} \times D_1.$$

On each leaf $\mathbb{T}^k \times \mathbb{R}^{n-k} \times \{y\}$ the flow is again given by straight line motion, but now the winding takes place on a cylinder rather than a torus.

The Liouville-Arnold-Jost Theorem describes qualitatively the behavior of an integrable system. In each case the serious and essential task remains of determining the action-angle variables explicitly.

Sketch of the proof of the Theorem:

As $\{\phi_i, \phi_j\} = 0$, $[v_{\phi_i}, v_{\phi_j}] = v_{\{\phi_i, \phi_j\}} = 0$ which shows that the Hamiltonian vector fields v_{ϕ_i} induced by the ϕ_i's commute. This means that we can immerse $\mathbb{R}^{2n}$ into M^{2n} as follows. Let $\psi_i^{t_i}(m) = \psi_i(t_i, m)$, $\psi_i(0, m) = m$, be the flow induced on M^{2n} by ϕ_i. Then, fixing $m_0 \in \phi^{-1}(0)$, the map

$$\mathbb{R}^n \ni t = (t_1, \ldots, t_n) \longmapsto \Gamma(t) \equiv \psi_1^{t_1} \circ \cdots \circ \psi_n^{t_n}(m_0)$$

takes $\mathbb{R}^n$ into the level set $N_0 = \{m \colon \phi_i(m) = \phi_i(m_0),\ i = 1, \ldots, n\}$. This is because $\frac{d}{dt_j}\phi_i(\psi_j^{t_j}(m)) = \{\phi_i, \phi_j\}(\psi_j^{t_j}(m)) = 0$, so that ϕ_i is a constant of the motion for the ϕ_j-flow, $1 \le i, j \le n$. Simple arguments show that Γ is onto N_0. Let $\Lambda = \{t \in \mathbb{R}^n \colon \Gamma(t) = m_0\}$. Now as the flows commute it follows that Λ is a lattice in $\mathbb{R}^n$. On the other hand $\mathbb{R}^n/\Lambda$ is mapped diffeomorphically onto N_0. But N_0 is compact by assumption. It follows that Λ *must* have n generators and hence $\mathbb{R}^n \backslash \Lambda$ is a torus, and hence N_0 is a torus. Then one has to "thicken" things ....

Elementary examples:

(i) *Harmonic oscillator*
 $M = \mathbb{R}^2, H = \frac{1}{2}(p^2 + w^2 q^2) = \phi_1$.

The system is integrable as ϕ_1 is conserved and $n = 1$ (clearly all Hamiltonian systems on 2-dimensional manifolds are integrable!). $N_0 = \{(q, p) \colon H = \phi_1 = C > 0\} = \{(q, p) \colon p^2 + w^2 q^2 = 2C\}$, which is clearly a torus. The equations of motion are

$$\dot{q} = H_p = p, \qquad \dot{p} = -Hq = -w^2 q,$$

with solution

$$q = \frac{\sqrt{2C}}{w}\sin(wt + \alpha), \qquad p = \sqrt{2C}\,\cos(wt + \alpha).$$

Note $p^2 + w^2 q^2 = 2C$.

The map ψ in the Theorem is constructed as follows. We can take $D_1 = \mathbb{R}^+$. Then

$$\mathbb{R}^+ \times \mathbb{T}^1 \ni (y, x) \longmapsto \psi(y, x) = (q(y, x), p(y, x)) \equiv \left(\sqrt{\frac{y}{\pi w}}\sin 2\pi x, \sqrt{\frac{wy}{\pi}}\cos 2\pi x\right).$$

We have

$$
\begin{aligned}
H \circ \psi(y, x) &= \frac{1}{2}\left(\left(\sqrt{\frac{wy}{\pi}}\cos 2\pi x\right)^2 + w^2\left(\sqrt{\frac{y}{\pi w}}\sin 2\pi x\right)^2\right) \\
&= \frac{w}{2\pi}y
\end{aligned}
$$

and

$$\psi^*(dq \wedge dp) = \left(\frac{1}{2}\frac{1}{\sqrt{\pi w y}}\sin 2\pi x\, dy + \sqrt{\frac{y}{\pi w}}\cos 2\pi x\, 2\pi\, dx\right) \wedge$$

$$\left(\frac{1}{2}\sqrt{\frac{w}{\pi y}}\cos 2\pi x\, dy - \sqrt{\frac{wy}{\pi}}\sin 2\pi x\, 2\pi\, dx\right)$$

$$= dx \wedge dy.$$

In the (x, y) variables the flow becomes

$$\dot{x} = \frac{\partial}{\partial y}H \circ \psi = \frac{w}{2\pi}, \qquad \dot{y} = -\frac{\partial}{\partial x}H \circ \psi = 0$$

so that $x(t) = \frac{wt}{2\pi} + x_0$, $y(t) = y_0$, which implies

$$q(t) = \sqrt{\frac{y_0}{\pi w}}\sin(wt + 2\pi x_0), \qquad p(t) = \sqrt{\frac{wy_0}{\pi}}\cos(wt + 2\pi x_0),$$

as it should.

Exercise: Analyze $H = \frac{1}{2}\sum\limits_{i=1}^{n} p_i^2 + w_i^2 q_i^2$ in $\mathbb{R}^{2n}$.

(ii) *Simple pendulum*

Here $M^2 = (\mathbb{T} \times \mathbb{R}, w = dq \times dp)$ and $H = \frac{1}{2}p^2 + 1 - \cos 2\pi q$, which gives rise to the equations of motion, $\dot{q} = H_p = p$, $\dot{p} = -H_q = -2\pi \sin 2\pi q$, or $\ddot{q} + 2\pi \sin 2\pi q = 0$. Here $2\pi q$ is the angle that the pendulum makes with the vertical

Exercises:

(a) The motion of the pendulum depends on the value of $H = C > 0$. Show that there are three different cases, $C < 2$, $C = 2$, $C > 2$. If $C < 2$, the pendulum oscillates back and forth with $|2\pi q(t)| < \pi$. If $C > 2$, the pendulum rotates "over the top". If $C = 2$, the pendulum moves from $2\pi q = -\pi$ to $2\pi q = +\pi$ as t runs from $-\infty$ to $+\infty$; this case is the so-called *separatrix* for the system.

(b) Describe $N_0 = \{(q, p)\colon H(q, p) = C\}$ in the above three cases and draw a picture of M^2 foliated by the invariance sets $N_0 = N_0(C)$ for all values of $C > 0$.

(c) Construct the map ψ of the Theorem.

Lecture 2

The Liouville-Arnold-Jost Theorem describes the flow of an integrable system, once we have determined that the system is indeed integrable. The big question is, how do we determine whether a given system is integrable? The true analog of this question already arose in freshman calculus. You are given a list of twenty integrals to do and you struggle with one of them, say, over the entire weekend. Then someone tells you to replace the variable x by $y = e^{x^2+x} \log(1 + x^4)$, say, and then the whole problem pops out! The truth is that there is no *systematic* way to determine whether a given system is integrable. You have to guess and be lucky. You can analyze large classes of systems which have similar features and free parameters and which are known to be integrable, and see whether a given system fits into the class. You can also test a system to see if it behaves like an integrable system. Does it have "soliton-like" solutions, for example. But to really discover the structure that makes it integrable, you have to guess — you just have to guess that "$y = e^{x^2+x} \log(1 + x^4)$" will do the job for you!

Now in the nineteenth century many systems were shown to be integrable. For example, in a famous paper Jacobi showed that geodesic flow on an ellipsoid (he considered only the 3-dimensional case) was integrable. This example generated great interest at the time because he showed that the functions he needed to solve the flow were hyperelliptic functions and that the flow was linearized by mapping certain divisors on a Riemann surface onto the associated Jacobi variety via the Abel map (for a recent analysis of the flow see [M2]). Of course the pride of 19th century mathematics was the Theory of Riemann surfaces, and the fact that this theory, recently developed, was exactly what was needed to explicitly solve some dynamical system, created great excitement in the mathematical world. Other examples, amongst many others, were the Neumann system of n harmonic oscillators $\ddot{x}_i + w_i^2 x_i = 0$, $1 \le i \le n$, constrained to lie on a sphere $\sum_{i=1}^{n} x_i^2 = 1$ (this system was shown to be integrable in the case $n = 3$ by C. Neumann — for a recent analysis of the general case, see, for example, [M2]), a variety of spinning tops, and of course Sonia Kovalevskaya's extraordinary discovery of her integrable top with its deep connections to complex function theory and the origins of Painlevé theory. Geometric symmetries of a system give rise to integrals of the motion, a fact most clearly understood by Emmy Noether — for example, translation invariance gives rise to conservation of momentum, etc. But this rarely gives us enough integrals: in general we must find other integrals which people like to think of as arising from "hidden symmetries". The whole exuberant development of the theory of integrable systems came to an abrupt halt at the end of the last century, however, as a result of a theorem of Poincaré, who showed for the 3-body problem that, apart from the geometric integrals, there are no other conserved quantities. At this point the explicit solution of dynamical systems declined as an area of mathematical activity, and the so-called qualitative theory of dynamical systems, pioneered by Poincaré himself, came to the fore.

This was the situation for almost 80 years, until the remarkable discovery of Kruskal and his collaborators (see [GGKM]). In 1967 they showed that the Korteweg de Vries (KdV) equation

(4) $$q_t - 6qq_x + q_{xxx} = 0$$

of shallow water wave theory was "integrable" in a precise sense and they showed
how to linearize the flow explicitly. The extraordinary message of their work was
that the integrals for the flow were contained in spectral and scattering theoretic
quantities of an associated operator

$$H(t) = -\frac{d^2}{dx^2} + q(x, t).$$

The fundamental significance of this work was quickly grasped by P. Lax [L],
who showed that the KdV equation was equivalent to an *isospectral deformation* of
$H(t)$. Indeed he showed that

$$(4) \quad \equiv \quad \frac{dH}{dt} = [B(t), H(t)] = B(t)H(t) - H(t)B(t),$$

$$\text{where } B(t) = -4\frac{d^3}{dx^3} + 3\left(q(x, t)\frac{d}{dx} + \frac{d}{dx}q(x, t)\right).$$

To see that the RHS is indeed an isospectral deformation of $H(t)$ let $Q(t)$ be the
solution of

$$\frac{dQ}{dt} = -QB, \qquad Q(t = 0) = I.$$

Then $\frac{dQ^*}{dt} = BQ^*$, as $B = -B^*$, and hence

$$\frac{d}{dt}(QQ^*) = -QBQ^* + QBQ^* = 0,$$

so that $QQ^* = I$. A similar argument shows that $Q^*Q = I$, and hence Q is unitary.
Set $L(t) = Q^*(t)H(0)Q(t)$. Then

$$\frac{d}{dt}L = BQ^*H(0)Q + Q^*H(0)(-QB) = [B, L].$$

As $L(0) = H(0)$, we conclude that $H(t) = L(t) = Q^*(t)H(0)Q(t)$, i.e., $H(t)$ is
unitarily equivalent to $H(t = 0)$. In particular, if $\lambda(H(t))$ is an eigenvalue of $H(t)$.
Then $\lambda(H(t)) = \text{const.}$, etc. Thus the spectrum of $H(t)$ gives integrals for the KdV
equation. These are the integrals that arise from "hidden symmetries".

   If the matter would have stayed there, we would just have been looking at some
glorious, but isolated, example. But now we know, some 27 years later, and due
to the efforts of many people, that there are literally hundreds of systems of prime
physical interest that are integrable via a spectral transform. More precisely, at-
tached to the dynamical system of interest, there is an associated linear operator,
and the motion of the system is equivalent to an isospectral deformation of the
operator, and hence the eigenvalues, etc., of the operator provide integrals for the
flow. And indeed many people in the "integrable business" believe that a system
is integrable, if and only if it is equivalent to the isospectral deformation of some
associated operator.

   In the rest of this lecture I am going to tell the story of just one such integrable
system. I hope you won't mind my focusing on this system, but its story contains the
elements of all other integrable systems. I am going to consider the so-called *Toda
Lattice*, consisting of n particles on the line interacting with exponential forces. Thus

we consider the flow generated on $\left(\mathbb{R}^{2n}, w = \sum_{i=1}^{n} dx_i \wedge dy_i \right)$ by the Hamiltonian

$H_T = \frac{1}{2} \sum_{i=1}^{n} y_i^2 + \sum_{i=1}^{n-1} (e^{(x_i - x_{i+1})}$. A general reference for the approach that we will follow is [DLTo] and the references therein.

The so-called Toda lattice equations generated by H_T have the form

$$(5) \qquad \ddot{x}_i = e^{x_{i-1} - x_i} - e^{x_i - x_{i+1}}, \qquad 1 \le i \le n,$$

where $e^{x_0 - x_1} = e^{x_n - x_{n+1}} \equiv 0$. This system has the standard geometric integrals $H_T = $ energy and $P = \sum_{i=1}^{n} y_i = $ total momentum, but it was clear from the numerical calculations of J. Ford in the mid 70's that something very special and "KdV-like" was going on with these equations. And indeed in 1975 Flaschka, and independently Manakov, showed that the Toda lattice was equivalent to an isospectral deformation, as follows.

Introduce *Flaschka's variables*,

$$a_k = -y_k/2, \qquad k = 1, \ldots, n,$$
$$b_k = \frac{1}{2} e^{\frac{1}{2}(x_k - x_{k+1})}, \qquad k = 1, \ldots, n - 1,$$

and consider the Jacobi matrix

$$L = \begin{pmatrix} a_1 & b_1 & & & \\ b_1 & a_2 & b_2 & & \bigcirc \\ & b_2 & \ddots & \ddots & \\ & & \ddots & \ddots & b_{n-1} \\ \bigcirc & & & b_{n-1} & a_n \end{pmatrix} = L^T$$

and the associated matrix

$$B = B(L) = \begin{pmatrix} 0 & b_1 & & & \bigcirc \\ -b_1 & 0 & b_2 & & \\ & -b_2 & \ddots & \ddots & \\ & & \ddots & \ddots & -b_{n-1} \\ \bigcirc & & -b_{n-1} & 0 \end{pmatrix} = -B^T.$$

Note that $B = L_+ - L_+^T$ where L_+ denotes the upper part of L. A straightforward calculation shows that

$$(5) \equiv \frac{dL}{dt} = [B, L].$$

Thus the eigenvalues $\lambda_1, \ldots, \lambda_n$ of the associated matrix L give n integrals for the Toda flow. Moreover (**Exercise**) one can show that

$$\{\lambda_i, \lambda_j\} = 0 \qquad 1 \le i, j \le n$$

in $\left(\mathbb{R}^{2n}, w = \sum_{i=1}^{n} dx_i \wedge dy_i \right)$. Hence the system is completely integrable.

According to the Liouville-Arnold-Jost Theorem, in appropriate variables, the flow becomes a straight line motion on some cylinder $\mathbb{R}^k \times \mathbb{T}^{n-k}$. What are these variables? Associated to each λ_i, there is a normalized eigenvector v_i, $\sum_{j=1}^{n} v_i^2(j) = 1$. All eigenvalues are simple (**Exercise**) and without loss of generality (**Exercise**) we can and do specify v_i uniquely by choosing $v_i(1) > 0$. Label the eigenvalues such that $\lambda_1 < \cdots < \lambda_n$. Differentiating the eigenvalue relation $(L - \lambda_i)v_i = 0$ with respect to t, we obtain

$$(\dot{L} - \dot{\lambda}_i)v_i + (L - \lambda_i)\dot{v}_i = 0,$$

or, as $\dot{\lambda}_i = 0$,

$$(BL - LB)v_i + (L - \lambda_i)\dot{v}_i = 0,$$

which implies that

$$(L - \lambda_i)(\dot{v}_i - Bv_i) = 0.$$

Hence $\dot{v}_i - Bv_i = \mu v_i$ for some scalar $\mu = \mu(t)$. But $(v_i, v_i) = 1$, and so $(v_i \dot{v}_i) = 0$. Thus $\mu = \mu(v_i, v_i) = \mu[((v_i, \dot{v}_i) - (v_i, Bv_i)] = 0$, as $B = -B^T$. This proves that $\dot{v}_i = Bv_i$. In particular

$$\dot{v}_i(1) = \left(e_1, \left(\begin{pmatrix} 0 & b_1 & \\ -b_1 & 0 & b_2 \\ & & \ddots \end{pmatrix} \begin{pmatrix} v_i(1) \\ v_i(2) \\ \vdots \end{pmatrix} \right) \right) = b_1 v_i(2).$$

But $(L - \lambda_i)v_i = 0$ implies, in particular, that $(a_1 - \lambda_i)v_i(1) + b_1 v_i(2) = 0$, and we obtain the relation

$$\dot{v}_i(1) = (\lambda_i - a_1(t))v_i(1).$$

In turn, for $i \neq j$, $\frac{d}{dt}\log(v_i(1)/v_j(1)) = \lambda_i - \lambda_j$, which leads to the formula

$$v_i(1,t) = \frac{v_i(1,0)e^{\lambda_i t}}{\left(\sum_{j=1}^{n} v_j^2(1,0)e^{2\lambda_j t} \right)^{1/2}}.$$

On the other hand, $\frac{d}{dt}\sum_{i=1}^{n} x_i = \sum y_i = P = \text{const.}$ This shows that

$$x_{cm} = \text{centre of mass} = \frac{1}{n}\sum_{i=1}^{n} x_i, \qquad \log\frac{v_i(1)}{v_1(1)}, \qquad 2 \leq i \leq n$$

are the variables which move linearly under (5).

From the point of view of the Liouville-Arnold-Jost Theorem, where is the cylinder $\mathbb{R}^k \times \mathbb{T}^{n-k}$? We have the map χ,

$$J = n \times n \text{ Jacobi matrices} \ni L \longmapsto \chi(L) = (\lambda_1, \ldots, \lambda_n, v_1(1), \ldots, v_n(1))$$

$$\in Y \equiv \left\{ (\alpha, \beta) \in \mathbb{R}^{2n} \colon \alpha_1 < \cdots < \alpha_n, \beta_i > 0, \sum_{i=1}^{n} \beta_i^2 = 1 \right\}.$$

One can show (**Exercise**) that χ is a bijection of J onto Y. Consider the invariant set

$$\{\alpha\} \times \mathbb{R} \times \left\{ \beta \in \mathbb{R}^n \colon \beta_i > 0, \sum_{i=1}^{n} \beta_i^2 = 1 \right\}.$$

Observe that $\log \frac{\beta_i}{\beta_1}$, $i = 2, \ldots, n$, are clearly coordinates for

$$\left\{ \beta \in \mathbb{R}^n \colon \beta_i > 0, \sum_{i=1}^{n} \beta_i^2 = 1 \right\},$$

which shows that this set is $\cong \mathbb{R}^{n-1}$. Thus there is no torus and the invariance set is $\mathbb{R}^n \times \mathbb{T}^{n-n} = \mathbb{R}^n$. The variables $\{\lambda_i\}$, $1 \le i \le n$, x_{cm}, $\log \frac{v_j(1)}{v_1(1)}$, $2 \le j \le n$, are, essentially, the action-angle variables for the Toda flow.

The above solution of the Toda lattice is due to Moser [M1]. Furthermore, Moser showed that as $t \to \infty$,

$$b_i(t) = \frac{1}{2} e^{\frac{1}{2}(x_i - x_{i+1})} \to 0$$
$$\dot{x}_i = y_i = -2a_i(t) \to \text{const.}$$

Physically this means that as $t \to \infty$, particle x_1 moves out to the left at a steady speed, followed by x_2, followed by x_3, etc.

In another important development in 1980, Bill Symes made a wonderful observation. Let $L_0 = L(t = 0)$ and let $e^{tL_0} = Q(t)R(t)$ be the (unique) QR factorization of e^{tL_0} i.e. Q is orthogonal and R is upper triangular with $R_{ii} > 0$. This is just the Gram-Schmidt procedure applied to the columns of e^{tL_0}.

Then, remarkably,

$$L(t) = Q^T(t)L_0 Q(t)$$

is the solution of the Toda lattice equations (5) with $L(0) = L_0$.

Exercise: Verify this by direct calculation.

Observe also that $e^{L(t=1)} = Q^T(1)e^{L_0}Q(1) = Q^T(1)Q(1)R(1)Q(1) = R(1)Q(1)$. Thus (the exponential of) $L(t)$ at time $t = 1$ can be obtained from (the exponential of) L_0 by simply permuting the factors $Q(1)$ and $R(1)$!

At this point Tomei, Nanda, L.C. Li and I became interested in the problem from the following point of view: the Toda lattice in fact gives an algorithm to compute the eigenvalues of a matrix L_0 (here, for the moment $L_0 = L_0^T = $ tridiagonal). Indeed, solve (5) with $L(0) = L_0$. As $t \to \infty$, $b_i(t) \to 0$. When $|b_i(t)| < \varepsilon \ll 1$, the spectrum of $L(t)$ is essentially given by the diagonal elements of $L(t)$, and hence by isospectrality, this gives the eigenvalues of the original matrix L_0! Try it out on your hand calculator: it works!

Our main question, however, was whether this *Toda algorithm* has anything to do with, and perhaps is competitive with, the standard methods for computing the eigenvalues of a matrix. Suppose we have a matrix L_0. How do we compute its eigenvalues? In $98/100$ cases, if you go to a computer and use an eigenvalue package, it will be based on the so-called QR *algorithm*. The heart of the algorithm, which is probably the most successful algorithm in numerical analysis, is the following.

Factor

$$L_0 = Q_0 R_0.$$

Write

$$L_1 \equiv R_0 Q_0 = Q_0^T L_0 Q_0 \text{ (so that } \text{spec}(L_1) = \text{spec}(L_0))$$

and factor

$$L_1 = Q_1 R_1.$$

Write

$$L_2 = R_1 Q_1 \quad \text{(so that } \text{spec}(L_2) = \text{spec}(L_1) = \text{spec}(L_0))$$

and factor

$$L_2 = Q_2 R_2$$

etc. In this way we produce a sequence $\{L_k\}$ of isospectral (in fact unitarily equiv-alent) matrices. In many, many cases L_k converges as $k \to \infty$ to a diagonal matrix L_∞, whose diagonal entries must therefore give the eigenvalues of the original matrix L_0!

Question: does this method have anything to do with the Toda lattice?

We made the following observation. Let $H_{QR}(L) = \text{tr}(L \log L - L)$. Through Flaschka's transformation, in the tridiagonal case, this is a function on $\mathbb{R}^{2n}$. Hence it gives rise to a flow $\phi_{QR}(t, L_0)$, $\phi_{QR}(0, L_0) = L_0$.

Then we showed that

$$\phi_{QR}(k; L_0) = L_k, \text{ the } k^{\text{th}} \text{ step in the } QR \text{ algorithm.}$$

In other words, we proved the following result.

Stroboscope Theorem (tridiagonal case)

The QR algorithm is the time 1 map of a completely integrable Hamiltonian system which commutes with the Toda flow. Furthermore the flow can be solved explicitly by a matrix factorization.

Note: $H_{QR} = \sum_{i=1}^{n} (\lambda_i \log \lambda_i - \lambda_i)$, $H_T = 2 \text{ tr } L^2 = 2 \sum_{i=1}^{n} \lambda_i^2$. Thus

$$\{H_{QR}, H_T\}, \ \{H_{QR}, \lambda_j\} = 0, \quad \text{as} \quad \{\lambda_i, \lambda_j\} = 0.$$

At this point we must introduce some information from Lie group theory. You cannot tell a Lie theorist that some dynamic is taking place on the space of tridi-agonal matrices without eliciting the following response (B. Kostant, $\sim$ '79): the tridiagonal matrices form a (minimal) co-adjoint orbit of the action of the lower triangular group $\mathcal{L}$ on its dual Lie algebra ℓ^*, presented as the symmetric matrices.

This remark opens up the whole game. As we saw in Lecture 1, dual Lie alge-bras are carriers of a (Kostant-Kirillov) Poisson bracket and co-adjoint orbits are the symplectic leaves of this Poisson manifold. In particular we can lift the Toda lattice to a flow on general symmetric matrices. So we can take $\frac{1}{4} H_T = \frac{1}{2} \text{ tr } L^2$ as a Hamiltonian on ℓ^*. With the Kostant-Kirillov bracket this generates a flow on gen-eral symmetric matrices which has the same form (**Exercise**) as in the tridiagonal case

$$\frac{dL}{dt} = [B(L), L], \qquad B(L) = L_+ - L_+^T.$$

Remark. The push forward, under Flaschka's transformation, of the form $\sum_{i=1}^{n} dx_i \wedge dy_i$ to the Jacobi matrices J, induces a 2-form which is a non-trivial multiple of the 2-form that J inherits as a co-adjoint orbit of $\mathcal{L}$. This explains why we must take $\frac{1}{4}H_T$, rather than H_T.

Now the QR algorithm also applies to general symmetric matrices L_0 (as long as $\det L_0 \neq 0$). We generate as before $L_0, L_1, L_2, \ldots, L_k, \ldots$, with $L_k \rightarrow L_\infty =$ diagonal as $k \rightarrow \infty$.

Question: Is the Stroboscope Theorem still true?

On generic co-adjoint orbits, which have dimension $2[n^2/4]$ (**Exercise**), the answer is yes! What are the new integrals? The eigenvalues of L are certainly still conserved. This gives $\sim n$ conserved quantities. But we need $\sim [n^2/4] - n$ more.

Consider a 4×4 example. From $\det(L - \lambda) = 0$ we obtain four eigenvalues $\lambda_{10}, \lambda_{20}, \lambda_{30}, \lambda_{40}$, say. Now consider the 3×3 matrix $(L - \lambda)_1$, obtained by chopping the first row and last column from $L - \lambda$

$$
\begin{pmatrix}
* - \lambda & * & * & * \\
\hline
* & * - \lambda & * & * \\
* & * & * - \lambda & * \\
* & * & * & * - \lambda
\end{pmatrix}.
$$

Then we consider $\det(L - \lambda)_1 = 0$, which has two solutions $\lambda_{11}, \lambda_{21}$. Now it turns out that a generic co-adjoint orbit is given by $\left\{ L = L^T: \sum_{i=1}^{4} \lambda_{i0} = \text{const}, \sum_{i=1}^{2} \lambda_{i1} = \text{const} \right\}$ which has dimension 10-2, which equals $2[4^2/4] = 8$, as it should. Thus we need four integrals. Fortunately $\lambda_{11}, \lambda_{21}$ turn out to be additional integrals for the flow and moreover, all the integrals $\lambda_{10}, \lambda_{20}, \lambda_{30}, \lambda_{40}, \lambda_{11}, \lambda_{21}$ Poisson commute with each other. Thus we may take, for example, the four independent integrals to be $\lambda_{10}, \lambda_{20}, \lambda_{30}$ and λ_{11}. Again the associated angles are the first components of the associated, appropriately normalized, eigenvectors. If $n > 4$, then we consider $\det(L - \lambda)_2 = 0$, where $(L - \lambda)_2$ is obtained by chopping the first two rows and the last two columns from $L - \lambda$, etc.

Now what about general real matrices $L \in M(n, \mathbb{R})$? The equation $\frac{dL}{dt} = [B(L), L]$, $B = L_+ - L_+^T$ still makes sense and generates a global flow (as $B = -B^T$). We need a symplectic structure, and in light of the foregoing discussion, we understand that what we really need is a Lie group and its dual Lie algebra. Consider invertible matrices L, L'. Factor $L = QR$, $L' = Q'R'$ and define a product

$$
L \circ L' \equiv QQ'R'R.
$$

It turns out that with this product the invertible matrices form a Lie group $G = G_{QR}$ with dual Lie algebra $\mathcal{G} = \mathcal{G}_{QR}$. Generic co-adjoint orbits have dimension $n^2 - n$ and if we look at the flow generated by $\frac{1}{4}H_T = \frac{1}{2}\operatorname{tr} L^2$ through the Kostant-Kirillov form on $\mathcal{G}_{QR}^*$ we again get the generalized Toda flow $\dot{L} = [B, L]$. Also the QR algorithm still applies.

Question: Is the Stroboscope Theorem still true?

On generic orbits, the answer is again yes!

We now need $(n^2 - n)/2$ integrals. Approximately $[n^2/4]$ integrals can be obtained by chopping $L - \lambda$ as before. Now consider the polynomial in two variables,

$$P(h, z; L) \equiv \det(L + hL^T - z) = \sum_{i,j} \gamma_{ij}(L)h^i z^j.$$

It turns out that the $\gamma_{ij}(L)$'s give the remaining approximately $[n^2/4]$ integrals. In other words the whole Riemann surface

$$P(h, z; L) = 0$$

is conserved by the flow. Moreover the flows linearize on an appropriately singularized Jacobi variety associated with the curve. Again the QR algorithm is the time 1 map of the flow generated by $H_{QR}(L) = \mathrm{tr}(L \log L - L)$ through the Kostant-Kirillov structure on $\mathcal{G}^*_{QR}$, etc.

Not only are the ideas and constructions introduced in this lecture of theoretical interest, but they can also be used to analyze the performance of eigenvalue and singular value algorithms. However, that's another story . . . .

Lecture 3

In the previous lecture we described how the QR algorithm can be viewed as the time 1 map of a Hamiltonian flow. Moreover the associated symplectic structure is the Kostant-Kirillov 2-form on co-adjoint orbits of a specific Lie group G_{QR}. As a set $\mathcal{G}_{QR}$, the Lie algebra of G_{QR}, is simply $M(n, \mathbb{R})$, the set of all real $n \times n$ matrices, but using the definition of the multiplication rule on G_{QR}, it is easy to see (**Exercise**) that the Lie bracket on $\mathcal{G}_{QR}$ is given by

$$[M, M']_{QR} = [\pi_\ell M, \pi_\ell M'] - [\pi_s M, \pi_s M']$$

where $\pi_s M = M_+ - M_+^T$, $\pi_\ell M = (1 - \pi_s)M = M_- + M_0 + M_+^T$. Here M_-, M_0, M_+ are the strictly lower, diagonal and strictly upper parts of M respectively, $M = M_- + M_0 + M_+$, and $[\cdot, \cdot]$ denotes the standard commutator. The dual Lie algebra $\mathcal{G}_{QR}^*$ can be identified with $\mathcal{G}_{QR}$ through the non-degenerate pairing

$$\langle M, N \rangle = \operatorname{tr} MN.$$

In the previous lecture we mentioned the basic "conjecture", strongly believed by many people, that every integrable system is an isospectral deformation of some associated linear operator. There is a more refined conjecture that I now want to describe.

Given a Lie algebra $(\mathcal{G}[\cdot, \cdot])$, a linear operator $R \colon \mathcal{G} \to \mathcal{G}$ is said to be a *classical R-matrix* (reference [STS]) if the bracket on $\mathcal{G}$ given by

$$[x, y]_R = \frac{1}{2}([Rx, y] + [x, Ry])$$

is again a Lie bracket. As $[\cdot, \cdot]_R$ is bilinear and skew, the only issue is whether it satisfies the Jacobi identity. Thus if R is a classical R-matrix on $\mathcal{G}$ we obtain two Lie algebras $(\mathcal{G}, [\cdot, \cdot])$ and $(\mathcal{G}_R \equiv \mathcal{G}, [\cdot, \cdot]_R)$, and two associated Lie-Poisson brackets (see Lecture 2),

$$\begin{aligned} \{F_1, F_2\}(\alpha) &= \alpha([dF_1(\alpha), dF_2(\alpha)]), & \alpha \in \mathcal{G}^*, \\ \{F_1, F_2\}_R(\alpha) &= \alpha([dF_1(\alpha), dF_2(\alpha)]_R), & \alpha \in \mathcal{G}_R^* \equiv \mathcal{G}^*. \end{aligned}$$

We call the first Lie algebra/bracket the *free* Lie algebra/bracket and the second the *interacting* Lie algebra/bracket.

An interesting and useful sufficient condition for $[\cdot, \cdot]_R$ to be a Lie bracket is the so-called *modified Yang-Baxter equation* (mYB),

(mYB) $\qquad [Rx, Ry] - 2R([x, y]_R) = -[x, y], \quad \text{for all} \quad x, y \in \mathcal{G}.$

Thus if R satisfies (mYB), $[\cdot, \cdot]_R$ is a Lie bracket. Now it is a fundamental observation (**Exercise**) the (mYB) can be reformulated, equivalently, as

(6) $\qquad [(R \pm 1)x, (R \pm 1)y] = 2(R \pm 1)[x, y]_R, \text{ for all } x, y \in \mathcal{G}.$

This shows immediately that R gives rise to a (not necessarily direct) decomposition of $\mathcal{G}$,

$$\mathcal{G} = \mathcal{G}_+ + \mathcal{G}_-,$$

into a sum of *subalgebras* $\mathcal{G}_\pm = (1 \pm R)\mathcal{G}$. Set $\pi_\pm = \frac{1}{2}(1 \pm R)$ so that

$$I = \pi_+ + \pi_-, \qquad R = \pi_+ - \pi_-.$$

Exercise 1: If the decomposition is direct, then $\pi_\pm$ are complementary projections. Conversely, if $\mathcal{G} = \mathcal{G}_+ \oplus \mathcal{G}_-$ is a direct decomposition of $\mathcal{G}$ into subalgebras with associated projections $\pi_\pm$, show that $R = \pi_+ - \pi_-$ solves (mYB).

Exercise 2: For $\mathcal{G} = gl(n, \mathbb{R})$, $R \equiv \pi_\ell - \pi_s$ solves (mYB) and $[\cdot, \cdot]_R$ is precisely the Lie algebra $[\cdot, \cdot]_{QR}$ above. Here $\mathcal{G}_+ = \{$lower triangular matrices$\}$ and $\mathcal{G}_- = \{$skew matrices$\}$.

Let G be a Lie group with Lie algebra $\mathcal{G}$. We say that $F\colon \mathcal{G}^* \to \mathbb{R}$ is Ad^*-*invariant* if

$$F(Ad_g^* \alpha) = F(\alpha)$$

for all $\alpha \in \mathcal{G}^*, g \in G$.

Example. $\mathcal{G} = Gl(n, \mathbb{R})$, $F(\alpha) = \pi\alpha^k$, $\alpha \in \mathcal{G}^* \cong \mathcal{G}$. We showed in Lecture 1 that $Ad_g^* \alpha = g^{-1}\alpha g$. Hence $F(Ad_g^* \alpha) = \operatorname{tr}(g^{-1}\alpha g)^k = \operatorname{tr} g^{-1}\alpha^k g = \operatorname{tr} \alpha^k = F(\alpha)$.

Finally we consider an *invariant, nondegenerate pairing* $(\cdot, \cdot)$ on $\mathcal{G}$, i.e.

$$(x, [y, z]) + ([x, y], z) = 0, \quad \text{for all} \quad x, y, z \in \mathcal{G}.$$

Exercise: Show that $(x, y) = \operatorname{tr} xy$ is an invariant, non-degenerate pairing on $\mathcal{G} = gl(n, \mathbb{R})$.

It is straightforward to see that Ad^*-invariant Hamiltonians H generate trivial flows through $\{\cdot, \cdot\}$, but in the interacting structure $\{\cdot, \cdot\}_R$, H gives rise to the differential equation

$$(7) \qquad\qquad \frac{d\alpha}{dt} = [\pi_- dH(\alpha), \alpha]$$

on $\mathcal{G}_R^*$, which is identified with $\mathcal{G}_R \equiv \mathcal{G}$, through the invariant pairing.

Exercise: For $H = \frac{1}{4}H_T = \frac{1}{2} \operatorname{tr} L^2$, $dH(L) = L$ and

$$\pi_-(dH(L)) = \pi_s L = L_+ - L_+^T = B(L).$$

This verifies that H generates the generalized Toda flow $\frac{dL}{dt} = [B(L), L]$ on $\mathcal{G}_R^* = \mathcal{G}_{QR}^*$.

We are interested in the above definitions and calculations primarily because of the following basic result.

Theorem 1 Ad^*-*invariant functions Poisson commute on* $(\mathcal{G}_R^*, \{\cdot, \cdot\}_R)$.

Theorem 2 *Suppose H is Ad^*-invariant and that the decomposition $\mathcal{G} = \mathcal{G}_+ \oplus \mathcal{G}_-$ is direct. Let $\alpha_0 \in \mathcal{G}^*$ be given. Then for some $0 < T = T(\alpha_0) \le \infty$, there exists a unique decomposition*

$$e^{t\,dF(\alpha_0)} = g_+(t)g_-(t)$$

for $0 \le t < T$, where $g_\pm(t) \in \exp \mathcal{G}_\pm$, and

$$\alpha(t) = Ad_{g_+(t)}^* \alpha_0 = Ad_{g_-(t)^{-1}}^* \alpha_0$$

solves (7) for $0 \le t < T$, with $\alpha(0) = \alpha_0$.

Example. In the QR case,

$\exp \mathcal{G}_+$ = identity component of the group of lower triangular, invertible matrices,

$\exp \mathcal{G}_-$ = identity component of the orthogonal matrices.

Now for $H = \frac{1}{2} \operatorname{tr} L^2$, $dH(L) = L$ and

$$e^{tdH(L_0)} = e^{tL_0} = g_+ g_-, \quad g_+ \text{ orthogonal}, \ g_- \text{ lower triangular},$$

and $L(t) = Ad^*_{g_+} L_0 = g_+^{-1} L_0 g_+ = g_+^T L_0 g_+$. This is exactly the Symes procedure described in Lecture 2 (with upper triangularity replaced by lower triangularity).

Proof of Theorem 1. Differentiating $F_i(Ad^*_{e^{tx}} \alpha) = F_i(\alpha)$, $i = 1, 2$, with respect to t at $t = 0$, we obtain for any $x \in \mathcal{G}$, $\alpha([x, dF_i(\alpha)]) = 0$. Thus

$$
\begin{aligned}
\{F_1, F_2\}_R(\alpha) &= \alpha([dF_1(\alpha), dF_2(\alpha)]_R) = \frac{1}{2}\alpha([RdF_1(\alpha), dF_2(\alpha)]) \\
&\quad + \frac{1}{2}\alpha([dF_1(\alpha), RdF_2(\alpha)]) \\
&= \frac{1}{2}\alpha([RdF_1(\alpha), dF_2(\alpha)]) - \frac{1}{2}\alpha([RdF_2(\alpha), dF_1(\alpha)]) \\
&= 0, \text{ by the invariance of } F_2 \text{ and } F_1.
\end{aligned}
$$

In particular, if $\{\lambda_i\}$ are the eigenvalues of $M \in M(n, \mathbb{R})$ in the QR case, then the $\{\lambda_i\}$ are clearly Ad^*-invariant and hence $\{\lambda_i, \lambda_i\}(M) = 0$, which shows that the eigenvalues give n commuting integrals for the Toda flow.

Exercise: Prove Theorem 2 by direct calculation.

We are now ready to consider infinite dimensional systems. Of course, an infinite-dimensional system, the KdV equation, was responsible for all the developments that we have been discussing so far. But as is often the case, the motivating example in a new theory is often not the simplest and most basic example. Now we know, for many reasons, some of which will become clear during this and the next lecture, that the Nonlinear Schrödinger (NLS) equation is the simplest example of an infinite dimensional, integrable Hamiltonian system and without further apologies I am going to restrict my attention to this equation. The discovery that the NLS equation is integrable is due to Zakharov and Shabat [ZS].

The NLS equation has the form

$$iy_t + y_{xx} - 2\kappa|y|^2 y = 0,$$

where $\kappa = 1$ is the defocusing case and $\kappa = -1$ is the focusing case. Both cases are of great interest, but again, without any further apologies, I am going to restrict myself in this and the next lecture to the simpler case when $\kappa = 1$, i.e. the defocusing case. So from now on we only consider

NLS $$iy_t + y_{xx} - 2|y|^2 y = 0.$$

This equation can be written in generalized Lax-pair form as

$$(8) \qquad \frac{d}{dt}\left(P - \frac{d}{dx}\right) = \left[Q, P - \frac{d}{dx}\right],$$

where

$$P = -iz\sigma_3 + \begin{pmatrix} 0 & y \\ \bar{y} & 0 \end{pmatrix}, \qquad \sigma_3 = \begin{pmatrix} 1 & 0 \\ 0 & -1 \end{pmatrix},$$

and

$$Q = -2iz^2\sigma_3 + 2z\begin{pmatrix} 0 & y \\ \bar{y} & 0 \end{pmatrix} + i\begin{pmatrix} -|y|^2 & y_x \\ -\bar{y}_x & |y|^2 \end{pmatrix}.$$

Exercise: Check this! Also verify that

$$\frac{d}{dt}\left(\tilde{P} - \frac{d}{dt}\right) = \left[\tilde{Q}, \tilde{P} - \frac{d}{dt}\right],$$

where

$$\tilde{P} = -iz\sigma_3 - i\begin{pmatrix} 0 & y \\ \bar{y} & 0 \end{pmatrix}$$

and

$$\tilde{Q} = -2iz^2\sigma_3 - 2iz\begin{pmatrix} 0 & y \\ \bar{y} & 0 \end{pmatrix} + i\begin{pmatrix} |y|^2 & -iy_x \\ i\bar{y}_x & -|y|^2 \end{pmatrix}$$

is the generalized Lax-pair form for the focusing NLS equation. This information will be needed in Dave McLaughlin's lecture.

Remarks

Equation (8) is slightly different from the Lax-pair for the KdV equation described in Lecture 2. Here the spectral parameter z is present in P and in Q and, in particular, the dependence of Q on z is quadratic. However, arguing as before, we see that for any fixed z,

$$P(t, z) - \frac{d}{dt} = U(t)^{-1}\left(P(0, z) - \frac{d}{dx}\right)U(t)$$

for some invertible operator $U(t)$. In particular the spectrum of the generalized eigenvalue equation

$$\frac{d}{dx}\psi = P(t, z)\psi = \left(-iz\sigma_3 + \begin{pmatrix} 0 & y \\ \bar{y} & 0 \end{pmatrix}\right)\psi$$

is conserved by the NLS equation.

Observe that equation (8) can be, and usually is, written in the form

$$\frac{\partial P}{\partial t} = \frac{\partial Q}{\partial x} + [Q, P],$$

which is the so-called *zero-curvature* form of the equation.

Now I want to draw your attention to a basic difference between finite and infinite dimensional systems. That is, the answer to a given infinite-dimensional problem depends, far more than in the finite-dimensional case, on the context of the problem.

For example, the NLS equation can be thought of as describing solutions which are periodic in x, $y(x,t) = y(x+1,t)$, say, or on the other hand, solutions which decay as $|x| \to \infty$, $y(x,t) \to 0$ as $|x| \to \infty$, the so-called scattering situation. The behavior of the system is radically different in these two situations. Moreover, even within a given context, a particular question can have different answers, depending on the detail required. For example, in the scattering situation for NLS, we may ask whether $y(x,t) \to 0$ as $t \to \infty$. If we are speaking about compacta in $\mathbb{R}$, the answer is yes, but considering the system more globally, $\int_{-\infty}^{\infty} |y(x,t)|^2 dx$ cannot go to zero as $t \to \infty$: in fact, it must be constant.

In the finite-dimensional case, there is usually only one reasonable answer to a question. But in the infinite-dimensional case, the situation is far more subtle, with different answers depending on the analytical nature of the question. These remarks must be kept in mind, in particular, when we try to apply dynamical system methods to PDE's. These methods give us a guide what to look for, but now each question can have many different answers. So, again without any apologies, I am going to restrict my attention to the scattering case for the defocusing NLS equation,

$$y(x) \to 0 \quad \text{sufficiently rapidly as} \quad |x| \to \infty.$$

The Hamiltonian character of the defocusing NLS equation in the scattering situation is discussed, for example, in [FT].

In the scattering situation (see, for example, [BC]), there exists for all $z \notin \mathbb{R}$, a unique solution $\Psi = \Psi(x,z)$ of the eigenvalue equation $\frac{\partial}{\partial x}\Psi = P\Psi$ with the properties

$$m \equiv \Psi e^{ixz\sigma_3} \to I = \begin{pmatrix} 1 & 0 \\ 0 & 1 \end{pmatrix} \quad \text{as} \quad x \to +\infty,$$

$$m(x,z) \quad \text{is bounded as} \quad x \to -\infty.$$

Using the (generalized) Lax-pair form for NLS, we find, as in the case of Toda, that $\frac{\partial}{\partial t}\Psi = Q\Psi + \Psi C$ for some matrix C. Letting $x \to \pm\infty$, and using the given asymptotics of m, we obtain $C = 2iz^2\sigma_3$. Thus $\frac{\partial \Psi}{\partial t} = Q\Psi + 2iz^2\Psi\sigma_3$.

Now it turns out that for all x and t,

$$\Psi_\pm(x,t,z) \equiv \lim_{\varepsilon \downarrow 0} \Psi(x,t,z \pm i\varepsilon)$$

exist for all $z \in \mathbb{R}$. Also $\Psi_\pm$ are clearly both solutions of $\frac{\partial \Psi_\pm}{\partial x} = P\Psi_\pm$: hence there exists a matrix $v = v(t,z)$, $z \in \mathbb{R}$, such that

$$(9) \qquad \Psi_+(x,t,z) = \Psi_-(x,t,z)v(t,z).$$

For obvious reasons, $v = v(t,z)$ is called the *jump matrix* for $\Psi(x,t,\cdot)$. Differentiating (9) with respect to t, we find $\Psi_{+t} = \Psi_{-t}v + \Psi_- v_t$, i.e. $2iz^2\Psi_+\sigma_3 + Q\Psi_+ = 2iz^2\Psi_-\sigma_3 v + Q\Psi_- v + \Psi_- v_t$, which implies

$$(10) \qquad v_t = 2iz^2[v,\sigma_3].$$

This is, explicitly, the *linearization of the flow*! Indeed inverse scattering theory (see, for example, [FT] or [BC]) tells one that the map $y \mapsto v$ is 1-1 (and onto, for v in a suitable space). Thus the v's (or, more precisely, the r's from which the v's

are constructed — see below) are essentially the "angles" for the problem. Solving
(10) we obtain

$$v(z,t) = e^{-2iz^2 t \sigma_3} v(z) e^{2izt\sigma_3}, \qquad z \in \mathbb{R},$$

where $v(z)$ is the jump matrix for the scattering problem $\frac{\partial}{\partial x} \Psi = P\Psi$ at time $t = 0$.
One shows easily (see [FT] or [BC]) that v has the form

$$v(z) = \begin{pmatrix} 1 - |r(z)|^2 & -\overline{r(z)} \\ r(z) & 1 \end{pmatrix}, \qquad z \in \mathbb{R},$$

for some function r with $\|r\|_{L^\infty(\mathbb{R})} < 1$. If $y(x, t = 0) \in \mathcal{S}(\mathbb{R})$, the Schwartz space on
$\mathbb{R}$, then also $r(z) \in \mathcal{S}(\mathbb{R})$. And conversely if $r \in \mathcal{S}(\mathbb{R})$ is given with $\|r\|_{L^\infty(\mathbb{R})} < 1$,
inverse scattering theory tells one that there exists a unique function $y \in \mathcal{S}(\mathbb{R})$ with
associated jump matrix $v = v(z) = \begin{pmatrix} 1 - |r|^2 & -\bar{r} \\ r & 1 \end{pmatrix}$. The function $r = r(z)$ is
called the *reflection coefficient* for the scattering problem $\frac{d\Psi}{dx} = P\Psi$.

If we set $m = \Psi e^{izx\sigma_3}$ as above, then from $\frac{d}{dx}\Psi = P\Psi$ we obtain

$$\frac{d}{dx}m = -iz[\sigma_3, m] + \begin{pmatrix} 0 & y \\ \bar{y} & 0 \end{pmatrix} m.$$

Also it is a fact (see [FT] or [BC]) that, for fixed x,

$$m(x, z) = I + \frac{m_1(x)}{z} + O\left(\frac{1}{z^2}\right) \quad \text{as } z \to \infty.$$

Substituting this asymptotic expansion into the above differential equation for m,
we obtain $i[\sigma_3, m_1] = \begin{pmatrix} 0 & y \\ \bar{y} & 0 \end{pmatrix}$, which establishes the basic formula

(11) $$y(x) = 2i(m_1(x))_{12}.$$

The preceding calculations can be summarized as follows. Let $y_0(x) \in \mathcal{S}(\mathbb{R})$
be given and let $r = r(z)$ be the corresponding reflection coefficient. Set $v_{x,t}(z) =$
$e^{-i(2tz^2 + xz)\sigma_3} v(z) e^{i(2tz^2 + xz)\sigma_3}$, where $v(z) = \begin{pmatrix} 1 - |r(z)| & -\overline{r(z)} \\ r(z) & 1 \end{pmatrix}$.

Theorem 3 *Let $m(z) = m(x, t, z)$ solve the* Riemann-Hilbert *(RH) problem*

- $m = m(\cdot) = m(x, t, \cdot)$ *is a 2×2 matrix valued analytic function in $\mathbb{C} \backslash \mathbb{R}$,*

- $m_+(z) = m_-(z)v_{x,t}(z), z \in \mathbb{R}$,

- $m(z) \to I$ *as* $z \to \infty$.

Then if $m = I + \frac{m_1(x,t)}{z} + O\left(\frac{1}{z^2}\right)$ as $z \to \infty$,

(12) $$y(x, t) \equiv 2i(m_1(x, t))_{12}$$

solves the NLS equation with initial data $y(x, 0) = y_0(x)$.

Thus we see that to analyze the solution of NLS we must solve a Riemann-Hilbert problem and control the solution of the problem as x, t become large. We will see how to do this in the next lecture.

Now the point I want to make here is that the above RH formulation should be understood in the classical R-matrix framework (see [FT] for details). Indeed we saw that if we are in an R-matrix situation, the solution of the equation generated by an invariant Hamiltonian can be solved by a factorization problem. For example, in the case of the Toda lattice on $\mathcal{G}^*_{QR}$, the abstract factorization problem guaranteed by the R-matrix formalism takes the concrete form of a finite-dimensional matrix factorization, $e^{tL0} = QR$. For NLS, as we see above, the abstract factorization takes the form of a Riemann-Hilbert factorization.

This theme, I believe, is a unifying concept in the theory of integrable systems. The more refined conjecture mentioned above is the following: not only is every integrable system integrable because there is an associated Lax-pair, but even more, in the background, there is an R-matrix, and hence a factorization problem, whose solution gives the solution of the basic dynamical system.

Remark 1. The first person to recognize that an inverse scattering problem was equivalent to a Riemann-Hilbert problem was Shabat [S].

Remark 2. In the infinite-dimensional setting, questions arise about the meaning of "complete integrability". In the finite-dimensional case, of dimension $2n$, say, we need n (commuting) integrals to integrate the system. In the infinite-dimensional case, what does it mean to have "half" the integrals? In what sense can we add one integral in or take one out? When do we have enough? These questions, and more, were considered in the case of KdV with periodic boundary conditions, for example, by McKean and Trubowitz, in their beautiful paper [MT]. In the scattering situation, however, these are in general only a finite number of L^2-eigenvalues, and these must be complemented with a "continuum of integrals" constructed from scattering theoretic quantities (see [FT]). In these four lectures we unfortunately do not have the time to develop these elegant considerations: rather, we take the practical point of view that a(n infinite-dimensional) system is integrable if we can construct, by hook or by crook, a one-to-one change of variables which linearizes the flow. For example for NLS, we have

$$y(x, t) \mapsto v_{x,t} \equiv r_{x,t} = re^{2i(2tz^2 + xz)} \mapsto \log r_{x,t} = \log r + 2i(2tz^2 + xz).$$

Analyzing the inverse map, $r_{x,t} \mapsto y(x, t)$, we are able to describe the behavior, and in particular, the asymptotic behavior of the flow.

Lecture 4

The Liouville-Arnold-Jost Theorem teaches us that the flow generated by a finite-dimensional integrable system is equivalent to a straight line motion on a cylinder $\mathbb{T}^k \times \mathbb{R}^{n-k}$. If $k = n$, say, the cylinder is a torus and the motion is almost periodic in time, etc. In the integrable, infinite-dimensional case, we still have straight line motion on a (now infinite-dimensional) cylinder, but when we invert the linearizing map, the implications of the straight line motion for the flow in physical x-space will depend on the topology we impose (see the discussion in the middle of Lecture 3). Special tools are needed to analyze the inverse map and again, without apologies, we restrict our attention to the defocusing NLS equation.

We are interested, in particular, in the long-time behavior of the solution $y(x, t)$ of NLS with initial data $y(x, 0) = y_0(x) \in \mathcal{S}(\mathbb{R})$. By Theorem 3, this reduces to controlling the large x, t behavior of a Riemann-Hilbert problem. Now a straightforward calculation (**Exercise**) shows that if $\|r\|_{L^\infty(\mathbb{R})} \ll 1$, then

$$
\begin{aligned}
y(x, t) &= 2i(m_1(x, t))_{12} \\
&\sim \text{const} \int_{-\infty}^{\infty} \overline{r(z)} e^{-2i(2tz^2 + xz)} dz
\end{aligned}
$$

and the long-time behavior of $y(x, t)$ can be computed to leading order using the standard method of stationary phase. In other words, if we think functorially, then there exists some (highly nonlinear) functional $\mathcal{F}$, $y(x, t) = \mathcal{F}(r(\cdot)e^{i(2t(\cdot)^2 + x(\cdot))})$, which reduces when r is small to the ordinary Fourier transform, to which the standard methods of stationary phase then apply.

This raises the following **question**! Does there exist some kind of "nonlinear" steepest-descent/stationary phase type method which can be applied directly to (11), even when $\|r\|_{L^\infty(\mathbb{R})}$ is not small?

More generally, we are looking for a steepest-descent/stationary phase-type method that applies directly to oscillatory Riemann-Hilbert problems (of the type occurring in Theorem 3 above, for example) as the external parameters (x and t in our case) become large.

What features should the method have? First let us review the *classical method of stationary phase* to determine the asymptotic behavior as $t \to \infty$ of an integral $F(t) = \int_C f(z)e^{itg(z)} dz$ on a contour C, say, where $f(z)$ and $g(z)$ are analytic.

Step 1 Locate the points z_j of stationary phase, $g'(z_j) = 0$.

Step 2 Determine paths of steep(est)-descent in the neighborhood of the z_j's: as one moves along these paths away from the stationary phase points, $|e^{it(z)}|$ should decrease exponentially.

Step 3 Using Cauchy's Theorem, deform C to coincide with these paths of steep(est)-descent near the z_j's.

Step 4 As $t \to \infty$, the problem is localized to the part of the deformed contour that lies in an ε-neighborhood of each z_j; the contribution from the remainder of the contour is exponentially small.

Step 5 The localized integrals can be computed explicitly: indeed, after scaling $z - z_j \to (z - z_j)\sqrt{t}$, the integrals reduce to Gaussian integrals on $\mathbb{R}$. Thus the leading asymptotics of $F(t)$ can be computed explicitly.

Recently Xin Zhou and I [DZ1] have discovered a method for RH problems that is the analog of the above classical method for integrals. In addition to the NLS equation, the method has now been used to obtain the asymptotics of a wide variety of problems ranging from integrable wave equations such as Modified KdV [DZ1] and KdV [DVZ1] etc., to "integrable" ode's such as the Painlevé equations [DZ3], to integrable discrete models such as the Toda lattice [K], to integrable statistical mechanical models such as the transverse Ising chain at the critical magnetic field [DZ2] and also the so-called XY model [DIZ2], to small dispersion problems [DVZ2], and finally, most recently, to the theory of random matrix models [DIZ2].

The application of the method to the NLS equation, which is the main subject of this lecture, was developed together with Alexander Its and X. Zhou [DIZ1]; for a more detailed exposition, refer to [DZ4], which is also intended as a pedagogic introduction to the general steepest-descent method for RH problems.

So how does the method work? Set $\theta(z) = 2z^2 + \frac{x}{t}z$. **Step 1** is the same. Setting $\theta'(z_0) = 4z_0 + \frac{x}{t} = 0$, we obtain the stationary phase point $z_0 = -x/4t$. To describe the analog of **Step 3**, we must explain what it means to "deform" a RH problem. What we mean is that we can replace a given RH problem on one contour by an equivalent RH problem on another contour — the one problem is solvable if and only if the other is solvable.

To be explicit, we consider more generally RH problems on oriented contours $\Sigma \subset \mathbb{C}$

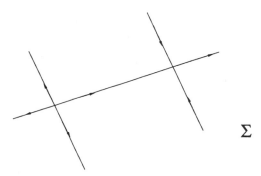

with given jump matrices $v \colon \Sigma \to Gl(n, \mathbb{C})$. If Σ is unbounded, we always assume $v(z) \to I$ sufficiently rapidly as $|z| \to \infty$. Recall that the orientation on an arc is equivalent to specifying the $\pm$ sides of the arc, e.g.,

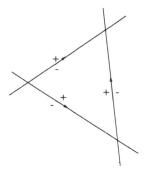

By convention, the $+$ side is taken to lie on the left as we traverse the arc in the direction of the orientation.

Given (Σ, v) we seek an $n \times n$ matrix-valued function m which is

* analytic in $\mathbb{C} \backslash \Sigma$,

and satisfies,

* $m_+(z) = m_-(z)v(z)$ for $z \in \Sigma$,

* $m(z) \to I$ as $z \to \infty$.

This is the RH problem determined by (Σ, v). It is not at all clear that, given Σ and v, the RH problem has a (unique) solution, and indeed, a beautiful and profound mathematical theory has been developed over the years to consider such questions (see, e.g., [CG]). Fortunately our RH problem in Theorem 3 has a unique solution (see e.g., [DZ4] for details).

Now suppose we have a RH problem (Σ, v)

and suppose that there are two distinguished points z_1 and z_2 on Σ which support a circular disk D intersecting Σ as follows,

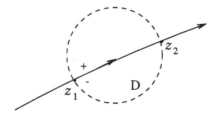

Suppose that within the disk $v(z)$ has a factorization $v(z) = b_-^{-1}(z)v_1(z)b_+(z)$, $z \in \Sigma \cap D$, where $b_\pm(z)$ have analytic continuations to the $\pm$ segments of the disk, respectively: $v_1(z)$, however, has no analyticity properties. Let m be the solution of the RH problem determined by (Σ, v).

Define $\widetilde{m}(z)$ on the complement of the extended contour $\widetilde{\Sigma} = \Sigma \cup \partial D$, as follows:

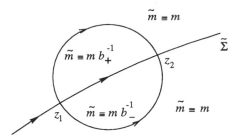

Observe that

- $\widetilde{m}$ is analytic in $\mathbb{C}\backslash\widetilde{\Sigma}$,

- satisfies $\widetilde{m}_+ = \widetilde{m}_- \widetilde{v}$ on $\widetilde{\Sigma}$,

- $\widetilde{m}(z) \to I$ as $z \to \infty$,

where $\widetilde{v}$ is defined on $\widetilde{\Sigma}$ by

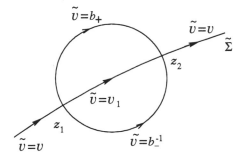

The RH problem corresponding to $(\widetilde{\Sigma}, \widetilde{v})$ is the *deformed RH* for (Σ, v); clearly the two problems are equivalent in the sense stated above.

For the RH problem, what do we mean by paths of steep(est)-descent as in **Step 2**? Somehow we feel that the main contribution to the solution of the RH problem should be coming from the stationary phase point z_0. The key to the problem is the signature table of $\operatorname{Re} i\theta = \operatorname{Re} i(2z^2 - 4z_0 z)$. In the z-plane we have

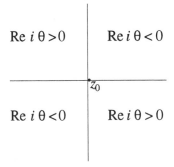

What we would like is to "put" the $e^{2it\theta}$ factor in $v_{x,t}$ in the region where $\operatorname{Re} i\theta < 0$, and the $e^{-2it\theta}$ factor in the region where $\operatorname{Re} i\theta > 0$: in that way the RH

problem would clearly localize to a neighborhood of z_0, as $t \to \infty$. This would then be the RH analog of **Step 2**. But how can we do it? Algebraically the two factors are locked together in the same expression. The basic observation is that we can factor $v_{x,t}(z)$,

$$v_{x,t}(z) = \begin{pmatrix} 1 - |r(z)|^2 & -\bar{r}e^{-2it\theta} \\ re^{2it\theta} & 1 \end{pmatrix} = \begin{pmatrix} 1 & -\bar{r}e^{-2it\theta} \\ 0 & 1 \end{pmatrix} \begin{pmatrix} 1 & 0 \\ re^{2it\theta} & 1 \end{pmatrix}$$

$$\equiv b_-^{-1} b_+.$$

Assuming for the moment that $r(z)$ has an analytic continuation to Im $z > 0$, and hence $\overline{r(z)} = r(\bar{z})$ has an analytic continuation to Im $z < 0$, we see that for $z > z_0$ in particular, this gives rise to a deformed RH problem in the sense described above (here $z_1 \leftrightarrow z_0, z_2 \leftrightarrow \infty$),

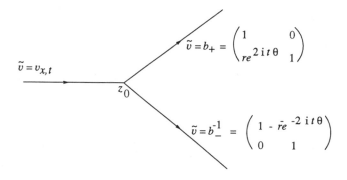

Clearly this is what we want as $b_+(z) \to I$ for z on the upper leg and $b_-(z) \to I$ for z on the lower leg, as $t \to \infty$ (see signature table of Re $i\theta$). But if we try the same trick for $z < z_0$, we obtain

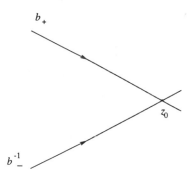

which is clearly **not** what we want as b_+, b_-^{-1} now blow up exponentially. What we want to do somehow is to reverse the order of appearance of the multipliers $e^{\pm 2it\theta}$ in the factorization of $v_{x,t}(z)$. Fortunately, this can be achieved by performing a lower/upper, as opposed to an upper/lower, factorization of $v_{x,t}$,

$$v_{x,t}(z) = \begin{pmatrix} 1 & 0 \\ \frac{r}{1-|r|^2}e^{2it\theta} & 1 \end{pmatrix} \begin{pmatrix} (1-|r(z)|^2) & 0 \\ 0 & (1-|r(z)|^2)^{-1} \end{pmatrix} \begin{pmatrix} 1 & \frac{-\bar{r}}{1-|r|^2}e^{-2it\theta} \\ 0 & 1 \end{pmatrix}.$$

(Recall that $\|r\|_{L^\infty(\mathbb{R})} < 1$ so that $1 - |r(z)|^2 \neq 0$.) To remove the diagonal factor set

$$m^{(1)} = m \begin{pmatrix} \delta^{-1} & 0 \\ 0 & \delta \end{pmatrix}$$

where δ is an as yet undetermined function, which we will choose below to be analytic and invertible in $\mathbb{C} \setminus (-\infty, z_0]$.

Simple algebra shows that $m^{(1)}$ solves a RH problem on $\Sigma^{(1)} = \Sigma = \mathbb{R}$ with

$$
\begin{aligned}
v^{(1)} &= \begin{pmatrix} \delta_-(z) & 0 \\ 0 & \delta_-^{-1}(z) \end{pmatrix} v_{x,t}(z) \begin{pmatrix} \delta_+^{-1} & 0 \\ 0 & \delta_+ \end{pmatrix}, \\
&= \begin{pmatrix} 1 & -\bar{r}\delta^2 e^{-2i\theta t} \\ 0 & 1 \end{pmatrix} \begin{pmatrix} 1 & 0 \\ r\delta^{-2} e^{2it\theta} & 1 \end{pmatrix} \text{ for } z > z_0, \\
&= \begin{pmatrix} 1 & 0 \\ \frac{r}{1-|r|^2} e^{2it\theta} \delta_-^{-2} & 1 \end{pmatrix} \begin{pmatrix} \frac{(1-|r|^2)\delta_-}{\delta_+} & 0 \\ 0 & \frac{\delta_+}{(1-|r|^2)\delta_-} \end{pmatrix} \begin{pmatrix} 1 & -\frac{\bar{r}e^{-2i\theta t}\delta_+^2}{1-|r|^2} \\ 0 & 1 \end{pmatrix} \\
&\qquad \text{for } z < z_0.
\end{aligned}
$$

We choose δ to solve the following scalar RH problem,

- $\delta(z)$ is analytic in $\mathbb{C} \setminus (-\infty, z_0]$,

- $\delta_+(z) = \delta_-(z)(1 - |r(z)|^2), z < z_0$,

  $= \delta_-(z), z > z_0$

- $\delta(z) \to 1$ as $z \to \infty$.

Such scalar problems can be solved explicitly by formula,

$$\delta(z) = e^{\frac{1}{2\pi i} \int_{-\infty}^{z_0} \frac{\log(1-|r(s)|^2)}{s-z} ds}.$$

With δ as above, $v^{(1)}$ takes the form

$$
\begin{aligned}
(13) \qquad v^{(1)}(z) &= \begin{pmatrix} 1 & -\bar{r}\delta^2 e^{-2it\theta} \\ 0 & 1 \end{pmatrix} \begin{pmatrix} 1 & 0 \\ r\delta^{-2} e^{2it\theta} & 1 \end{pmatrix}, z > z_0, \\
&= \begin{pmatrix} 1 & 0 \\ \frac{r}{1-|r|^2} \frac{e^{2it\theta}}{\delta_-^2} & 1 \end{pmatrix} \begin{pmatrix} 1 & -\frac{\bar{r}}{1-|r|^2} \delta_+^2 e^{-2it\theta} \\ 0 & 1 \end{pmatrix}, z < z_0 \\
&\equiv b_-^{-1} b_+.
\end{aligned}
$$

Deforming now leads to the RH problem

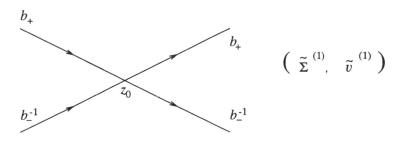

with jump matrices $\tilde{v}^{(1)}$ as indicated above. Referring to the signature table for Re $i\theta$, we see that all the jump matrices converge exponentially to the identity, uniformly on the complement of ε-neighborhoods of z_0 in $\widetilde{\Sigma}^{(1)}$.

What if r is not analytic? In this case, for $z > z_0$, we factor $r(z)e^{2it\theta(z)} = \hat{r}(z,t) + r_1(z,t)$, where $\hat{r}(\cdot,t)$ is analytic and $r_1(z,t)$ is small as $t \to \infty$, and for $z < z_0$ we factor $r(1 - |r(z)|^2)^{-1}e^{2it\theta(z)}$ similarly. Instead of (12), this leads to a factorization $v^{(1)} = b_-^{-1}v_1b_+$, where $v_1 \to I$ rapidly as $t \to \infty$. Under deformation this, in turn leads to a RH problem of the form

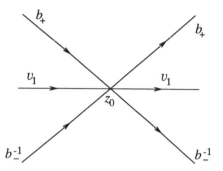

But as $v_1 \to I$ rapidly, we again end up, to any desired order, with a RH problem on a cross of type $\left(\widetilde{\Sigma}^{(1)}, \tilde{v}^{(1)}\right)$ as before. Either way, for any $\varepsilon > 0$, the RH problem localizes to a RH problem on a small cross of size ε, as $t \to \infty$

This completes Steps 1, 2, 3 and 4.

In the analog of **Step 5**, we again scale $z \to (z - z_0)\sqrt{t}$ and this leads to a RH problem on an infinite cross, but now the jump matrices have constant coefficients. At this stage all the analytical tricks have been used up and there are no small parameters left to scale. Like the Gaussian integral that you just have to compute in the classical, scalar case, this is a problem you just have to solve: either the heavens are with you or against you. Fortunately, in this problem they are with you and it turns out that this local RH problem can be solved explicitly in terms of parabolic cylinder functions. We thus obtain the leading order behavior of the solution, with controlled error terms. Furthermore, the method is algorithmic and we do not need to make an a priori ansatz for the asymptotic form of the solution.

The kind of result that we obtain is the following.

Let $y_{as}(x,t) = t^{-1/2}\alpha(z_0)e^{i\frac{x^2}{4t} - i\nu(z_0)\log 8t}$, $z_0 = -x/4t$, where

$$\nu(z_0) = -\frac{1}{2\pi}\log(1 - |r(z_0)|^2) > 0,$$

$$|\alpha(z_0)|^2 = \frac{1}{2}\nu(z_0),$$

$$\arg \alpha(z_0) = \frac{1}{\pi} \int_{-\infty}^{z_0} \log(z_0 - s) d(\log(1 - |r(s)|^2))$$
$$+ \frac{\pi}{4} + \arg \Gamma(i\nu(z_0)) - \arg r(z_0)$$

(Γ = gamma function). Fix $M > 1$.

Theorem 4 *Let $y(x,t)$ solve NLS with $y(x, t = 0) = y_0(x) \in \mathcal{S}(\mathbb{R})$. Then as $t \to \infty$*

(i) $y(x,t) = y_{as}(x,t) + O\left(\frac{\log t}{t}\right)$ *for* $|z_0| = \left|-\frac{x}{4t}\right| < M$,

(ii) *for any j,*

$$y(x,t) = y_{as}(x,t) + O\left(\frac{1}{|x|}j + c_j(z_0)\frac{\log |x|}{|x|}\right), |z_0| > \frac{1}{M},$$

where $c_j(z_0)$ decreases rapidly as $|z_0| \to \infty$.

Note that the asymptotics match in the common region $M^{-1} < \left|-\frac{x}{4t}\right| < M$.
Remarks
(1) The asymptotic form of the solution $y \sim y_{as}$ for NLS was first obtained by Zakharov and Manakov [ZM], using a so-called "isomonodromic-type" technique. The calculations in [ZM], however, contain no error bounds.
(2) The scaled RH problem arising in Step 5 above, first arose, and was solved, in the work of Its and Novokshenov (see [IN]).

The steepest-descent/stationary phase type method presented above for RH problems is a kind of nonlinear Fourier integral method, and we may reasonably raise the following **question**: do the analogs of singular optical phenomena, such as caustics, occur in the non-linear problem? The NLS equation is indeed the simplest case because there are no caustics i.e. the pair of equations

$$\frac{d}{dz}\left(2z^2 + \frac{x}{t}z\right) = 0, \qquad \frac{d^2}{dz^2}\left(2z^2 + \frac{x}{t}z\right) = 0$$

has no solutions. In the case of the Modified KdV equation, however, there is a similar RH to NLS, but now θ is replaced by $\theta_{\mathrm{MKdV}} = \psi z^3 + \frac{x}{t}z$. For this problem there are indeed caustics:

$$\frac{d}{dz}\left(4z^3 + \frac{x}{t}z\right) = 0 \Rightarrow z = \pm z_0 = \pm\sqrt{-\frac{x}{12t}}, \quad x < 0,$$

$$\frac{d^2}{dz^2}\left(4z^3 + \frac{x}{t}z\right) = 0 \Rightarrow z = \pm z_0 = 0.$$

Thus $x/t \to 0$ is a "caustic" region, and indeed there are **two** main regions for the asymptotic behavior of MKdV (see [SA], [DZ1]). In the first region, $M^{-1} < \sqrt{-\frac{x}{12t}} < M$, $x < 0$, the stationary phase points $\pm z_0$ are separated and do not interact to leading order and the solution is a sum of two terms of the same type as the asymptotic form y_{as} occurring in NLS. In the *caustic region*, $\left|\frac{x}{t^{1/3}}\right| \leq c$, however, the solution $y(x,t)$ of Modified KdV now takes a new and different form, $y \sim$

$\frac{1}{t^{1/3}} u \left(\frac{x}{t^{1/3}} \right)$ where $u = u(s)$ solves the Painlevé II equation $u''(s) = su(s) + 2u^3(s)$. Again there are overlapping regions, etc.

A problem which has two stationary phase points but no caustic region arises in evaluating the long-time asymptotics for the autocorrelation function $\chi(t) = \langle \sigma_0^x(t)\sigma_0^x \rangle_T$ at temperature T for the transverse Ising chain at the critical magnetic field with Hamiltonian $H = -\frac{1}{2} \sum_{\ell \in \mathbb{Z}} (\sigma_\ell^x \sigma_{\ell+1}^x + \sigma_\ell^z)$. In [DZ2], Zhou and I show that

$$\chi(t) = e^{\frac{t}{\pi} \int_{-1}^{1} \log|\tanh(s/T)| ds + O(\log t)}.$$

Now, so far, we have presented the RH method as a nonlinear analog of the classical method of stationary phase. But it turns out that there are *genuinely nonlinear* phenomena which can arise. For example for KdV, which is closely related to the Modified KdV equation, a *third* region, called the *collisionless shock region* and discovered by Ablowitz and Segur [AS], appears between the regions $M^{-1} < \sqrt{-\frac{x}{12t}} < M$, $|x/t^{1/3}| < c$, in describing the long-time behavior of the solution. In this region, which is described by the inequalities $c_1^{-1} < \frac{-x}{t^{1/3}(\log t)^{2/3}} < c_1$, the solution of KdV behaves like a modulated cnoidal wave. As is shown in [DVZ1], this region arises from the fact that the full line joining the stationary phase points

$$-z_0 \bullet\!\!\!-\!\!\!\bullet\, z_0$$

becomes "electrified" and each point on the line contributes equally to the long-time behavior of KdV. Thus in the steepest-descent/stationary phase method for RH problems we can have "lines of stationary-phase", and not just points of stationary phase.

Remarks A complete description of the leading asymptotics of the solution of the Cauchy problem for KdV, with connection formulae between different regions (including the collisionless shock region, which is their discovery!) was presented by Ablowitz and Segur [AS], but without precise information on the phase (and without error estimates). In a later development, Segur and Ablowitz [SA] used a modification of the method in [ZM] to derive the leading asymptotics of the solutions of the MKdV, KdV (but not in the collisionless shock region) and Sine-Gordon equations, including full information on the phase. For more historical information on the contributions of many authors to the calculation of the asymptotic behavior of integrable systems, see, for example, [DIZ1].

Acknowledgements

These notes were prepared with the support of NSF Grant No. DMS–9203771 and NSF Grant No. DMS–9500867.

References

[A]   V. Arnold, *Mathematical Methods of Classical Mechanics*, Springer-Verlag, New York, 1978.

[AM] R. Abraham and J.E. Marsden, *Foundations of Mechanics*, Second Edition, Benjamin/Cummings, Reading, Massachusetts, 1978.

[AS] M.J. Ablowitz and H. Segur, Asymptotic solutions of the Korteweg de Vries equation, *Stud. Appl. Math.*, **57**, No. 1 (1977), 13–44.

[BC] R. Beals and R. Coifman, Scattering and inverse scattering for first order operators, *Comm. Pure Appl. Math.*, **37** (1984), 39–90.

[CG] K. Clancey and I. Gohberg, *Factorization of matrix functions and singular integral operators*, Operator Theory, Vol. 3, Birkhäuser Verlag, Basel, 1981.

[DIZ1] P.A. Deift, A.R. Its and X. Zhou, *Long-time asymptotics for integrable non-linear wave equations, Important developments in soliton theory*, A.S. Fokas and V.E. Zakharov, eds., Berlin, Heidelberg, New York, Springer, 1993.

[DIZ2] P.A. Deift, A.R. Its and X. Zhou, A Riemann-Hilbert approach to asymptotic problems arising in the theory of random matrix models, and also in the theory of integrable statistical mechanics, Preprint, 1995.

[DLTo] P.A. Deift, L.C. Li and C. Tomei, *Symplectic aspects of some eigenvalue algorithms*, in Important developments in soliton theory, eds. A.S. Fokas and V.E. Zakharov, Springer-Verlag, Berlin, Heidelberg, 511–536, 1993.

[DLTr] P. Deift, F. Lund and E. Trubowitz, Nonlinear wave equations and constrained harmonic motion, *Comm. Math. Phys.*, **74** (1980), 141–188.

[DVZ1] P.A. Deift, S. Venakides and X. Zhou, The collisionless shock region for the long-time behavior of solutions of the KdV equation, *Comm. Pure Appl. Math.*, **47** (1994), 199–206.

[DVZ2] P.A. Deift, S. Venakides and X. Zhou, New results in small dispersion KdV by an extension of the steepest descent method for Riemann-Hilbert problems, Part I, Preprint, 1995.

[DZ1] P.A. Deift and X. Zhou, A steepest descent method for oscillatory Riemann-Hilbert problems, Asymptotics for the MKdV equation, *Ann. of Math.*, **137** (1993), 245–308.

[DZ2] P.A. Deift and X. Zhou, *Long-time asymptotics for the autocorrelation function of the transverse Ising chain at the critical magnetic field, Singular limits of dispersive waves*, N.M. Erolani, I.R. Grabitov, C.D. Levermore, and D. Serre, eds., NATO ASI Series B, Physics Vol. 320, Plenum Press, New York and London, 1994.

[DZ3] P.A. Deift and X. Zhou, Asymptotics for the Painlevé II equation, Announcement in *Adv. Stud. Pure Math.*, **23** (1994), 17–26. Full paper in *Comm. Pure Appl. Math.* **48** (1995), 277–337.

[DZ4] P.A. Deift and X. Zhou, *Long-time behavior of the non-focusing nonlinear Schrödinger equation — a case study*, Lectures in Math. Sciences, University of Tokyo, 5, 1994.

[FT]   L. Faddeev and L. Takhtajan, *Hamiltonian methods in the theory of solitons*, Springer-Verlag, Berlin, Heidelberg, 1987.

[G]   H. Goldstein, *Classical Mechanics*, Addison-Wesley, 1950.

[GGKM] C.S. Gardner, J.M. Greene, M.D. Kruskal and R.M. Miura, Korteweg de Vries equation and generalizations, *Phys. Rev. Letters* **19** (1967), 1095–1097.

[IN]   A.R. Its and V.Yu. Novokshenov, *The isomonodromic method in the theory of Painlevé equations*, Lecture Notes in Math., 1191, Springer-Verlag, Berlin, Heidelberg, 1986.

[K]   S. Kamvissis, Long-time behavior of the Toda lattice under initial data decaying at infinity, *Comm. Math. Phys.*, **153** (1993), 479–519.

[L]   P. Lax, Integrals of nonlinear equations of evolution and solitary waves, *Comm. Pure Appl. Math.*, **21** (1968), 467–490.

[M1]   J. Moser, *Finitely many mass points on the line under the influence of an exponential potential — an integrable system*, Lecture Notes in Physics **38**, ed. J. Moser, Springer-Verlag, New York, 467–497, 1975.

[M2]   J. Moser, *Geometry of quadrics and spectral theory*, The Chern Symposium, Berkeley, 1979, Springer-Verlag, New York, Heidelberg, Berlin, 147–188, 1980.

[MT]   H.P. McKean and E. Trubowitz, Hill's operator and hyperelliptic function theory in the presence of infinitely many branch points, *Comm. Pure Appl. Math.*, **29** (1976), 143–226.

[S]   A.B. Shabat, *One dimensional perturbations of a differential operator and the inverse scattering problem*, in Problems in Mechanics and Mathematical Physics, Nauka, Moscow, 1976.

[SA]   H. Segur and M.J. Ablowitz, Asymptotic solutions of nonlinear evolution equations and a Painlevé transcendent, *Physica*, **3D**, 1+2, (1981), 165–184.

[STS]   M.A. Semenov-Tian-Shanskii, What is a classical R-matrix? *Func. Anal. Appl.*, **17** (1984), 259–272 (English version).

[ZM]   V.E. Zakharov and S.V. Manakov, Asymptotic behavior of nonlinear wave systems integrated by the inverse method, *Zh. Eksp. Teor. Fiz.*, **71** (1976), 203–215 (Russian); *Sov. Phys. - JETP*, **44**, No. 1 (1976), 106–112 (English).

[ZS]   V.E. Zakharov and A.B. Shabat, Exact theory of two-dimensional self-focusing and one-dimensional self-modulation of waves in nonlinear media, *Zh. Eksp. Teor. Fiz.*, **61** (1971), 118–134; *Soviet Physics JETP* **34** (1972), 62–69.

Section III

Amplitude Equations

Lectures in Applied Mathematics
Volume **31**, 1996

The Complex Ginzburg-Landau Equation
as a Model Problem

C. David Levermore and Marcel Oliver

ABSTRACT. The generalized complex Ginzburg-Landau (CGL) equation describes the evolution of a complex-valued field $u = u(x,t)$ by

$$\partial_t u = Ru + (1 + i\nu)\Delta u - (1 + i\mu)|u|^{2\sigma} u \,.$$

It has a long history in physics as a generic amplitude equation near the onset of instabilities that lead to chaotic dynamics in fluid mechanical systems, as well as in the theory of phase transitions and superconductivity. In this role it and equations like it have had remarkable success in describing evolution phenomena in a broad range of physical systems, from fluids to optics. More recently, it has been proposed and studied as a model for "turbulent" dynamics in nonlinear partial differential equations. It is a particularly interesting model in this respect because it is a dissipative version of the nonlinear Schrödinger equation, a Hamiltonian equation which can possess solutions that form localized singularities in finite time.

This article summarizes the status of well-posedness and regularity questions for the CGL equation subject to periodic boundary conditions in any spatial dimension. It also discusses the relationship of these results to the above investigations and, in particular, to analogous issues for Navier-Stokes fluid turbulence. Appropriately defined weak solutions exist globally in time and unique classical solutions are found locally. Conditions are given under which these classical solutions exist globally in time and have bounds on all derivatives that are uniform over large times, thereby providing the setting in which ideas from finite dimensional dynamical systems can be applied to the CGL equation. Moreover, by using a characterization of Gevrey classes in terms of decay of Fourier coefficients, these solutions are shown to be analytic for positive times. Refinements in both some of the theorems and some of the proofs are introduced.

1991 *Mathematics Subject Classification.* Primary 35-02, 35Q99; Secondary 35A05, 35B65.

This article expands the lecture presented by the first author on 20 June 1994 during the AMS-SIAM Summer School on *Dynamical Systems and Probabilistic Methods for Nonlinear Waves* held at the Mathematical Sciences Research Institute (MSRI) in Berkeley. It was prepared while the authors were visiting the MSRI, which is supported in part by the NSF under grant DMS-9022140, the CNLS at Los Alamos National Laboratory, the Institute for Advanced Study (IAS) under a grant from the Alfred P. Sloan Foundation and the Ecole Normale Supérieure de Cachan.

1. Introduction

The description of chaotic dynamics or, more dramatically, turbulence in continuum systems by nonlinear evolution equations (often partial differential equations) is one of the central challenges facing theoretical physics and applied mathematics. We would like to be able to apply ideas from the studies of dynamical systems or statistical physics to these evolution equations in order to gain a deeper understanding of these phenomena and, as a byproduct, to hopefully achieve simpler or more tractable descriptions. One of the main mathematical stumbling blocks to studying these questions often lies in our lack of understanding of even the most basic issues of existence, uniqueness and regularity for solutions of the underlying evolution equations. This situation has led mathematicians and physicists either to derive, usually through systematic asymptotics, or to propose, usually in an ad hoc manner based on intuition and symmetries, simpler equations for which these mathematical issues can be addressed with greater success.

One such equation is the generalized complex Ginzburg-Landau (CGL) equation, which describes the evolution of a complex-valued field $u = u(x,t)$ over a d-dimensional spatial domain by

$$(1.1) \qquad \partial_t u = Ru + (1 + i\nu)\Delta u - (1 + i\mu)|u|^{2\sigma}u.$$

Here the parameters R and σ are positive while ν and μ are real. Without loss of generality, the equation has been normalized so that the coefficients of the linear and nonlinear dissipation (damping) terms are unity. The R is the coefficient of the linear driving term, without which all solutions would decay to zero. Clearly, σ sets the degree of the nonlinearity. The ν and μ are the coefficients of the linear and nonlinear dispersive terms respectively.

The CGL equation with a cubic nonlinearity ($\sigma = 1$) has a long history as a generic amplitude equation derived asymptotically near the onset of instabilities in fluid dynamical systems. The case with real coefficients was first derived by Newell and Whitehead [65] and Segel [74] to describe Bénard convection. The case with complex coefficients was put forth in a general setting by Newell and Whitehead [66] and DiPrima, Eckhaus, and Segel [20], and was shown by Stewartson and Stuart [76] to apply to plane Poiseuille flow. The reader may also want to consult more comprehensive reviews of the application of amplitude equations, such as the CGL equation to chemical reactions [51] and to pattern formation [17, 64]. Here we remove it from any particular physical context and investigate it as a mathematical model of a variety of possible phenomena in nonlinear partial differential equations. Considered over the d-dimensional torus $\mathbb{T}^d$, the CGL equation is both theoretically and numerically tractable, and has proved fruitful for illustration of connections between infinite dimensional and finite dimensional dynamical systems [3, 24, 73]. We shall therefore focus our study on the spatially periodic CGL equation (1.1) and point out those results that extend to $\mathbb{R}^d$.

This lecture will present to nonspecialists one overview of how the generalized CGL equation can be viewed as a dynamical system. We have therefore tried not to presume familiarity with material beyond that from standard graduate level analysis and functional analysis courses. The style will be expository. We will give a big picture along with enough detail to clarify the strategies that are used to prove many basic theorems. Without any pretense of being exhaustive, we review several of the avenues of investigation taken in the literature, all of which have application far beyond the context of the CGL equation.

The material is organized as follows. Section 2 begins to paint the CGL phase portrait by identifying some special solutions, examining their stability, and describing how chaotic behavior might arise. Section 3 presents the attributes of the CGL equation that allow it to be a focus for general studies of turbulent behavior. In particular its relation to the nonlinear Schrödinger (NLS) equation,

$$(1.2) \qquad \partial_t u = i\nu\Delta u - i\mu|u|^{2\sigma}u \,,$$

is discussed as an analogue to the relation of the driven Navier-Stokes equations to the Euler equations in fluid dynamics. The existence of global (in time) weak solutions to the CGL equation for L^2 initial data is proved in Section 4. Such solutions exist in any spatial dimension and degree of nonlinearity. The result is similar in spirit to that of Leray's existence theorem for global weak solutions to the Navier-Stokes equations. Also similarly, the question of uniqueness for weak solutions remains open. Section 5 discusses the existence and uniqueness of local (in time) classical solutions—even C^∞ solutions provided σ is a positive integer. Initial data in the classes L^p and H^q are considered. In cases when s can be negative this will include initial data that are distributions. Section 6 presents local estimates which show that when σ is an integer the solutions are not only C^∞, but also contained in a subclass of the real analytic functions. The derivation of these estimates employs a characterization of Gevrey classes in terms of decay of Fourier coefficients. Our inability to unconditionally extend the local solutions to global ones is a manifestation of the instability in the inviscid dynamics. One cannot generally exclude the possibility of a finite-time blow-up, even in dissipative systems. Section 7 establishes global classical solutions through *a priori* bounds of L^p norms and does not restrict σ to integer values. When σ is a positive integer, direct global bounds on every H^n norm are derived in Section 8. The role of these results in applying dynamical systems methods to the study of the long-time behavior of CGL solutions is discussed.

2. The CGL Equation as a Model for the Onset of Chaos

This investigation has two main themes. The first theme is to gain as complete an understanding as possible of the CGL "phase portrait" in any infinite dimensional phase space in which the flow is defined globally in time. The second theme is to justify the asymptotic derivation of the CGL equation as an amplitude equation and to determine whether the CGL phase portrait can be related to the onset of chaos in solutions of the original dynamical system. This question will be taken up in the lecture of A. Mielke [63] and is also reviewed in [17, 51], so it will not be discussed here.

Crucial to the understanding of the phase portrait of any dynamical system is the nature of the long-time asymptotics. Our main focus will be to establish existence and properties of a globally attracting set. Such a set, the *global attractor*, is compact, invariant under the flow, and attracts all orbits uniformly over bounded sets of initial data [29, 78]. Naturally, the attractor contains the ω-limit set and the *unstable manifold* of the ω-limit set of every orbit. The nature of orbits on and about the attractor provides valuable information for understanding the mechanism for the onset of chaos.

Our investigation begins by identifying classes of solutions that are known explicitly and which are therefore amenable to local analysis using linear and nonlinear methods. First, the number of linearly unstable directions gives a lower bound on the dimension of the unstable manifold, thereby providing a lower bound on the dimension of the attractor. This analysis was first carried out in detail for the cubic CGL equation for $d = 1$ by Doering, Gibbon, Holm and Nicolaenko [24] and for $d = 1, 2$ by Ghidaglia and Héron [35]. Second, we will indicate the nonlinear techniques used to paint a picture of the bifurcations from these solutions as R increases.

To illustrate the procedure, we first linearize the CGL equation (1.1) about the zero solution and obtain

$$(2.1) \qquad \partial_t \tilde{u} = R\tilde{u} + (1 + i\nu)\Delta\tilde{u}.$$

Then a small perturbation with a pure Fourier mode $\tilde{u}(x, t) = \tilde{a}(t) \exp(i\xi \cdot x)$ where $\xi \in 2\pi\mathbb{Z}^d$ $(= (2\pi\mathbb{Z})^d)$ evolves according to

$$(2.2) \qquad \partial_t \tilde{a} = R\tilde{a} - (1 + i\nu)|\xi|^2 \tilde{a}.$$

This shows that with respect to this perturbation the zero solution is linearly unstable when $|\xi|^2 < R$, linearly asymptotically stable when $|\xi|^2 > R$, and linearly neutrally stable when $|\xi|^2 = R$. Hence, the dimension of the subspace of linear instability, which gives a lower bound on the dimension of the globally attracting set, grows like $R^{d/2}$ as R becomes large.

The simplest class of nontrivial spatially periodic solutions is comprised of rotating wave solutions, which have the form

$$(2.3) \qquad u(x, t) = a \exp\left(i(\xi \cdot x - \omega t)\right),$$

where $a \in \mathbb{C}$, $\omega \in \mathbb{R}$, and $\xi \in 2\pi\mathbb{Z}^d$ satisfy

$$(2.4) \qquad \omega = \mu R + (\nu - \mu)|\xi|^2, \qquad |a|^{2\sigma} = R - |\xi|^2.$$

The largest of these, corresponding to $\xi = 0$, is the spatially homogeneous solution, sometimes referred to as the Stokes solution. For each wave vector ξ such that $|\xi|^2 < R$, formulas (2.3) and (2.4) define a circle of rotating wave solutions that is parameterized by the phase of a. These solutions that play an important role in many physical applications and their existence is one of the reasons that the CGL initial-value problem posed in $L^p(\mathbb{T}^d)$ is more interesting than that posed in $L^p(\mathbb{R}^d)$. Notice that, as R increases, the circle of rotating waves corresponding to wave vector ξ bifurcates off the zero solution at $R = |\xi|^2$, thereby corresponding to the addition of two real dimensions to the subspace of linear instability of the zero solution for each such ξ as R increases beyond $|\xi|^2$. There are on the order of $R^{d/2}$ such circles as R becomes large.

Due to the simple form of the rotating waves (2.3), their linear stability can be analyzed exactly. This is not something one can do for any nontrivial solution of any partial differential equation because the associated linearized equation will generally have coefficients that depend on both space and time. Indeed, a naive linearization of the CGL equation (1.1) about a rotating wave (2.3) leads to a variable coefficient equation. However, if one introduces a perturbation of the form

$$(2.5) \qquad u(x, t) = a \exp\left(i(\xi \cdot x - \omega t)\right)(1 + \epsilon \tilde{u}(x, t)),$$

then the resulting linearization is

$$(2.6) \qquad \partial_t \tilde{u} = 2(1 + i\nu)i\xi \cdot \nabla\tilde{u} + (1 + i\nu)\Delta\tilde{u} - (1 + i\mu)\sigma|a|^{2\sigma}(\tilde{u} + \tilde{u}^*) \, .$$

This equation can be viewed as a real linear system for the real and imaginary parts of $\tilde{u}$ (alternatively, for $\tilde{u}$ and $\tilde{u}^*$) with constant coefficients, which can therefore be analyzed using Fourier modes. Due to the multiplicative form of the perturbation (2.5), the Fourier mode of $\tilde{u}$ with wave vector $\tilde{\xi}$ corresponds to the Fourier mode of the perturbation with wave vector $\xi + \tilde{\xi}$. Hence, this procedure is often called a side-band analysis and $\tilde{\xi}$ a side-band wave vector.

   This general analysis will not be carried out here (see [24] for details), but some of its features will be illustrated by analyzing the linear stability of the homogeneous solution. In that case (2.6) simplifies to

$$(2.7) \qquad \partial_t \tilde{u} = (1 + i\nu)\Delta\tilde{u} - (1 + i\mu)\sigma R(\tilde{u} + \tilde{u}^*) \, .$$

When it is written as a real linear system for $\tilde{u}_R$ and $\tilde{u}_I$, the real and imaginary parts of $\tilde{u}$, this equation becomes

$$(2.8) \qquad \partial_t \begin{pmatrix} \tilde{u}_R \\ \tilde{u}_I \end{pmatrix} = \begin{pmatrix} \Delta - 2\sigma R & -\nu\Delta \\ \nu\Delta - 2\mu\sigma R & \Delta \end{pmatrix} \begin{pmatrix} \tilde{u}_R \\ \tilde{u}_I \end{pmatrix} \, .$$

It is easily checked that the Fourier mode with wave vector $\tilde{\xi}$ will decay exponentially provided

$$(2.9) \qquad (1 + \nu^2)|\tilde{\xi}|^4 + 2\sigma R(1 + \mu\nu)|\tilde{\xi}|^2 > 0 \, .$$

This will certainly be satisfied for every nonzero $\tilde{\xi}$, no matter what values the degree σ and forcing R may have, whenever μ and ν lie in the region of the $\mu\nu$-plane where $(1 + \mu\nu) \geq 0$, commonly referred to as the modulationally stable region (illustrated in Figure 7.1 on page 39).

   A slightly more general class than the rotating wave solutions are the harmonic (or monochromatic) solutions, which have the form

$$(2.10) \qquad u(x, t) = a(t)\exp(i\xi \cdot x) \, ,$$

where the complex-valued amplitude $a = a(t)$ satisfies

$$(2.11) \qquad \frac{da}{dt} = \big(R - (1 + i\nu)|\xi|^2\big)a - (1 + i\mu)|a|^{2\sigma}a \, .$$

Multiplying this equation by a^* and separating real and imaginary parts gives

$$(2.12) \qquad \frac{1}{2}\frac{d}{dt}|a|^2 = \big(R - |\xi|^2\big)|a|^2 - |a|^{2\sigma+2} \, ,$$

$$(2.13) \qquad \frac{1}{2i}\left(a^*\frac{da}{dt} - a\frac{da^*}{dt}\right) = -\nu|\xi|^2|a|^2 - \mu|a|^{2\sigma+2} \, .$$

It is easily seen that when $R - |\xi|^2 < 0$ these solutions decay exponentially to zero. In this case, these solutions are part of the stable manifold of the trivial solution. When $R - |\xi|^2 = 0$ these solutions decay algebraically to zero, reflecting the marginal stability of the zero solution at the given value of R to perturbations with wave vector ξ. When $R - |\xi|^2 > 0$ these solutions are attracted exponentially to a rotating wave solution corresponding to the wave vector ξ. In this case, these solutions lie on the intersection of the unstable manifold of the zero solution with the stable manifold of the rotating wave solution of wave vector ξ.

The rotating waves are also part of the more general class of traveling wave solutions, each of which has the form

$$(2.14) \qquad u(x,t) = \exp(-i\alpha t)\, v(x - ct)\,,$$

where $\alpha \in \mathbb{R}$ is the phase frequency, $c \in \mathbb{R}^d$ is the wave velocity and $v = v(z)$ is the complex-valued profile function over $z = x - ct \in \mathbb{T}^d$ that satisfies

$$(2.15) \qquad (R + i\alpha)v + (1 + i\nu)\Delta_z v - c \cdot \nabla_z v - (1 + i\mu)|v|^{2\sigma}v = 0\,.$$

Within the class of traveling waves are the so-called standing waves, for which $c = 0$, and which possess a nontrivial spatial structure in both amplitude and phase. A rotating wave (2.3) will solve (2.15) provided

$$(2.16) \qquad \omega = \alpha + \xi \cdot c\,.$$

More general traveling wave solutions will bifurcate off these rotating wave solutions as R increases [77]. In particular, standing waves bifurcate off the spatially homogeneous solution [67]. Traveling wave solutions have also been shown to exist when $d = 1$ near the NLS limit [18]. However, a complete understanding of all solutions of (2.15) is still lacking.

Stability is not as easy to assess for these traveling waves as it was for the rotating waves. For example, to study their linear stability, introduce a perturbation of the form

$$(2.17) \qquad u(x,t) = \exp(-i\alpha t)\big(v(x - ct) + \epsilon\,\tilde{v}(x - ct, t)\big)\,,$$

into the CGL equation and retain only terms of order ϵ to obtain

$$(2.18) \qquad \begin{aligned} \partial_t \tilde{v} &= (R + i\alpha)\tilde{v} + (1 + i\nu)\Delta_z \tilde{v} - c \cdot \nabla_z \tilde{v} \\ &\quad - (1 + i\mu)|v|^{2\sigma}\tilde{v} - \sigma(1 + i\mu)|v|^{2\sigma-2}v(v^*\tilde{v} + v\tilde{v}^*)\,. \end{aligned}$$

The operator on the right side is independent of t, so the linear stability analysis is thereby reduced to studying the semigroup generated by that operator, which is not easy because the operator generally has coefficients with a complicated dependence on z.

Harder still is analytically determining the bifurcation sequence of solutions in the attractor. Some insight into these questions has been gained through numerical studies [57]. For the parameter choices studied in [57], as R is increased, the standing wave solutions undergo supercritical Hopf bifurcations, creating two stable periodic orbits. Further increase in R gradually deforms these orbits into a pair of homoclinic orbits which connect the homogeneous solution to itself (the homogeneous solution is really a fixed point of the CGL system). When this homoclinic orbit is broken, a single stable periodic orbit appears which looks like a combination of the two periodic orbits that formed the homoclinic orbit. This bifurcation sequence is known as a "gluing bifurcation" (see [33] and the references therein); it is not a period doubling bifurcation, but rather a length doubling that can be understood in a three dimensional context. At this homoclinic connection, a chaotic set is not created because of the stable nature of the periodic orbits involved: the orbits bring their attracting neighborhoods to the homoclinic orbit, which is consequently a *stable* object in the phase space, and therefore cannot produce a hyperbolic chaotic set of trajectories when the homoclinic connection is broken by a change in R [57]. As R is increased beyond this initial bifurcation, a series of unstable homoclinic orbits occur that connect the homogeneous solution to itself.

These homoclinic orbits lead to chaotic attractors through so-called "homoclinic explosions", similar to those found in the Lorenz equations.

A special case of the CGL equation for which the attractor can, in principle, be completely characterized is when $\mu = \nu$. In that case it is natural to introduce the new complex-valued field $v = v(x,t)$ by

$$(2.19) \qquad u(x,t) = \exp(-iR\mu t)\, v(x,t)\,.$$

The evolution of v is then governed by

$$(2.20) \qquad \partial_t v = -(1+i\mu)\frac{\delta G}{\delta v^*}\,,$$

where

$$(2.21) \qquad \frac{\delta G}{\delta v^*} = -\Delta v + |v|^{2\sigma} v - Rv\,,$$

which is the classical variational partial derivative with respect to v^* of the so-called Ginzburg-Landau functional G defined by

$$(2.22) \qquad G = \int_{\mathbb{T}^d} |\nabla v|^2 + \frac{1}{\sigma+1}|v|^{2\sigma+2} - R|v|^2 \, dx\,.$$

Computing the time derivative of this functional under the flow (2.20) gives

$$(2.23) \qquad \frac{dG}{dt} = \int_{\mathbb{T}^d} \frac{\delta G}{\delta v}\partial_t v + \frac{\delta G}{\delta v^*}\partial_t v^* \, dx = -2\int_{\mathbb{T}^d} \left|\frac{\delta G}{\delta v}\right|^2 dx\,,$$

which shows that G is a monotonically decreasing function of time except at points where $\delta G/\delta v^*$ vanishes. Such points are stationary points of the flow (2.20) and, hence, are standing waves of the CGL flow. They include as a special case the circle of stationary harmonic waves given by

$$(2.24) \qquad v(x) = b\exp(i\xi{\cdot}x)\,, \qquad |b|^{2\sigma} = R - |\xi|^2 > 0\,,$$

for some $b \in \mathbb{C}$ and $\xi \in 2\pi\mathbb{Z}^d$. When $d = 1$, the stationary points generally satisfy the ordinary differential equation

$$(2.25) \qquad -\partial_{xx}v + |v|^{2\sigma}v - Rv = 0\,,$$

so their existence may be determined through a phase plane analysis. Furthermore, these solutions may be expressed explicitly in terms of elliptic functions when $\sigma = 1, 2$ and in terms of hyperelliptic functions when $\sigma > 2$ is an integer.

In general, (2.23) shows that the functional G formally acts as a global Lyapunov functional for the flow, a fact that will be made rigorous once sufficient regularity is established to justify the formal manipulations. This will be done in Section 5. The stability of any stationary point v can be examined through the spectrum of $\mathcal{H}(v)$, the Hessian operator of the functional G at the point v, which is given by

$$(2.26) \qquad \begin{aligned} \mathcal{H}(v) &\equiv \begin{pmatrix} \dfrac{\delta^2 G}{\delta v^*\delta v} & \dfrac{\delta^2 G}{\delta v^*\delta v^*} \\[2mm] \dfrac{\delta^2 G}{\delta v\delta v} & \dfrac{\delta^2 G}{\delta v\delta v^*} \end{pmatrix} \\[3mm] &= \begin{pmatrix} -\Delta + (\sigma+1)|v|^{2\sigma} - R & \sigma|v|^{2\sigma-2}v^2 \\[2mm] \sigma|v|^{2\sigma-2}v^{*2} & -\Delta + (\sigma+1)|v|^{2\sigma} - R \end{pmatrix}\,. \end{aligned}$$

The Hessian is the self-adjoint linear operator associated with the real quadratic form that is the second variation of G at v. Zero is always in the spectrum of $\mathcal{H}(v)$ with multiplicity at least one because $(iv, -iv^*)^T$ is always the null vector associated with changing v by a constant phase. The analysis of the remaining spectrum of $\mathcal{H}(v)$ is generally difficult because of its nonconstant coefficients.

In the case when v is a stationary harmonic wave (2.24) the Hessian operator (2.26) is unitarily similar to the constant coefficient self-adjoint operator

$$(2.27) \qquad \tilde{\mathcal{H}}(v) = \begin{pmatrix} -\Delta - 2i\xi \cdot \nabla + \sigma|b|^{2\sigma} & \sigma|b|^{2\sigma} \\ \sigma|b|^{2\sigma} & -\Delta + 2i\xi \cdot \nabla + \sigma|b|^{2\sigma} \end{pmatrix},$$

whose spectrum can be analyzed through Fourier analysis. For example, the dimension of the unstable manifold of the circle of stationary harmonic waves (2.24) is given by the number of nonzero $\tilde{\xi} \in 2\pi\mathbb{Z}^d$ such that

$$(2.28) \qquad |\tilde{\xi}|^4 + 2\sigma|b|^{2\sigma}|\tilde{\xi}|^2 - 4(\xi\cdot\tilde{\xi})^2 < 0,$$

assuming there are no such $\tilde{\xi}$ for which the left side of this equation vanishes. In particular, the circle of stationary harmonic waves (2.24) will be stable whenever $(\sigma + 2)|\xi|^2 < \sigma R$.

When $d = 1$ and $\sigma = 1$ the unstable manifold of each stationary point determined by (2.25) can also be analyzed through the Hessian (2.26) by using the machinery of the integrable NLS equation [44]. However, in all other cases our knowledge of the phase portrait for even this simplest of all CGL equations (2.20) is far from complete.

3. The CGL Equation as a Model for Turbulent Phenomena

Much work on the theory of turbulent phenomena has focused on simple fluids, where the phenomena are presumably described by the incompressible Navier-Stokes equations [5]. One of the major obstacles in this area of study is that fact that, as of this writing, the existence of smooth, or *classical*, or *strong* solutions to the Navier-Stokes initial value problem in three spatial dimensions has never been demonstrated for arbitrarily large (albeit smooth) initial data, or arbitrarily strong (albeit smooth) applied forces [14]. Leray established the existence, although not the uniqueness, of *weak* solutions to the Navier-Stokes equations over half a century ago [55], and interpreted the implied loss of regularity in the general case as an indication of the onset of turbulent behavior. This would mean that viscous dissipation in three spatial dimensions is unable to inhibit singularities driven by instabilities in the underlying inviscid system, the Euler equations. There is, however, an obvious logical gap associated with connecting our inability to prove regularity and uniqueness with an actual failure of those properties. Moreover, the necessity of a breakdown of regularity in the Euler equations remains an open area of investigation which is discussed in the lecture by P. Constantin [12]. Uniqueness of these weak solutions is not established, and this fact carries its own implications for the question of the self consistency of any hydrodynamic description at all! These issues remain today in an unsatisfactory state of nonresolution.

There are two key attributes of the CGL equation that allow us to seriously consider it as a worthwhile focus for general studies of turbulent behavior. First, in the absence of driving ($R = 0$) it has an inviscid limit that is a Hamiltonian system, namely, when $\mu\nu \neq 0$ it is the nonlinear Schrödinger (NLS) equation

$$(3.1) \qquad i\partial_t u = -\nu\Delta u + \mu|u|^{2\sigma}u \,,$$

which is essentially the usual Schrödinger equation of quantum mechanics with its potential V replaced by $\mu|u|^{2\sigma}$. Second, this inviscid limit possesses a self-focusing mechanism capable of provoking spatially localized singularities in the solutions, within a finite time, starting from arbitrarily smooth initial data. This effect can be thought of as a rather violent cascade process whereby energy concentrated in low wavenumber modes is rapidly transported to high wavenumber modes. These two properties are in direct analogy to the problem of incompressible fluid turbulence, at least so far as the conventional wisdom is concerned. The Euler equations constitute the underlying Hamiltonian dynamics of the dissipative Navier-Stokes equations, and instabilities in the inviscid equations can create a cascade of energy between widely separated length scales. We will examine these two attributes in more detail below.

The NLS equation can be naturally recast as a Hamiltonian system in the form

$$(3.2) \qquad i\nu\partial_t u = \frac{\delta H}{\delta u^*} \,,$$

where the Hamiltonian H is given by

$$(3.3) \qquad H \equiv \int_{\mathbb{T}^d} \nu^2|\nabla u|^2 + \frac{\mu\nu}{\sigma+1}|u|^{2\sigma+2}\, dx \,.$$

Defining the Poisson bracket of any two functionals F and G to be

$$(3.4) \qquad \{F,G\} \equiv \frac{1}{i\nu}\int_{\mathbb{T}^d}\left(\frac{\delta F}{\delta u}\frac{\delta G}{\delta u^*} - \frac{\delta F}{\delta u^*}\frac{\delta G}{\delta u}\right)dx \,,$$

the evolution of any functional F under the NLS flow (3.2) is governed by

$$(3.5) \qquad \frac{dF}{dt} = \{F,H\} \,.$$

At least formally, the Hamiltonian H is conserved by the NLS flow. Besides H, the NLS flow also generally conserves the mass M and the momentum P, which are given by

$$(3.6) \qquad M \equiv \int_{\mathbb{T}^d}|u|^2\,dx \,, \qquad P \equiv \frac{\nu}{2i}\int_{\mathbb{T}^d}\left(u^*\nabla u - u\nabla u^*\right)dx \,.$$

These quantities play an important role in the general study of the NLS equation, but such a digression would take us to far afield. However, it should be noted that the one-dimensional cubic NLS equation ($d = \sigma = 1$) is completely integrable and therefore possesses an infinite number of conserved quantities in involution with respect to the Poisson bracket (3.4) [82, 83].

Ginibre and Velo [36] showed that the nonlinear Schrödinger equation is a locally well-posed initial-value problem in the Hilbert space $H^1(\mathbb{R}^d)$ provided

$$(3.7) \qquad d < 2 + \frac{2}{\sigma} \,.$$

More precisely, if (3.7) is satisfied then for every $\rho > 0$ there exists a time $T(\rho) > 0$ such that for every initial data $u^{\text{in}} \in H^1(\mathbb{R}^d)$ with $\|u^{\text{in}}\|_{H^1} \leq \rho$ there exists a unique $u \in C([0,T], H^1(\mathbb{R}^d))$ satisfying a mild form of the NLS initial-value problem. Moreover, the quantities M, P, and H are all conserved by this solution. Extending this result to $\mathbb{T}^d$ is not direct, and requires some different ideas. Recently, Bourgain has developed new techniques that allow the NLS dynamics over periodic domains to be extended to larger spaces [8, 9], but here we shall confine ourselves to H^1 dynamics. We will leave imprecise the exact nature of what it means to be a mild solution of the NLS initial-value problem—it is not important for the following considerations. However, it is important to understand that whenever (3.7) is satisfied, the NLS equation will make sense as a dynamical system in H^1, at least locally in time, with M, P and H conserved.

Condition (3.7) arises from the need to control the contribution of the nonlinear term of the NLS equation to the Hamiltonian H (or equivalently, to control the $L^{2\sigma+2}$ norm of u) with the H^1 norm of u, which we take to be

$$(3.8) \qquad \|u\|_{H^1} \equiv \left(\int_{\mathbb{T}^d} \kappa^2 |u|^2 + |\nabla u|^2 \, dx \right)^{\frac{1}{2}},$$

where $-\kappa^2$ is the largest negative eigenvalue of the Laplacian. This control is achieved through a Gagliardo-Nirenberg type of Sobolev inequality [2, 32, 68], which states that there exists a positive nondimensional constant $C(\sigma, d) < \infty$ such that the $L^{2\sigma+2}$ norm of u can be bounded by a geometric average of the L^2 and H^1 norms of u as

$$(3.9) \qquad \begin{aligned} \|u\|_{L^{2\sigma+2}} &= \left(\int_{\mathbb{T}^d} |u|^{2\sigma+2} \, dx \right)^{\frac{1}{2\sigma+2}} \\ &\leq C(\sigma, d) \left(\int_{\mathbb{T}^d} \kappa^2 |u|^2 + |\nabla u|^2 \, dx \right)^{\frac{\theta}{2}} \left(\int_{\mathbb{T}^d} |u|^2 \, dx \right)^{\frac{1-\theta}{2}} \\ &= C(\sigma, d) \|u\|_{H^1}^{\theta} \|u\|_{L^2}^{1-\theta}, \end{aligned}$$

whenever the geometric interpolation parameter θ satisfies

$$(3.10) \qquad \theta \equiv \frac{\sigma d}{2\sigma + 2} < 1.$$

The above value of θ is determined by balancing units of length in (3.9), thereby rendering $C(\sigma, d)$ nondimensional, which leads to the relation

$$(3.11) \qquad \frac{1}{2\sigma+2} d = \frac{\theta}{2}(d-2) + \frac{1-\theta}{2} d.$$

The inequality in (3.10) is equivalent to condition (3.7).

These local solutions can be extended to be global solutions if one can find an upper bound on $\|u\|_{H^1}$ that is uniform in time. When $\mu\nu > 0$ such a bound follows directly from the estimate

$$(3.12) \qquad \nu^2 \|u\|_{H^1}^2 \leq H + \nu^2 \kappa^2 M.$$

On the other hand, when $\mu\nu < 0$ the Gagliardo-Nirenberg inequality (3.9) yields the estimate

$$(3.13) \qquad \begin{aligned} H + \nu^2 \kappa^2 M &= \nu^2 \|u\|_{H^1}^2 + \frac{\mu\nu}{\sigma+1} \|u\|_{L^{2\sigma+2}}^{2\sigma+2} \\ &\geq \nu^2 \|u\|_{H^1}^2 + \frac{\mu\nu}{\sigma+1} C(\sigma, d)^{2\sigma+2} \|u\|_{H^1}^{(2\sigma+2)\theta} \|u\|_{L^2}^{(2\sigma+2)(1-\theta)}, \end{aligned}$$

where θ is given by (3.10). Because $\|u\|_{L^2}$ is conserved, the above estimate gives a uniform upper bound for $\|u\|_{H^1}$ whenever the degree of $\|u\|_{H^1}$ in the second term is less that in the first term, which is the case if and only if $\sigma d < 2$. For the marginal case $\sigma d = 2$, the above estimate gives a uniform upper bound for $\|u\|_{H^1}$ provided $\|u\|_{L^2}$ is small enough.

The above argument shows that necessary conditions for the possible formation of finite-time singularities are $\nu\mu < 0$ and $\sigma d \geq 2$. This is simply understood by considering the NLS equation to be a "mean field" quantum mechanical description of locally interacting particles. Without loss of generality, taking $\nu > 0$, the potential in (3.1) is $V = \mu|u|^{2\sigma}$. When $\mu > 0$ the interaction is repulsive, and consequently there is no tendency for the particles to concentrate; this is referred to as the defocusing case. On the other hand, when $\mu < 0$ the interaction is attractive, providing the so-called self-focusing mechanism, and the particles tend to concentrate; this is referred to as the focusing case. When $\sigma d < 2$ this attraction is not strong enough that it can induce a collapse of the matter. However, as will be argued below, when $\sigma d \geq 2$ this attraction is strong enough that collapse may occur. The term *critical* indicates that the condition for blow-up is satisfied as an equality, $\sigma d = 2$. The terms *supercritical* and *subcritical* indicate $\sigma d > 2$ and $\sigma d < 2$ respectively.

That the self-focusing mechanism can produce finite time blow-up for solutions of the NLS equation when $\sigma d \geq 2$ was first established over $\mathbb{R}^d$ [**38, 81, 84**] by the following argument. Given any $u \in H^1(\mathbb{R}^d)$, consider the vector-valued quantity $\bar{x}$ and scalar-valued quantity I defined by

$$(3.14) \qquad \bar{x} \equiv \frac{1}{M}\int_{\mathbb{R}^d} x\,|u|^2\,dx\,, \qquad I \equiv \int_{\mathbb{R}^d} \tfrac{1}{8}|x - \bar{x}|^2\,|u|^2\,dx\,,$$

which, by analogy with classical mechanics, can be interpreted as the center of mass and the moment of inertia of $|u|^2$ respectively. These quantities will be finite for any $u \in H^1(\mathbb{R}^d)$ for which

$$(3.15) \qquad \int_{\mathbb{R}^d} |x|^2\,|u|^2\,dx < \infty\,.$$

Moreover, I will be positive for any nonzero $u \in H^1(\mathbb{R}^d)$. Now consider any initial data $u^{\mathrm{in}} \in H^1(\mathbb{R}^d)$ which satisfies (3.15). When (3.7) is satisfied it can be shown that, so long as the corresponding H^1 solution $u(t)$ of the NLS equation exists, condition (3.15) remains satisfied. The quantities $\bar{x}$ and I of (3.14) can then be differentiated and written in terms of the conserved quantities M, P and H defined in (3.6) and (3.3), so that

$$
\begin{aligned}
\frac{d\bar{x}}{dt} &= \frac{2P}{M}\,, \\[4pt]
\frac{dI}{dt} &= \frac{\nu}{4i}\int_{\mathbb{R}^d}(x - \bar{x})\cdot\left(u^*\nabla u - u\nabla u^*\right)dx\,, \\[4pt]
\frac{d^2 I}{dt^2} &= \int_{\mathbb{R}^d}\nu^2|\nabla u|^2 + \frac{\sigma d}{2}\frac{\mu\nu}{\sigma + 1}|u|^{2\sigma+2}\,dx - \frac{1}{2}\frac{d\bar{x}}{dt}\cdot P \\[4pt]
&= H - \frac{|P|^2}{M} + \left(\frac{\sigma d}{2} - 1\right)\frac{\mu\nu}{\sigma + 1}\int_{\mathbb{R}^d}|u|^{2\sigma+2}\,dx\,.
\end{aligned}
$$

(3.16)

If we assume that $\nu\mu < 0$ and $\sigma d \geq 2$, which are necessary conditions for a possible singularity to develop, then

$$(3.17) \qquad \frac{d^2 I}{dt^2} \leq H - \frac{|P|^2}{M}\,.$$

This inequality can be integrated to obtain the bound

$$(3.18) \qquad I(t) \leq I(0) + \frac{dI}{dt}(0)\, t + \frac{1}{2} \left(H - \frac{|P|^2}{M} \right) t^2 \,.$$

It is clear that if u^{in} is chosen so that $H - |P|^2/M$ is negative then $I(t)$ would have to become zero at a finite time T_*, in contradiction with (3.14), which shows it to be positive. The H^1 solution must therefore develop a singularity no later than the time T_*, an upper bound that could be far larger than the actual blow-up time. By exploiting the local nature of the self-focusing mechanism, this argument has been extended to bounded domains for various kinds of boundary conditions (including periodic) in [48] by introducing a cut-off version of the $|x - \bar{x}|^2$ weight in (3.14).

The nature of the NLS singularity has been studied extensively both asymptotically and numerically [53, 54, 71], and is understood to behave self-similarly in the supercritical case and nearly so in the critical case. In this sense, our knowledge about the possible breakdown of classical solutions is much better for the NLS equation than for the three-dimensional Euler equations, where it is not even known whether or not singularities develop. Numerical simulations show that the self-focusing mechanism persists in the CGL equation inside the modulationally unstable regime $1 + \nu\mu < 0$ and when $\sigma d \geq 2$ is capable of provoking many spatially localized near NLS-singularities in solutions, starting from arbitrarily smooth initial data. Each near singularity produces a rather violent cascade process whereby energy concentrated in low wavenumber modes is rapidly transported to high wavenumber modes, after which it is damped by the CGL dissipation and then is dispersed in a somewhat less violent inverse cascade process whereby some undissipated energy returns from high wavenumber modes to low wavenumber modes [39, 75]. Admittedly the mechanisms by which energy transfer is generated for the CGL equation (self-focusing) is very different than that for the 3-dimensional Euler equations (vortex stretching), and we make no claims as to a direct physical analogy. Self-focusing blow-up in the NLS equation plays a more direct role in describing strong turbulence in plasma (the NLS equation is a limit of the Zakharov equations [40, 81]), but dissipative mechanisms in plasma do not directly correspond to the dissipative terms in the CGL equation. Rather, we regard the CGL equation as a convenient nonlinear dissipative partial differential equation displaying intrinsic inviscid instabilities that we may exploit to test and hone our mathematical tools.

Finally, there are several distinct advantages to considering the CGL equation rather than, say, the Navier-Stokes equations or the Zakharov equations. First, there is the obvious point that there are simply fewer dependent variables in the CGL equation, so we may in general expect the analysis to be simpler. Indeed, the basic techniques that have been developed for the study of semilinear parabolic equations [80], and the NLS equation over $\mathbb{R}^d$ [10, 24, 79] and, just recently, over $\mathbb{T}^d$ [8, 9], can be applied to the CGL equation. Second, the self-focusing mechanism and its resulting finite time blow-up for solutions of the NLS equation is quite well understood both physically and mathematically (also see [54, 61, 62, 71] and references therein). Third, from a numerical viewpoint it is much faster and more convenient to run simulations in low spatial dimensions, and the self-focusing blow-up behavior is already present in the 1-dimensional NLS equation for $\sigma \geq 2$.

4. Global Weak Solutions

For many nonlinear evolution equations of mathematical physics, classical solutions are either known to break down or not known to exist after a finite time. The best example of the former case is the formation of shocks in solutions of the compressible Euler equations of gas dynamics, while the best example of the later case is the open problem as to whether temporally global classical solutions exist for the incompressible Navier-Stokes equations of hydrodynamics. One approach around this problem is to enlarge the notion of a solution to that of a so-called *weak solution* by only requiring that the equation be satisfied in a "weak" sense, namely, after integrating it against a member of a prescribed class of test functions. In a seminal paper of 1934, Leray [55] proved the existence of a temporally global weak solution to the incompressible Navier-Stokes equations over the whole space $\mathbb{R}^3$ for any initial data with finite energy. Subsequently, the strategy used by Leray has been applied to prove the existence of global weak solutions for a number of other equations, most notably for the Boltzmann equation by DiPerna and Lions [19]. This section gives the analogue of this result for the CGL equation (1.1) over $\mathbb{T}^d$, the d-dimensional periodic box.

The technical setup and notation for this section is as follows. Throughout this section it is more convenient to use $\gamma = \sigma + 1$. We consider the evolution of a complex valued field $u = u(t, x)$ governed by the generalized CGL equation

$$(4.1) \qquad \partial_t u = Ru + (1 + i\nu)\Delta u - (1 + i\mu)|u|^{2(\gamma-1)}u,$$

with initial condition

$$(4.2) \qquad u(0, x) = u^{\mathrm{in}}(x).$$

As before, we assume $R > 0$ and $\gamma > 1$ ($\sigma > 0$), while ν and μ can take any real value. This arbitrariness contrasts sharply with the theory of classical solutions presented in later sections where the admissible parameter values of γ, ν, and μ and the dimension d are interrelated.

Without loss of generality the units of length may be chosen so that

$$(4.3) \qquad \int_{\mathbb{T}^d} dx = 1.$$

We employ the notation

$$(4.4) \qquad \langle f \rangle = \int_{\mathbb{T}^d} f \, dx,$$

which by (4.3) denotes the mean value of f over the torus.

In order to state the main result of this section we must first introduce the principal spaces involved. For $1 \leq p \leq \infty$, the classical Lebesgue L^p-space over $(\mathbb{T}^d, dx)$ will be denoted $L^p(\mathbb{T}^d)$. The Sobolev space of functions in $L^2(\mathbb{T}^d)$ with partial derivatives in $L^2(\mathbb{T}^d)$ will be denoted $H^1(\mathbb{T}^d)$. Given any Banach space $\mathbb{X}$ with norm $\|\cdot\|_{\mathbb{X}}$ and $1 \leq p \leq \infty$, the space of (equivalence classes of) measurable functions $v = v(t)$ from $[0, \infty)$ into $\mathbb{X}$ such that $\|v\|_{\mathbb{X}} \in L^p([0, T])$ for every $T > 0$ will be denoted $L^p_{\mathrm{loc}}([0, \infty), \mathbb{X})$. And finally $C([0, \infty), \mathrm{w}\text{-}L^2(\mathbb{T}^d))$ will denote the space of continuous functions from $[0, \infty)$ into $\mathrm{w}\text{-}L^2(\mathbb{T}^d)$, which denotes $L^2(\mathbb{T}^d)$ equipped with its weak topology. This means that $v \in C([0, \infty), \mathrm{w}\text{-}L^2(\mathbb{T}^d))$ if for every $\psi \in L^2(\mathbb{T}^d)$ the function $t \mapsto \langle \psi^* v(t) \rangle$ is in $C([0, \infty))$ endowed with the usual topology of uniform convergence over compact intervals. We remark that $L^p_{\text{loc}}([0,\infty), \mathbb{X})$ and $C([0,\infty), \text{w-}L^2(\mathbb{T}^d))$ are Frechét spaces rather than Banach spaces. As such, their topologies are completely determined by the class of convergent sequences.

The main result of this section, established in [**25**], is the following global existence theorem for L^2 initial data; it is the CGL analogue of Leray's result for the Navier-Stokes equation.

THEOREM 4.1. *Given* $u^{\text{in}} \in L^2(\mathbb{T}^d)$, *there exists a function*

$$(4.5) \quad u \in C([0,\infty), \text{w-}L^2(\mathbb{T}^d)) \cap L^2_{\text{loc}}([0,\infty), H^1(\mathbb{T}^d)) \cap L^{2\gamma}_{\text{loc}}([0,\infty), L^{2\gamma}(\mathbb{T}^d)),$$

that satisfies the initial condition (4.2) *and the weak form of the CGL equation* (4.1):

$$(4.6) \quad \begin{aligned} 0 &= \langle \psi^* u(t_2) \rangle - \langle \psi^* u(t_1) \rangle - R \int_{t_1}^{t_2} \langle \psi^* u \rangle \, dt' \\ &\quad + \int_{t_1}^{t_2} \langle (1+i\nu) \nabla \psi^* \cdot \nabla u \rangle \, dt' + \int_{t_1}^{t_2} \langle (1+i\mu) \psi^* |u|^{2(\gamma-1)} u \rangle \, dt', \end{aligned}$$

for every $[t_1, t_2] \subset [0,\infty)$ *and every test function* $\psi \in C^\infty(\mathbb{T}^d)$. *Moreover, it satisfies the energy relation*

$$(4.7) \quad \begin{aligned} \tfrac{1}{2}\|u(t)\|_{L^2}^2 &+ \int_0^t \|\nabla u\|_{L^2}^2 \, dt' + \int_0^t \|u\|_{L^{2\gamma}}^{2\gamma} \, dt' \\ &\leq \tfrac{1}{2}\|u^{\text{in}}\|_{L^2}^2 + R \int_0^t \|u\|_{L^2}^2 \, dt', \end{aligned}$$

for every $t \in [0,\infty)$.

REMARK 4.1. The notion of weak solution given above is stronger in two ways than the notion of weak solution in the sense of distributions. Recall that u is a weak solution of the CGL equation in the sense of distributions when it satisfies the equation one obtains formally by integrating (4.1) over $[0,\infty) \times \mathbb{T}^d$ against a test function $w = w(t,x) \in C^\infty_c([0,\infty) \times \mathbb{T}^d)$ and integrating by parts so as to put all the derivatives onto the test function. It can be seen directly that if u satisfies (4.6) then it satisfies the distribution form of the CGL equation for test functions in the factored form $w = \phi(t)\psi^*(x)$ where $\phi \in C^\infty_c([0,\infty))$ and $\psi \in C^\infty(\mathbb{T}^d)$. Indeed, set $t_1 = 0$ in (4.6), multiply by $\partial_t \phi(t_2)$ and integrate t_2 over $[0,\infty)$, and integrate by parts to put the spatial derivatives onto ψ^*. As linear combinations of factored test functions are dense in $C^\infty_c([0,\infty) \times \mathbb{T}^d)$, it is clear that (4.6) alone implies that u is a weak solution of the CGL equation in the sense of distributions. In addition to (4.6), the weak solutions given in Theorem 4.1 are required to satisfy the energy relation (4.7).

REMARK 4.2. The strategy of the proof follows that introduced by Leray in the context of the Navier-Stokes equations, as well as to that of many other existence proofs for weak solutions of other equations. For this reason, the proof will be given in sufficient detail so as to clearly illustrate this strategy. Roughly, the idea is to construct a sequence of solutions to equations that approximate (4.6), then show the sequence is relatively compact in a topology that is strong enough to allow us pass from the approximate equation to the limit (4.6) for any converging subsequence. This involves striking a balance between the facts that compactness is easier to establish for weaker topologies, while convergence is easier to prove in stronger topologies. Uniqueness can never be asserted by such a compactness argument, but often requires the knowledge of additional regularity of the solution. Since we do not limit ourselves to any particular spatial dimension, or to any particular nonlinearity, the results of this section are equally applicable to the supercritical, critical, and subcritical CGL equations.

PROOF. The proof proceeds in four distinct steps.

STEP 1. *Construct a family of approximate solutions u_ϵ constructed by any method that yields a consistent weak formulation and an energy relation—for example, the Fourier-Galerkin method. Let P_ϵ denote the L^2-orthogonal projection onto the span of all Fourier modes of wave vectors ξ with $|\xi| \leq 1/\epsilon$. Define $u_\epsilon^{\mathrm{in}} = P_\epsilon u^{\mathrm{in}}$ and let $u_\epsilon = u_\epsilon(t)$ be the unique solution of the ordinary differential initial-value problem*

(4.8)
$$\partial_t u_\epsilon = R u_\epsilon + (1 + i\nu)\Delta u_\epsilon - (1 + i\mu)P_\epsilon\big(|u_\epsilon|^{2(\gamma-1)}u_\epsilon\big)\,,$$
$$u_\epsilon(0) = u_\epsilon^{\mathrm{in}} \in P_\epsilon L^2(\mathbb{T}^d) \subset C^\infty(\mathbb{T}^d)\,.$$

The regularized initial data $u_\epsilon^{\mathrm{in}}(x)$ converges to u^{in} strongly in $L^2(\mathbb{T}^d)$ as ϵ tends to zero and is chosen so that $\|u_\epsilon^{\mathrm{in}}\|_{L^2} \leq \|u^{\mathrm{in}}\|_{L^2}$.

Furthermore, these solutions will satisfy the regularized version of the weak form (4.6) given by

(4.9)
$$0 = \langle \psi^* u_\epsilon(t_2)\rangle - \langle \psi^* u_\epsilon(t_1)\rangle - R \int_{t_1}^{t_2} \langle \psi^* u_\epsilon \rangle\, dt'$$
$$+ \int_{t_1}^{t_2} \langle (1+i\nu)\nabla\psi^* \cdot \nabla u_\epsilon \rangle\, dt' + \int_{t_1}^{t_2} \langle (1+i\mu)\psi_\epsilon^* |u_\epsilon|^{2(\gamma-1)}u_\epsilon \rangle\, dt'\,,$$

for every $[t_1, t_2] \subset [0, \infty)$ and $\psi \in C^\infty(\mathbb{T}^d)$. Here $\psi_\epsilon \equiv P_\epsilon \psi$ will converge to ψ in C^∞ as ϵ tends to zero. Moreover, these solutions will satisfy the regularized version the energy relation (4.7) as the equality

(4.10)
$$\tfrac{1}{2}\|u_\epsilon(t)\|_{L^2}^2 + \int_0^t \|\nabla u_\epsilon\|_{L^2}^2\, dt' + \int_0^t \|u_\epsilon\|_{L^{2\gamma}}^{2\gamma}\, dt'$$
$$= \tfrac{1}{2}\|u_\epsilon^{\mathrm{in}}\|_{L^2}^2 + R \int_0^t \|u_\epsilon\|_{L^2}^2\, dt'\,.$$

for every $t \in [0, \infty)$.

Step 1 follows from the standard Picard local existence theory for ordinary differential equations applied to (4.8) as posed in the finite dimensional space $P_\epsilon L^2(\mathbb{T}^d)$, and the fact that (4.10) provides a global L^2 bound on the solutions, ensuring that they are global. Hence, we omit all details of the proof and proceed under the premise of Step 1.

STEP 2. *Show that the sequence u_ϵ is a relatively compact set (has compact closure) in*

(4.11) $\quad C([0,\infty), \text{w-}L^2(\mathbb{T}^d)) \wedge \text{w-}L^2_{\mathrm{loc}}([0,\infty), H^1(\mathbb{T}^d)) \wedge \text{w-}L^{2\gamma}_{\mathrm{loc}}([0,\infty), L^{2\gamma}(\mathbb{T}^d))\,.$

REMARK 4.3. Here the notation "$\wedge$" indicates the intersection equipped with the weak topology induced by the inclusion maps; this means that a sequence in the intersection is convergent if and only if it converges in each space separately. The sense of convergence for these spaces individually is as follows. We have $v_n \to v$ in w-$L^2_{\text{loc}}([0,\infty), H^1(\mathbb{T}^d))$ when for every $T > 0$ and every $\psi \in L^2_{\text{loc}}([0,\infty), H^1(\mathbb{T}^d))$ we have

$$(4.12) \qquad \int_0^T \langle \psi^* v_n + \nabla \psi^* \cdot \nabla v_n \rangle \, dt \to \int_0^T \langle \psi^* v + \nabla \psi^* \cdot \nabla v \rangle \, dt \,.$$

Similarly, we have $v_n \to v$ in w-$L^{2\gamma}_{\text{loc}}([0,\infty), L^{2\gamma}(\mathbb{T}^d))$ when for every $T > 0$ and every $\psi \in L^{(2\gamma)^*}_{\text{loc}}([0,\infty), L^{(2\gamma)^*}(\mathbb{T}^d))$ we have

$$(4.13) \qquad \int_0^T \langle \psi^* v_n \rangle \, dt \to \int_0^T \langle \psi^* v \rangle \, dt \,,$$

where $(2\gamma)^* = 2\gamma/(2\gamma - 1)$. Finally, we have $v_n \to v$ in $C([0,\infty), \text{w-}L^2(\mathbb{T}^d))$ when for every $\psi \in L^2(\mathbb{T}^d)$ we have

$$(4.14) \qquad \langle \psi^* v_n(t) \rangle \to \langle \psi^* v(t) \rangle \,,$$

uniformly on compact subsets of $[0,\infty)$.

PROOF OF STEP 2. Equation (4.10), along with $\|u^{\text{in}}_\epsilon\|_{L^2} \leq \|u^{\text{in}}\|_{L^2}$, implies that

$$(4.15) \qquad \tfrac{1}{2}\|u_\epsilon(t)\|^2_{L^2} \leq \tfrac{1}{2}\|u^{\text{in}}\|^2_{L^2} + R \int_0^t \|u_\epsilon\|^2_{L^2} \, dt' \,,$$

from which the Gronwall Lemma then gives

$$(4.16) \qquad \|u_\epsilon(t)\|^2_{L^2} \leq \|u^{\text{in}}\|^2_{L^2} \, e^{2Rt} \,.$$

Inserting this into the right side of the regularized energy relation (4.10) yields the explicit bound, uniform in ϵ:

$$(4.17) \qquad \tfrac{1}{2}\|u_\epsilon(t)\|^2_{L^2} + \int_0^t \|\nabla u_\epsilon\|^2_{L^2} \, dt' + \int_0^t \|u_\epsilon\|^{2\gamma}_{L^{2\gamma}} \, dt' \leq \tfrac{1}{2}\|u^{\text{in}}\|^2_{L^2} \, e^{2Rt} \,.$$

This bound establishes that the sequence $\{u_\epsilon\}$ is contained in compact sets of both w-$L^2_{\text{loc}}([0,\infty), H^1(\mathbb{T}^d))$ and w-$L^{2\gamma}_{\text{loc}}([0,\infty), L^{2\gamma}(\mathbb{T}^d))$, because norm bounded sets are relatively compact in weak-$*$ topologies, which are the same as the weak topologies on these reflexive spaces. For the same reason, the uniform bound (4.16) also shows that $\{u_\epsilon(t)\}$ is a relatively compact set in w-$L^2(\mathbb{T}^d)$ for every $t \geq 0$.

In order to complete Step 2 it must be shown that $\{u_\epsilon\}$ is a relatively compact set in $C([0,\infty), \text{w-}L^2(\mathbb{T}^d))$. Compactness requires more than just boundedness here because of the strong topology over t. We appeal to the Arzela-Ascoli theorem [**7, 72**] which asserts that $\{u_\epsilon\}$ is a relatively compact set in $C([0,\infty), \text{w-}L^2(\mathbb{T}^d))$ if and only if

(i) $\{u_\epsilon(t)\}$ is a relatively compact set in w-$L^2(\mathbb{T}^d)$ for every $t \geq 0$;
(ii) $\{u_\epsilon\}$ is equicontinuous in $C([0,\infty), \text{w-}L^2(\mathbb{T}^d))$.

As was noted at the end of the last paragraph, condition (i) is satisfied. In order to establish (ii), we must show for every $\psi \in L^2(\mathbb{T}^d)$

(ii') $\{\langle \psi^* u_\epsilon \rangle\}$ is equicontinuous in $C([0,\infty))$.

This is done by first using the regularized weak form (4.9) of the CGL equation to establish (ii') for ψ in C^∞ and then using a density argument to extend (ii') to the general case of ψ in $L^2(\mathbb{T}^d)$ (see [**25**]). $\square$

STEP 3. *Show the sequence $\{u_\epsilon\}$ is relatively compact in $L^2_{\text{loc}}([0,\infty), L^2(\mathbb{T}^d))$ and $L^{2\gamma-1}_{\text{loc}}([0,\infty), L^{2\gamma-1}(\mathbb{T}^d))$ considered with their usual strong topologies.*

REMARK 4.4. This step is necessary because Step 2 allows us to assert the existence of a weakly convergent subsequence of $\{u_\epsilon\}$, say with a limit u, from which we may conclude only that

$$(4.18) \qquad \int_0^T \|u\|_{L^2}^2 \, dt \le \liminf_{\epsilon \to 0} \int_0^T \|u_\epsilon\|_{L^2}^2 \, dt \, .$$

However, the argument that recovers (4.7) from (4.10) will need

$$(4.19) \qquad \int_0^T \|u\|_{L^2}^2 \, dt = \lim_{\epsilon \to 0} \int_0^T \|u_\epsilon\|_{L^2}^2 \, dt \, ,$$

which requires strong L^2 convergence. Similarly, the argument that recovers the nonlinear term in (4.6) from that in (4.9) will require strong $L^{2\gamma-1}$ convergence.

PROOF OF STEP 3. The crucial point is to use the results of Step 2 along with the following embedding lemma, which is essentially due to Leray.

LEMMA 4.2. *The injection*

$$(4.20) \qquad C([0,\infty), \text{w-}L^2(\mathbb{T}^d)) \wedge \text{w-}L_{\text{loc}}^2([0,\infty), H^1(\mathbb{T}^d)) \hookrightarrow L_{\text{loc}}^2([0,\infty), L^2(\mathbb{T}^d))$$

is continuous.

This lemma is proved by demonstrating that sequences that converge to zero in both $C([0,\infty), \text{w-}L^2(\mathbb{T}^d))$ and $\text{w-}L_{\text{loc}}^2([0,\infty), H^1(\mathbb{T}^d))$ will also converge in the space $L_{\text{loc}}^2([0,\infty), L^2(\mathbb{T}^d))$. The proof is fairly direct, so we refer to [25] for details. The key tool used is the Rellich Theorem, which states that $H^1 \hookrightarrow L^2$ compactly.

Given the lemma, the L^2 portion of the assertion of Step 3 is argued as follows. Step 2 states that $\{u_\epsilon\}$ is a relatively compact set in both $C([0,\infty), \text{w-}L^2(\mathbb{T}^d))$ and $\text{w-}L_{\text{loc}}^2([0,\infty), H^1(\mathbb{T}^d))$, and because the continuous image of a compact set is compact, it follows that $\{u_\epsilon\}$ is a relatively compact set in $L_{\text{loc}}^2([0,\infty), L^2(\mathbb{T}^d))$. Therefore, any subsequence of $\{u_\epsilon\}$ that converges in both $C([0,\infty), \text{w-}L^2(\mathbb{T}^d))$ and $\text{w-}L_{\text{loc}}^2([0,\infty), H^1(\mathbb{T}^d))$ will be strongly convergent in $L_{\text{loc}}^2([0,\infty), L^2(\mathbb{T}^d))$.

The $L^{2\gamma-1}$ portion of the assertion of Step 3 is a direct consequence of the L^2 portion. We consider two cases. First, if $\gamma \le 3/2$ (so that $2\gamma - 1 \le 2$) then the injection

$$(4.21) \qquad L_{\text{loc}}^2([0,\infty), L^2(\mathbb{T}^d)) \hookrightarrow L_{\text{loc}}^{2\gamma-1}([0,\infty), L^{2\gamma-1}(\mathbb{T}^d)) \, ,$$

is continuous, so that strong convergence in $L_{\text{loc}}^2([0,\infty), L^2(\mathbb{T}^d))$ also gives strong convergence in $L_{\text{loc}}^{2\gamma-1}([0,\infty), L^{2\gamma-1}(\mathbb{T}^d))$, as was asserted. Second, if $\gamma > 3/2$ (so that $2\gamma - 1 > 2$) then the strong convergence in $L_{\text{loc}}^2([0,\infty), L^2(\mathbb{T}^d))$ and the weak convergence in $L_{\text{loc}}^{2\gamma}([0,\infty), L^{2\gamma}(\mathbb{T}^d))$ combine in a standard interpolation argument to yield the result [25]. $\square$

STEP 4. *Go to the limit. That is, the weak solution u in Theorem 4.1 is identified as the limit of a convergent subsequence of $\{u_\epsilon\}$. The fact that this subsequence converges in the various function spaces is used to verify the weak form (4.6) and energy relation (4.7).*

PROOF OF STEP 4. Step 2 ensures that there is a subsequence of $\{u_\epsilon\}$, which we also refer to as $\{u_\epsilon\}$, that simultaneously converges to a limit u in the spaces

$C([0,\infty), \text{w-}L^2(\mathbb{T}^d))$, w-$L^2_{\text{loc}}([0,\infty), H^1(\mathbb{T}^d))$ and w-$L^{2\gamma}_{\text{loc}}([0,\infty), L^{2\gamma}(\mathbb{T}^d))$. Thus, as was asserted in (4.5),

$$(4.22) \quad u \in C([0,\infty), \text{w-}L^2(\mathbb{T}^d)) \cap L^2_{\text{loc}}([0,\infty), H^1(\mathbb{T}^d)) \cap L^{2\gamma}_{\text{loc}}([0,\infty), L^{2\gamma}(\mathbb{T}^d)).$$

Step 3 then implies the strong convergence of u_ϵ to u in both $L^2_{\text{loc}}([0,\infty), L^2(\mathbb{T}^d))$ and $L^{2\gamma-1}_{\text{loc}}([0,\infty), L^{2\gamma-1}(\mathbb{T}^d))$. All that remains is to show that the limit u satisfies the weak form of the CGL equation (4.6) as well as the energy relation (4.7). Toward this end we check convergence of each term in the respective regularized versions, (4.9) and (4.10).

First consider the regularized weak form of the CGL equation (4.9) for an arbitrary test function $\psi \in C^\infty(\mathbb{T}^d)$ and interval $[t_1, t_2] \subset [0, \infty)$:

$$(4.23) \quad \begin{aligned} 0 = {}& \langle \psi^* u_\epsilon(t_2) \rangle - \langle \psi^* u_\epsilon(t_1) \rangle - R \int_{t_1}^{t_2} \langle \psi^* u_\epsilon \rangle \, dt' \\ & + \int_{t_1}^{t_2} \langle (1 + i\nu) \nabla \psi^* \cdot \nabla u_\epsilon \rangle \, dt' + \int_{t_1}^{t_2} \langle (1 + i\mu) \psi_\epsilon^* |u_\epsilon|^{2(\gamma-1)} u_\epsilon \rangle \, dt'. \end{aligned}$$

The convergence of u_ϵ to u in $C([0,\infty), \text{w-}L^2(\mathbb{T}^d))$ means (see (4.14)) that $\langle \psi^* u_\epsilon(t) \rangle$ converges to $\langle \psi^* u(t) \rangle$ uniformly over $[t_1, t_2]$, whereby

$$(4.24) \qquad \langle \psi^* u_\epsilon(t_1) \rangle \to \langle \psi^* u(t_1) \rangle, \qquad \langle \psi^* u_\epsilon(t_2) \rangle \to \langle \psi^* u(t_2) \rangle,$$

$$(4.25) \qquad \int_{t_1}^{t_2} \langle \psi^* u_\epsilon \rangle \, dt' \to \int_{t_1}^{t_2} \langle \psi^* u \rangle \, dt'.$$

When combined with (4.25), the convergence of u_ϵ in w-$L^2_{\text{loc}}([0,\infty), H^1(\mathbb{T}^d))$ (see (4.12)) then yields

$$(4.26) \qquad \int_{t_1}^{t_2} \langle (1+i\nu)\nabla\psi^* \cdot \nabla u_\epsilon \rangle \, dt' \to \int_{t_1}^{t_2} \langle (1+i\nu)\nabla\psi^* \cdot \nabla u \rangle \, dt'.$$

In order to pass to the limit in the nonlinear term of (4.23) we shall exploit the generally useful fact that if $\{v_n\}$ is a weakly convergent sequence in a Banach space $\mathbb{X}$ and $\{f_n\}$ is a strongly convergent sequence in the dual space $\mathbb{X}^*$ then the sequence $\{f_n(v_n)\}$ is convergent in $\mathbb{C}$. Here, the strong convergence of u_ϵ in $L^{2\gamma-1}_{\text{loc}}([0,\infty), L^{2\gamma-1}(\mathbb{T}^d))$ implies the weak convergence of $|u_\epsilon|^{2(\gamma-1)} u_\epsilon$ in $L^1_{\text{loc}}([0,\infty), L^1(\mathbb{T}^d))$ and, hence, in the Banach space $L^1([t_1, t_2] \times \mathbb{T}^d)$. The fact that ψ_ϵ converges to ψ uniformly over $\mathbb{T}^d$ implies that it converges strongly in $L^\infty([t_1, t_2] \times \mathbb{T}^d)$, which is the dual of $L^1([t_1, t_2] \times \mathbb{T}^d)$. It follows that

$$(4.27) \qquad \int_{t_1}^{t_2} \langle (1+i\mu)\psi_\epsilon^* |u_\epsilon|^{2(\gamma-1)} u_\epsilon \rangle \, dt' \to \int_{t_1}^{t_2} \langle (1+i\mu)\psi^* |u|^{2(\gamma-1)} u \rangle \, dt'.$$

By combining (4.25)–(4.27), one can pass to the limit in each term of (4.23) and conclude that u satisfies the weak form of the CGL equation (4.6).

To recover the energy relation (4.7) consider its regularized version (4.10):

$$(4.28) \quad \begin{aligned} \tfrac{1}{2} \|u_\epsilon(t)\|_{L^2}^2 &+ \int_0^t \|\nabla u_\epsilon\|_{L^2}^2 \, dt' + \int_0^t \|u_\epsilon\|_{L^{2\gamma}}^{2\gamma} \, dt' \\ &= \tfrac{1}{2} \|u_\epsilon^{\text{in}}\|_{L^2}^2 + R \int_0^t \|u_\epsilon\|_{L^2}^2 \, dt'. \end{aligned}$$

First examining the right side, the strong convergence of the initial data in $L^2(\mathbb{T}^d)$ implies

$$(4.29) \qquad \|u_\epsilon^{\mathrm{in}}\|_{L^2}^2 \to \|u^{\mathrm{in}}\|_{L^2}^2 \,,$$

while the strong convergence of u_ϵ in $L^2_{\mathrm{loc}}([0,\infty), L^2(\mathbb{T}^d))$ implies

$$(4.30) \qquad \int_0^t \|u_\epsilon\|_{L^2}^2 \, dt' \to \int_0^t \|u\|_{L^2}^2 \, dt' \,.$$

The right side of (4.28) is therefore convergent as ϵ tends to zero. Turning to the left side of (4.28), the convergence of u_ϵ in $C([0,\infty), \text{w-}L^2(\mathbb{T}^d))$, together with the general fact that the norm of the weak limit of a sequence is a lower bound for the limit inferior of the norms, yields

$$(4.31) \qquad \|u(t)\|_{L^2}^2 \le \liminf_{\epsilon \to 0} \|u_\epsilon(t)\|_{L^2}^2 \,.$$

Similarly, the convergence in $\text{w-}L^2_{\mathrm{loc}}([0,\infty), H^1(\mathbb{T}^d))$ and $\text{w-}L^{2\gamma}_{\mathrm{loc}}([0,\infty), L^{2\gamma}(\mathbb{T}^d))$ implies

$$(4.32) \qquad \int_0^t \|u\|_{H^1}^2 \, dt' \le \liminf_{\epsilon \to 0} \int_0^t \|u_\epsilon\|_{H^1}^2 \, dt' \,,$$

$$(4.33) \qquad \int_0^t \|u\|_{L^{2\gamma}}^{2\gamma} \, dt' \le \liminf_{\epsilon \to 0} \int_0^t \|u_\epsilon\|_{L^{2\gamma}}^{2\gamma} \, dt' \,.$$

Taken together, (4.30) and (4.32) give

$$(4.34) \qquad \int_0^t \|\nabla u\|_{L^2}^2 \, dt' \le \liminf_{\epsilon \to 0} \int_0^t \|\nabla u_\epsilon\|_{L^2}^2 \, dt' \,.$$

By combining (4.31), (4.33) and (4.34), one obtains a lower bound on the on the limit inferior of (4.28). The energy relation (4.7) is thereby satisfied and Theorem 4.1 is proved.   □

REMARK 4.5. The energy inequality (4.7) in Theorem 4.1 can be strengthened slightly. Indeed, it is readily seen that the approximate solutions defined by (4.8) of Step 1 will satisfy the energy relation

$$(4.35) \qquad \begin{aligned} \tfrac{1}{2}\|u_\epsilon(t_2)\|_{L^2}^2 + \int_{t_1}^{t_2} \|\nabla u_\epsilon\|_{L^2}^2 \, dt &+ \int_{t_1}^{t_2} \|u_\epsilon\|_{L^{2\gamma}}^{2\gamma} \, dt \\ &= \tfrac{1}{2}\|u_\epsilon(t_1)\|_{L^2}^2 + R \int_{t_1}^{t_2} \|u_\epsilon\|_{L^2}^2 \, dt \,, \end{aligned}$$

for every $[t_1, t_2] \subset [0,\infty)$. The convergence of u_ϵ to u in $L^2_{\mathrm{loc}}([0,\infty), L^2(\mathbb{T}^d))$ implies $\|u_\epsilon\|_{L^2}$ converges to $\|u\|_{L^2}$ in $L^2_{\mathrm{loc}}([0,\infty))$, whereby, upon also employing the Cantor diagonal method, one can pass to a subsequence of u_ϵ such that

$$(4.36) \qquad \|u_\epsilon(t)\|_{L^2} \to \|u(t)\|_{L^2} \qquad \text{for almost every } t \in [0,\infty) \,.$$

Call the set of points for which the limit holds E. When t_1 is restricted to lie in E one can argue starting from (4.35) as in (4.28–4.34) to show that u satisfies

$$(4.37) \qquad \begin{aligned} \tfrac{1}{2}\|u(t)\|_{L^2}^2 + \int_{t_1}^{t_2} \|\nabla u\|_{L^2}^2 \, dt &+ \int_{t_1}^{t_2} \|u\|_{L^{2\gamma}}^{2\gamma} \, dt \\ &\le \tfrac{1}{2}\|u(t_1)\|_{L^2}^2 + R \int_{t_1}^{t_2} \|u\|_{L^2}^2 \, dt \,, \end{aligned}$$

for every $[t_1, t_2] \subset [0, \infty)$ with $t_1 \in E$.

REMARK 4.6. Replacing $\mathbb{T}^d$ with $\mathbb{R}^d$ and $C^\infty(\mathbb{T}^d)$ with the compactly supported test functions $C_c^\infty(\mathbb{R}^d)$ everywhere in the statement of Theorem 4.1 gives the analogous $\mathbb{R}^d$ result. The strategy for its proof is embodied in the same four steps used above for the periodic case. Indeed, Step 2 and Step 4 have only superficial differences from those above. The Fourier-Galerkin method used in Step 1 to construct approximate classical solutions breaks down over the whole space, but can be replaced by so-called smoothing approximations. Specifically, a contraction mapping argument such as in [36], can be used to construct smooth solutions of the initial-value problem

$$(4.38) \quad \partial_t u_\epsilon = R u_\epsilon + (1 + i\nu)\Delta u_\epsilon - (1 + i\mu) j_\epsilon * \left(|j_\epsilon * u_\epsilon|^{2(\gamma-1)} j_\epsilon * u_\epsilon \right),$$
$$u_\epsilon(0, x) = u_\epsilon^{\mathrm{in}}(x) = j_\epsilon * u^{\mathrm{in}}(x),$$

where the regularization is achieved through convolution with a smooth, nonnegative mollifier j_ϵ. Finally, details of Step 3 must be modified to reflect differences in the Rellich Theorem for unbounded domains. These technical changes are straightforward, and so are omitted.

REMARK 4.7. One can show the existence of global weak solutions for the defocusing ($\mu\nu > 0$) NLS equation (1.2) with any $\sigma > 0$ for finite-energy initial data using similar arguments, but using the conservation laws of mass and energy in place of the dissipation relation (4.7) to obtain the weak compactness of Step 2.

5. Local Classical Solutions

The solutions found in the last section are too weak to obtain proofs of their uniqueness or regularity. In this section we establish the local existence, uniqueness, and regularity results for classical solutions of the generalized complex Ginzburg-Landau equation (1.1). Our objective here is to provide the basic results necessary to justify the formal manipulations necessary to establish the existence of global solutions in Section 7. More general results might be derived by adapting the important new techniques developed by Bourgain [8, 9] for the NLS equation over $\mathbb{T}^d$, however our needs here are less demanding. Indeed, our proofs mostly use standard techniques [32, 43, 70, 80] and some detail will be omitted.

Local existence and uniqueness of classical solutions of the generalized complex Ginzburg-Landau equation will be established by using the contraction mapping theorem. In the first part of this section we will describe—first as an abstract result, then applied to the CGL equation—the setting necessary to obtain a so-called mild solution from continuous initial data, and elevate it to a classical solution by a bootstrapping procedure. In the second part of this section the contraction mapping argument is reexamined in order to extend the class of initial data that evolve into local classical solutions.

Consider the evolution of $u = u(t)$ in a Banach space $\mathbb{X}$ to be governed by the abstract initial-value problem

$$(5.1) \quad \partial_t u = L u + N(u), \qquad u(0) = u^{\mathrm{in}} \in \mathbb{X}.$$

Here the linear operator L is assumed to be the infinitesimal generator of a strongly continuous semigroup $S(t)$ over $\mathbb{X}$ (so that the linear problem with $N = 0$ is well-posed, see [70]). The perturbation N is usually a nonlinear map over $\mathbb{X}$.

The well-posedness of the above initial-value problem can be established by a contraction mapping argument after formally recasting (5.1) in its so-called mild formulation

$$(5.2) \qquad u(t) = S(t)u^{\text{in}} + \int_0^t S(t - t')N(u(t'))\, dt'\,.$$

In order to employ the contraction mapping theorem we shall assume that the perturbation N is locally Lipschitz as a map from $\mathbb{X}$ into itself. More specifically, this means

(i) $\|N(u)\|_{\mathbb{X}} \le C_{\text{bd}}(\|u\|_{\mathbb{X}})$ for every $u \in \mathbb{X}$,

(ii) $\|N(u_1) - N(u_2)\|_{\mathbb{X}} \le C_{\text{Lip}}(\|u_1\|_{\mathbb{X}}, \|u_2\|_{\mathbb{X}})\,\|u_1 - u_2\|_{\mathbb{X}}$
     for every $u_1, u_2 \in \mathbb{X}$,

where $C_{\text{bd}}(\cdot)$ and $C_{\text{Lip}}(\cdot, \cdot)$ are nondecreasing functions of their arguments. We remark that N and L can always be chosen to satisfy $N(0) = 0$ in which case condition (ii) implies (i). Given such a perturbation N, one can prove the following basic result [**70**, page 184].

THEOREM 5.1 (Basic Local Existence). *For every $\rho > 0$ there exists a time $T(\rho) > 0$ such that for every initial data $u^{\text{in}} \in \mathbb{X}$ with $\|u^{\text{in}}\|_{\mathbb{X}} \le \rho$ there exists a unique $u \in C([0, T], \mathbb{X})$ satisfying the mild formulation (5.2). In addition, the map from u^{in} to u is a locally Lipschitz function from $\mathbb{X}$ to $C([0, T], \mathbb{X})$.*

Such a u is called a *mild solution* for the initial-value problem (5.1); it provides the starting point for an analysis demonstrating that it is in fact a classical solution.

REMARK 5.1. If $u \in C([0, T], \mathbb{X})$ is a mild solution of (5.1) then a direct calculation shows that it is also a weak solution of (5.1) in the sense that

$$(5.3) \qquad \begin{aligned} &\langle \psi | u(t_2) \rangle_{\mathbb{X}} - \langle \psi | u(t_1) \rangle_{\mathbb{X}} \\ &\qquad = \int_{t_1}^{t_2} \langle L^* \psi | u(t) \rangle_{\mathbb{X}}\, dt + \int_{t_1}^{t_2} \langle \psi | N(u(t)) \rangle_{\mathbb{X}}\, dt\,, \end{aligned}$$

for every $0 \le t_1 \le t_2 \le T$ and $\psi \in \mathcal{D}(L^*)$. Here $\langle \cdot | \cdot \rangle_{\mathbb{X}}$ is the usual bilinear duality between $\mathbb{X}$ and its dual space $\mathbb{X}^*$, while L^* is the usual dual adjoint of L with domain $\mathcal{D}(L^*)$ dense in $\mathbb{X}^*$.

We now apply this general theory to the initial-value problem for the spatially periodic generalized complex Ginzburg-Landau equation. In that case we choose

$$(5.4) \qquad Lu = (1 + i\nu)\Delta u + Ru\,, \qquad N(u) = -(1 + i\mu)|u|^{2\sigma}u\,.$$

The associated semigroup $S(t)$ acting on $u \in \mathbb{X}$ can be written as a convolution, $S(t)u = G_t * u$, with its Green function $G_t = G_t(x)$ for $t > 0$ given by

$$(5.5) \qquad \begin{aligned} G_t(x) &= \sum_{n \in \mathbb{Z}^d} g_t(x + n)\,, \\ g_t(x) &= \frac{1}{(4\pi(1 + i\nu)t)^{d/2}} \exp\left(-\frac{|x|^2}{4(1 + i\nu)t} + Rt\right). \end{aligned}$$

The integral equation (5.2) recast in terms of this Green function takes the form

$$(5.6) \qquad u(t) = G_t * u^{\text{in}} + \int_0^t G_{t-t'} * N(u(t'))\, dt'\,.$$

In order to apply the local existence theorem one only need identify the space $\mathbb{X}$ and verify the conditions (i) and (ii).

For $t > 0$ the Green function (5.5) satisfies the L^1-estimate

$$(5.7) \qquad \|G_t\|_{L^1} \leq \sum_{n \in \mathbb{Z}^d} \int_{\mathbb{T}^d} |g_t(x+n)| \, dx = \int_{\mathbb{R}^d} |g_t(x)| \, dx = (1+\nu^2)^{d/4} \, e^{Rt},$$

from which it follows that $S(t)$ is bounded over $L^p(\mathbb{T}^d)$ for every $1 \leq p \leq \infty$ with

$$(5.8) \qquad \|S(t)u\|_{L^p} = \|G_t * u\|_{L^p} \leq \|G_t\|_{L^1} \|u\|_{L^p} \leq (1+\nu^2)^{d/4} \, e^{Rt} \|u\|_{L^p}.$$

Moreover, it can be shown [32] that $S(t)$ is a strongly continuous semigroup over $C(\mathbb{T}^d)$ and over $L^p(\mathbb{T}^d)$ for every $1 \leq p < \infty$.

REMARK 5.2. The bounded operators $S(t)$ defined by $S(0) = I$ and $S(t)u = G_t * u$ for $t > 0$ form a semigroup over $L^\infty(\mathbb{T}^d)$ that is strongly continuous for every $t > 0$ but only weak-$*$ continuous at $t = 0$. (A sequence $\{v_n\}$ converges to v in the weak-$*$ topology on $L^\infty(\mathbb{T}^d)$ if it does so integrated against arbitrary functions in $L^1(\mathbb{T}^d)$.)

We first appeal to the local existence theorem with $\mathbb{X} = C(\mathbb{T}^d)$. The perturbation N given by (5.4) is clearly locally Lipschitz continuous as a map from $C(\mathbb{T}^d)$ into itself. A direct application of the local existence theorem then yields a unique mild solution $u = u(t)$ of the CGL equation over a time interval $[0, T]$ that depends only on the L^∞ norm of u^{in}. This solution is the limit of a sequence $\{u^{(n)}\}$ of successive iterates of (5.6), say that defined by

$$(5.9) \qquad \begin{aligned} u^{(0)}(t) &= G_t * u^{\text{in}}, \\ u^{(n+1)}(t) &= G_t * u^{\text{in}} + \int_0^t G_{t-t'} * N\big(u^{(n)}(t')\big) \, dt', \end{aligned}$$

that converge in $C([0,T], C(\mathbb{T}^d))$ for some T chosen sufficiently small that the sequence contracts.

Additional regularity must be demonstrated in order to elevate these mild solutions to classical solutions, specifically they ought to be C^2 in space and C^1 in time, so that the derivatives appearing in the equation are classical. This is done using the following standard bootstrapping argument. We use the regularity of G_t for $t > 0$ to estimate the gradient of the successive iterates (5.9) in the L^∞ norm to show that the sequence $\{u^{(n)}\}$ converges in $C((0,T], C^1(\mathbb{T}^d))$. This is accomplished as follows. Take the gradient of the integral equation (5.9),

$$(5.10) \qquad \nabla u^{(n+1)}(t) = \nabla G_t * u^{\text{in}} + \int_0^t \nabla G_{t-t'} * N\big(u^{(n)}(t')\big) \, dt',$$

and estimate

(5.11)

$$\|\nabla u^{(n+1)}(t)\|_{L^\infty} \leq \|\nabla G_t\|_{L^1} \|u^{\text{in}}\|_{L^\infty} + \int_0^t \|\nabla G_{t-t'}\|_{L^1} \|N(u^{(n)}(t'))\|_{L^\infty} \, dt'.$$

Then, use the L^1-estimate

$$\|\nabla G_t\|_{L^1} \le \sum_{n \in \mathbb{Z}^d} \int_{\mathbb{T}^d} |\nabla g_t(x+n)| \, dx$$

(5.12)

$$= \int_{\mathbb{R}^d} |\nabla g_t(x)| \, dx = C_d \, (1+\nu^2)^{d/4} \, \frac{e^{Rt}}{\sqrt{t}}$$

where $C_d > 0$ is a constant depending only on the dimension d, to find that

$$\|\nabla u^{(n+1)}(t)\|_{L^\infty} \le C_d \, (1+\nu^2)^{d/4} \, \frac{e^{Rt}}{\sqrt{t}} \, \|u^{\mathrm{in}}\|_{L^\infty}$$

(5.13)

$$+ C_d \, (1+\nu^2)^{d/4} \int_0^t \frac{e^{R(t-t')}}{\sqrt{t-t'}} \, \|N(u^{(n)}(t'))\|_{L^\infty} \, dt' \,.$$

This shows that each $\nabla u^{(n)}$ lies in $C((0,T], C(\mathbb{T}^d))$. By invoking the Lipschitz continuity of N, a similar estimate shows that the sequence is Cauchy in $C((0,T], C(\mathbb{T}^d))$, so that $u \in C((0,T], C^1(\mathbb{T}^d))$. If $u^{\mathrm{in}} \in C^1(\mathbb{T}^d)$, it is possible to rewrite the integral equation for ∇u without putting any derivatives on G_t, namely

$$\nabla u(t) = G_t * \nabla u^{\mathrm{in}} + \int_0^t G_{t-t'} * \big(DN(u(t')) \nabla u(t') \big) \, dt' \,,$$

(5.14)

where $DN(u)$ represents the derivative of $N(u)$ with respect to u, which, when $N(u)$ is given by (5.4), acts on an arbitrary function w by

$$DN(u)w = -(1+i\mu)\big((\sigma+1)|u|^{2\sigma}w + \sigma|u|^{2\sigma-2}u^2 w^*\big) \,.$$

(5.15)

Consequently, estimates analogous to (5.13) will not exhibit a singularity at $t=0$ and hence yield a $u \in C([0,T], C^1(\mathbb{T}^d))$. A repetition of the above regularity argument starting from (5.14) rather than (5.6) then shows that u is in $C((0,T], C^2(\mathbb{T}^d))$. Moreover, because the CGL equation relates the first time derivative to the second space derivative, the solution must also be in $C^1((0,T], C(\mathbb{T}^d))$ and is therefore a classical solution of the CGL equation so long as it is a mild solution. Summarizing, we have proved the following result.

THEOREM 5.2 (Local Classical Solutions for C^0 Initial Data). *For every $\rho > 0$ there exists a time $T(\rho) > 0$ such that for initial data $u^{\mathrm{in}} \in C(\mathbb{T}^d)$ with $\|u^{\mathrm{in}}\|_{L^\infty} \le \rho$ there exists a unique*

$$u \in C([0,T], C(\mathbb{T}^d)) \cap C((0,T], C^2(\mathbb{T}^d)) \cap C^1((0,T], C(\mathbb{T}^d)) \,,$$

(5.16)

satisfying the CGL initial-value problem. In addition, the map from u^{in} to u is a locally Lipschitz function from $C(\mathbb{T}^d)$ to $C([0,T], C(\mathbb{T}^d))$. Moreover, for initial data $u^{\mathrm{in}} \in C^2(\mathbb{T}^d)$ one has $u \in C([0,T], C^2(\mathbb{T}^d)) \cap C^1([0,T], C(\mathbb{T}^d))$.

In this generality one can not expect these solutions to be in $C((0,T], C^3(\mathbb{T}^d))$ because unbounded singularities will be introduced at zeros of u upon further differentiation of (5.15). However, additional regularity can be gain in some cases. For example, when σ is a positive integer the nonlinearity is a polynomial in u and u^* and one can freely differentiate (5.15) without introducing any singularities. Whence, continuing with the bootstrapping argument, it can be shown that for every t in $(0,T]$ the solution has at least one more spatial derivative than it had initially. But then repeating this argument implies that the solution is in $C((0,T], C^\infty(\mathbb{T}^d))$. Moreover, because the equation relates temporal derivatives to spatial derivatives, the solution must possess all temporal derivatives too and is therefore a smooth (C^∞) solution of the CGL equation (1.1) so long as it is a mild solution. More precisely, we have proved the following result.

THEOREM 5.3 (Local Smooth Solutions). *Let $\sigma > 0$ be an integer. Then for every $\rho > 0$ there exists a time $T(\rho) > 0$ such that for every initial data $u^{\mathrm{in}} \in C(\mathbb{T}^d)$ with $\|u^{\mathrm{in}}\|_{L^\infty} \leq \rho$ there exists a unique*

$$(5.17) \qquad u \in C([0,T], C(\mathbb{T}^d)) \cap C^\infty((0,T] \times \mathbb{T}^d),$$

satisfying the CGL initial-value problem. Moreover, given smooth initial data $u^{\mathrm{in}} \in C^\infty(\mathbb{T}^d)$ then $u \in C^\infty([0,T] \times \mathbb{T}^d)$.

REMARK 5.3. The above result justifies the formal manipulations we will carry out in Section 8, thereby enabling us to focus our attention on the establishment of global uniform Sobolev bounds. Moreover, in Section 6 we will extend Theorem 5.3 to show that when σ is a positive integer these solutions are real analytic.

Even when σ is not a positive integer, we can still advance the bootstrapping argument so long as the differentiation of (5.15) does not introduce unbounded singularities at the zeros of u. The lowest degree of homogeneity for the factors u and u^* appearing in a term of the $(n+1)^{\mathrm{st}}$ derivative of $|u|^{2\sigma}u$ will be $2\sigma - n$, and this can be controlled whenever $\sigma \geq n/2$. In that case the bootstrapping argument will gain an additional n spatial derivatives, showing that the solution is in $C((0,T], C^{n+2}(\mathbb{T}^d))$. This observation then leads to the following.

THEOREM 5.4 (Local C^k Solutions). *Let $\sigma \geq n/2$ for some positive integer n. Then for every $\rho > 0$ there exists a time $T(\rho) > 0$ such that for every initial data $u^{\mathrm{in}} \in C(\mathbb{T}^d)$ with $\|u^{\mathrm{in}}\|_{L^\infty} \leq \rho$ there exists a unique*

$$(5.18) \qquad u \in C([0,T], C(\mathbb{T}^d)) \cap C((0,T], C^{n+2}(\mathbb{T}^d)) \cap C^1((0,T], C^n(\mathbb{T}^d)),$$

satisfying the CGL initial-value problem. Moreover, for $u^{\mathrm{in}} \in C^{n+2}(\mathbb{T}^d)$ one has $u \in C([0,T], C^{n+2}(\mathbb{T}^d)) \cap C^1([0,T], C^n(\mathbb{T}^d))$.

REMARK 5.4. Of course, we can continue to trade two spatial derivatives for one time derivative. If $n = 2k$ is even this leads to $u \in C^{k+1}([0,T], C(\mathbb{T}^d))$, while if $n = 2k+1$ is odd one finds $u \in C^{k+1}([0,T], C^1(\mathbb{T}^d))$.

REMARK 5.5. Theorem 5.4 will be applied in Section 7 with $n = 1$ in order to justify some of the formal manipulations there, enabling us to focus attention on the question of global existence.

REMARK 5.6. Each of the above results have a version for initial data u^{in} in $L^\infty(\mathbb{T}^d)$. By Remark 5.2 the semigroup $S(t)$ acting on $L^\infty(\mathbb{T}^d)$ is generally not strongly continuous at $t = 0$ but rather is weak-$*$ continuous. The corresponding concession must be made regarding u in the conclusions of the L^∞ versions of the above theorems. For example, in the L^∞ version of Theorem 5.2 one must replace (5.15) with

$$(5.19) \qquad u \in C([0,T], \mathrm{w}^*\text{-}L^\infty(\mathbb{T}^d)) \cap C((0,T], C^2(\mathbb{T}^d)) \cap C^1((0,T], C(\mathbb{T}^d)),$$

where $\mathrm{w}^*\text{-}L^\infty(\mathbb{T}^d)$ denotes $L^\infty(\mathbb{T}^d)$ endowed with its weak-$*$ topology. Outside of that consideration, the proofs of the L^∞ versions are virtually identical to those given above [**25**].

The basic existence argument given above can be refined to greatly enlarge the class of initial data that evolve into classical solutions for positive times. We recall the mild formulation (5.2),

$$(5.20) \qquad u(t) = S(t)u^{\mathrm{in}} + \int_0^t S(t-t')N(u(t'))\,dt'\,.$$

In this formulation it is natural to distinguish three spaces. One aims at proving that for initial data in some space $\mathbb{X}$ the solution is in a "better" space $\mathbb{Y}$ on some interval $(0,T]$. In general, the nonlinearity maps $\mathbb{Y}$ into a third space we shall denote by $\mathbb{Z}$.

For suitable choices of $\mathbb{X}$, $\mathbb{Y}$ and $\mathbb{Z}$ the semigroup will be strongly continuous on $\mathbb{X}$ and also map $\mathbb{X}$ and $\mathbb{Z}$ back into the "better" space $\mathbb{Y}$ with a singularity at $t = 0$, so that near $t = 0$ one has nonnegative constants α and β such that

$$(5.21) \qquad \|S(t)w\|_{\mathbb{Y}} \le c\,t^{-\alpha}\,\|w\|_{\mathbb{X}} \quad \text{for every } w \in \mathbb{X}\,,$$

$$(5.22) \qquad \|S(t)w\|_{\mathbb{Y}} \le c\,t^{-\beta}\,\|w\|_{\mathbb{Z}} \quad \text{for every } w \in \mathbb{Z}\,.$$

Again, we will assume that N is locally Lipschitz, now as a map from $\mathbb{Y}$ to $\mathbb{Z}$. Without loss of generality one can assume that $N(0) = 0$, and for our convenience we will only consider the case where the Lipschitz condition can be written as

$$(5.23) \qquad \|N(u_1) - N(u_2)\|_{\mathbb{Z}} \le c\left(\|u_1\|_{\mathbb{Y}}^{2\sigma} + \|u_2\|_{\mathbb{Y}}^{2\sigma}\right)\|u_1 - u_2\|_{\mathbb{Y}}$$

for some $\sigma \ge 0$. The σ above will be identified with the σ of the CGL nonlinearity (5.4). This covers many applications, including the CGL equation. Modifications that embrace more general nonlinearities are not hard [43, 70].

THEOREM 5.5 (Extended Local Existence). *Given the above estimates (5.21), (5.22) and (5.23) where the exponents α, β and σ satisfy*

$$(5.24) \qquad\qquad 0 \le \beta < 1\,,$$

$$(5.25) \qquad\qquad 0 \le (2\sigma + 1)\alpha < 1\,,$$

$$(5.26) \qquad\qquad \beta + 2\sigma\alpha < 1\,,$$

for every $\rho > 0$ there exists a time $T(\rho) > 0$ such that for every initial data $u^{\mathrm{in}} \in \mathbb{X}$ with $\|u^{\mathrm{in}}\|_{\mathbb{X}} \le \rho$ there exists a unique

$$(5.27) \qquad\qquad u \in C([0,T],\mathbb{X}) \cap C((0,T],\mathbb{Y})$$

satisfying the mild formulation (5.20). In addition, the map from u^{in} to u is a locally Lipschitz function from $\mathbb{X}$ to $C([0,T],\mathbb{X})$.

PROOF. In order to indicate how to set up the contraction mapping argument, we first estimate the fixed point equation (5.20) in the $\mathbb{Y}$ norm,

$$(5.28) \qquad \|u(t)\|_{\mathbb{Y}} \le c\,t^{-\alpha}\,\|u^{\mathrm{in}}\|_{\mathbb{X}} + c\int_0^t (t-t')^{-\beta}\,\|u(t')\|_{\mathbb{Y}}^{2\sigma+1}dt'\,,$$

or
$$(5.29)$$

$$t^{\alpha}\,\|u(t)\|_{\mathbb{Y}} \le c\,\|u^{\mathrm{in}}\|_{\mathbb{X}} + c\,t^{\alpha}\int_0^t (t-t')^{-\beta}\,t'^{-(2\sigma+1)\alpha}\left(t'^{\alpha}\,\|u(t')\|_{\mathbb{Y}}\right)^{2\sigma+1}dt'$$

$$\le c\,\|u^{\mathrm{in}}\|_{\mathbb{X}} + c\,\sup_{t'\in[0,t]}\left(t'^{\alpha}\,\|u(t')\|_{\mathbb{Y}}\right)^{2\sigma+1}t^{\alpha}\int_0^t (t-t')^{-\beta}\,t'^{-(2\sigma+1)\alpha}\,dt'\,.$$

This suggests that we might be able to prove contractivity in the space $\mathbb{E}([0,T])$, defined as the completion of $C([0,T], \mathbb{Y})$ in the norm

$$(5.30) \qquad \|w\|_{\mathbb{E}} \equiv \sup_{t \in [0,T]} t^{\alpha} \|w(t)\|_{\mathbb{Y}}.$$

Indeed, (5.29) shows that provided

$$(5.31) \qquad t^{\alpha} \int_0^t (t-t')^{-\beta} t'^{-(2\sigma+1)\alpha} dt' \to 0 \quad \text{as } t \to 0,$$

an assumption that is directly seen to be equivalent to (5.24–5.26), there exists a $T > 0$ such that the right side of the fixed point equation (5.20) maps sufficiently large balls with respect to the $\mathbb{E}$ norm into themselves. Invoking the Lipschitz condition (5.23), one can show in a similar way that this mapping contracts under the same assumption (5.31) for some $T > 0$. Then by the contraction mapping theorem, there exists a unique solution $u \in \mathbb{E}([0,T]) \subset C((0,T], \mathbb{Y})$ to the fixed point problem (5.20).

Finally one can show that $u \in C([0,T], \mathbb{X})$. To this end we only need to check the continuity of u at $t = 0$. First subtracting u^{in} from both sides of the mild formulation (5.20), a direct estimate in $\mathbb{X}$, assuming a continuous embedding $\mathbb{Z} \hookrightarrow \mathbb{X}$, gives

$$
\begin{aligned}
(5.32) \quad \|u(t) - u^{\text{in}}\|_{\mathbb{X}} &\leq \|S(t)u^{\text{in}} - u^{\text{in}}\|_{\mathbb{X}} + c \int_0^t t'^{-(2\sigma+1)\alpha} \left(t'^{\alpha} \|u(t')\|_{\mathbb{Y}} \right)^{2\sigma+1} dt' \\
&\leq \|S(t)u^{\text{in}} - u^{\text{in}}\|_{\mathbb{X}} + c \|u\|_{\mathbb{E}}^{2\sigma+1} \int_0^t t'^{-(2\sigma+1)\alpha} dt'.
\end{aligned}
$$

The first term on the right converges to zero due to the strong $\mathbb{X}$ continuity of the linear semigroup. Since $\|u\|_{\mathbb{E}}$ is bounded, the second term converges to zero provided that $(2\sigma + 1)\alpha < 1$, a condition already contained in (5.31). $\qquad \square$

REMARK 5.7. Strictly speaking, here we have proved existence in a smaller space, namely $\mathbb{E}([0,T])$, than that asserted in (5.27).

Let us now apply Theorem 5.5 to the CGL equation where, as before, N and L are given by (5.4) and the associated linear semigroup by (5.5). We first consider the case where the initial data is in L^p, i.e. $\mathbb{X} = L^p(\mathbb{T}^d)$. We also take $\mathbb{Z} = L^p(\mathbb{T}^d)$ and let $\mathbb{Y} = L^r(\mathbb{T}^d)$ where r is still to be determined. It takes some relatively straightforward applications of Hölder and Young inequalities to verify that the Lipschitz condition (5.23) is satisfied whenever

$$(5.33) \qquad r \geq (2\sigma + 1)p.$$

In order to compute the exponent α of (5.21), the exponent of the "penalty" incurred by using $S(t)$ as a smoothing operator from L^p to L^r, we estimate

$$(5.34) \qquad \|G_t * w\|_{L^r} \leq \|G_t\|_{L^q} \|w\|_{L^p} \leq \|G_t\|_{L^1}^{1/q} \|G_t\|_{L^\infty}^{1-1/q} \|w\|_{L^p}$$

where

$$(5.35) \qquad 1 + \frac{1}{r} = \frac{1}{p} + \frac{1}{q}.$$

By using the L^1 estimate (5.7) and the L^∞ estimate

$$(5.36) \qquad \|G_t\|_{L^\infty} \leq \frac{(1+\nu^2)^{d/4} e^{Rt}}{(4\pi(1+\nu^2)t)^{d/2}} \{a + b[(1+\nu^2)t]^{d/2}\},$$

for positive absolute constants a and b, one finds that for small t

$$(5.37) \qquad \|G_t\|_{L^1} = O(t^0), \qquad \|G_t\|_{L^\infty} = O(t^{-d/2}).$$

This used with (5.34) and (5.35) gives

$$(5.38) \qquad \|S(t)w\|_{L^r} \le c\, t^{-\frac{d}{2}(\frac{1}{p}-\frac{1}{r})} \|w\|_{L^p},$$

i.e. $\alpha = \beta = \frac{d}{2}(\frac{1}{p} - \frac{1}{r})$. Then Theorem 5.5, first applied in the marginal case $r = (2\sigma + 1)p$, asserts the existence of a time $T > 0$ and a unique local solution

$$(5.39) \qquad u \in C([0,T], L^p(\mathbb{T}^d)) \cap C((0,T], L^r(\mathbb{T}^d))$$

subject to the constraint $(2\sigma + 1)\alpha < 1$, i.e. $\sigma d < p$.

Since $u(t) \in L^r(\mathbb{T}^d)$ for all $t > 0$, one can re-apply the argument to show that

$$(5.40) \qquad u \in C((0,T], L^{r_m}(\mathbb{T}^d)) \quad \text{where } r_m = (2\sigma + 1)^m p$$

for every $m \in \mathbb{N}$. Finally, for some $r_m > (\sigma + \frac{1}{2})d$ we can take $\mathbb{X} = \mathbb{Z} = L^{r_m}(\mathbb{T}^d)$ and $\mathbb{Y} = C(\mathbb{T}^d)$ endowed with the supremum norm to obtain $u \in C((0,T], C(\mathbb{T}^d))$. An application of Theorem 5.2 then yields the following.

THEOREM 5.6 (Local Classical Solutions for L^p Initial Data). *If p satisfies*

$$(5.41) \qquad 1 \le p < \infty, \qquad and \qquad \sigma d < p,$$

then for every $\rho > 0$ there exists a time $T(\rho) > 0$ such that for every initial data $u^{\mathrm{in}} \in L^p(\mathbb{T}^d)$ with $\|u^{\mathrm{in}}\|_{L^p} \le \rho$ there exists a unique

$$(5.42) \qquad u \in C([0,T], L^p(\mathbb{T}^d)) \cap C((0,T], C^2(\mathbb{T}^d)) \cap C^1((0,T], C(\mathbb{T}^d)),$$

satisfying the CGL initial-value problem. In addition, the map from u^{in} to u is a locally Lipschitz function from $L^p(\mathbb{T}^d)$ to $C([0,T], L^p(\mathbb{T}^d))$.

REMARK 5.8. For $p \ge 2$ the uniqueness of such classical solutions can be extended to hold within the class of weak solutions of Theorem 4.1. In other words, so long as a weak solution of the CGL equation is classical then it is unique. What is striking here though is that in subcritical cases ($\sigma d < 2$) Theorem 5.6 asserts the existence of solutions with L^p initial data for some $p < 2$—solutions which therefore lie outside the class of weak solutions of Theorem 4.1. In fact, as we will show below, for the subcritical case $\sigma = d = 1$ one can use the same technique to establish the existence of solutions for initial data in classes of distributions $H^q(\mathbb{T})$ for $-\frac{1}{2} < q < 0$.

REMARK 5.9. This result does more than extend to L^p the class of initial data that evolve into classical solutions for positive times. More importantly, it estimates the time interval over which any classical solution will exist in terms of its L^p norm. This means that in order to prove that any classical solution is global in time it suffices to control its L^p norm, where p satisfies (5.41). In Section 7 we present such temporally global *a priori* L^p bounds for an interval of ν values depending only on σ and d, but not μ. In that case one has global classical solutions (Theorem 7.1).

REMARK 5.10. The proof given here differs from that in [80] in that the space $\mathbb{E}([0,T])$ used here for the contraction mapping argument is larger than the space used there, but can be abstracted from [80] with minor modifications. Essentially the same methods lead to similar results for the CGL equation over $\mathbb{R}^d$. More sophisticated techniques lead to similar results even in the dissipationless case of the NLS equation, both over $\mathbb{R}^d$ [10, 24, 79] and over $\mathbb{T}$ where Bourgain [8, 9] has developed an $L^2(\mathbb{T})$ theory for $\sigma < 2$.

REMARK 5.11. An important special case is when $p = 2\sigma + 2$, in which case the above theorem yields local classical solutions whenever

$$(5.43) \qquad\qquad d < 2 + \frac{2}{\sigma}.$$

This condition coincides with condition (3.7) for the existence and uniqueness of local solutions for the NLS equation in $H^1(\mathbb{T}^d)$. In Section 7 we present temporally global *a priori* $L^{2\sigma+2}$ bounds for a region of the $\mu\nu$-plane depending on σ, but not d. In that case, whenever (5.43) is satisfied one has global classical solutions (Theorem 7.2).

In the following, we will use Theorem 5.5 in its full generality to prove existence of a local solution starting from H^q initial data. The Sobolev spaces $H^q(\mathbb{T}^d)$ can be defined for any $q \in \mathbb{R}$. To do so we will use $\widehat{w}(\xi)$ to denote the Fourier coefficients of a function $w \in L^2(\mathbb{T}^d)$, so that

$$(5.44) \qquad w(x) = \sum_{\xi \in 2\pi\mathbb{Z}^d} \widehat{w}(\xi)\, e^{i\xi \cdot x}, \qquad \widehat{w}(\xi) = \int_{\mathbb{T}^d} e^{-i\xi \cdot x}\, w(x)\, dx.$$

The H^q norm of a function w can then be defined [2] as

$$(5.45) \qquad \|w\|_{H^q} \equiv \left(\sum_{\xi \in 2\pi\mathbb{Z}^d} (1 + |\xi|^2)^q\, |\widehat{w}(\xi)|^2 \right)^{\frac{1}{2}}.$$

The existence result we are going to prove is relatively technical and will not be used in any of the subsequent sections, so that it may be skipped in a first reading. The motivation for this further inquiry comes from the observation that in the CGL equation $|u|^{2\sigma}$ scales dimensionally like Δ so that by formally trading powers for derivatives in condition (5.56), one transforms Theorem 5.6 into the following conjecture. Given $u^{\text{in}} \in H^q(\mathbb{T}^d)$ where $q > \frac{d}{2} - \frac{1}{\sigma}$, there exists a unique local classical solution. We will show under certain restrictions that this is indeed the case [56].

We first remark that it is very problematic to conduct the argument entirely in H^q spaces. The difficulty lies in obtaining a Lipschitz estimate (5.23), which is easily accomplished only when $\mathbb{Y} = \mathbb{Z} \equiv H^q(\mathbb{T}^d)$ is an algebra, i.e. when $q > d/2$. (See [2] for a review of Sobolev algebras.) In this case the basic local existence result, Theorem 5.1, immediately assures the existence of a local mild solution. This approach however does clearly not exhaust the conjectured range of admissible values for q.

Secondly, we observe that the only interesting case is $q < 0$. Namely, when $q > \frac{d}{2} - \frac{1}{\sigma}$ and $q > 0$, one can always find a $p > \sigma d$ so that the Sobolev embedding $H^q \hookrightarrow L^p$ holds, i.e. that $\frac{1}{p} > \frac{1}{2} - \frac{q}{d}$. Therefore in this case local classical solutions are already guaranteed by Theorem 5.5.

Assume therefore that $u^{\text{in}} \in H^q(\mathbb{T}^d) \equiv \mathbb{X}$ for $q < 0$. Take the two other spaces as in the L^p argument, i.e. $\mathbb{Y} = L^r(\mathbb{T}^d)$ and $\mathbb{Z} = L^p(\mathbb{T}^d)$. While $\beta = \frac{d}{2}(\frac{1}{p} - \frac{1}{r})$ and the restriction $r \geq (2\sigma + 1)p$ are as before, it still remains to find the exponent α of the "penalty" for using $S(t)$ as a smoothing operator from H^q to L^r. This can be done by going via the intermediate space L^2, i.e. by using (5.38) to infer that for $r \geq 2$,

$$(5.46) \qquad \|S(2t)w\|_{L^r} \leq c\, t^{-\frac{d}{2}(\frac{1}{2}-\frac{1}{r})} \|S(t)w\|_{L^2} .$$

The convolution $S(t)w = G_t * w$ is best estimated by deriving bounds on its Fourier transform. Specifically, we have

$$(5.47) \qquad
\begin{aligned}
\|G_t * w\|_{L^2}^2 &= \sum_{\xi \in 2\pi \mathbb{Z}^d} |\widehat{g}(\xi)\widehat{w}(\xi)|^2 \\
&\leq \sup_{\xi \in 2\pi \mathbb{Z}^d} \left\{ (1+|\xi|^2)^{-q} |\widehat{g}(\xi)|^2 \right\} \sum_{\xi \in 2\pi \mathbb{Z}^d} (1+|\xi|^2)^q\, |\widehat{w}(\xi)|^2 ,
\end{aligned}$$

where $\widehat{w}(\xi)$ denotes the Fourier coefficients of w and the Fourier coefficients of G_t are given by

$$(5.48) \qquad \widehat{g}(\xi) = e^{Rt}\, e^{-(1+i\nu)|\xi|^2 t} .$$

The supremum in (5.47) is easily estimated by simple optimization with respect to $|\xi|^2$, resulting in

$$(5.49) \qquad \|G_t * w\|_{L^2}^2 \leq \left(\frac{-q}{2et}\right)^{-q} e^{2(R+1)t} \|w\|_{H^q}^2 .$$

For sufficiently small t one has altogether

$$(5.50) \qquad \|S(t)w\|_{L^r} \leq c\, t^{-\frac{d}{2}(\frac{1}{2}-\frac{1}{r})-\frac{-q}{2}} \|w\|_{H^q} .$$

Again, we take the marginal case of the Lipschitz condition, $r = (2\sigma + 1)p$. Then condition $\beta + 2\sigma\alpha < 1$ of Theorem 5.5 is satisfied if and only if

$$(5.51) \qquad q > \frac{d}{2} - \frac{1}{\sigma} .$$

We still need to check the other two conditions of that theorem. Condition (5.25) imposes an upper bound on r, namely

$$(5.52) \qquad r < \frac{2(2\sigma + 1)d}{(2\sigma + 1)(d - 2q) - 4} .$$

Similarly, (5.24) imposes a lower bound on r,

$$(5.53) \qquad r > \sigma d .$$

Note first that given (5.51), the upper and lower bounds on r can be satisfied simultaneously. For technical reasons, we also have two additional lower bounds on r. In the derivation of (5.50) it was necessary to assume that $r \geq 2$, and we must also require $p \geq 1$, i.e. $r \geq 2\sigma + 1$. These combined with (5.52) translate into

$$(5.54) \qquad q > -\frac{2}{2\sigma + 1}$$

and

$$(5.55) \qquad q > \frac{d}{2} - \frac{d + 2}{2\sigma + 1}$$

respectively. Then finally Theorem 5.5 yields a unique mild solution $u \in C((0,T], \mathbb{Y})$ for some $T > 0$. Moreover, since $u(t) \in L^r(\mathbb{T}^d)$ for all $t \in (0,T]$, further regularity can be achieved by a direct application of Theorem 5.6. One thus has the following.

THEOREM 5.7 (Local Classical Solutions for H^q Initial Data). *Provided q satisfies* (5.54), (5.55) *and*

$$(5.56) \qquad q > \frac{d}{2} - \frac{1}{\sigma}$$

then for every $\rho > 0$ there exists a time $T(\rho) > 0$ such that for every initial data $u^{\text{in}} \in H^q(\mathbb{T}^d)$ with $\|u^{\text{in}}\|_{H^q} \leq \rho$ there exists a unique

$$(5.57) \qquad u \in C([0,T], H^q(\mathbb{T}^d)) \cap C((0,T], C^2(\mathbb{T}^d)) \cap C^1((0,T], C(\mathbb{T}^d)),$$

satisfying the CGL initial-value problem. In addition, the map from u^{in} to u is a locally Lipschitz function from $H^q(\mathbb{T}^d)$ to $C([0,T], H^q(\mathbb{T}^d))$.

REMARK 5.12. We believe conditions (5.54) and (5.55) are technical and can be removed by more sophisticated techniques. Moreover, by following [10], the use of Besov spaces should allow one to obtain the marginal case when (5.56) is satisfied as an equality. Note also that provided $\sigma \geq \frac{1}{2}$ and $\sigma d \geq 1$—thus in particular when σ is an integer—condition (5.56) dominates and we have proved local existence for the conjectured class of initial data.

REMARK 5.13. The results in this section extend to $\mathbb{R}^d$ with only minor modifications in their proofs.

6. Analyticity of Solutions

Let us now discuss Gevrey class regularity which is a stronger notion than the C^∞ regularity we have encountered before. It not only asserts that all derivatives of the solution u are bounded, but also that these bounds depend on the order of the derivatives in some prescribed way. Gevrey [34] used this notion as a setting in which to extend Cauchy-Kowalevski existence arguments to classes of functions that are not necessarily analytic (for a review of the analytic case see e.g. [45]). In fact, they are special cases of the *quasianalytic* classes which had been introduced earlier by Hadamard [41]. La Vallée Poussin [52] showed that, among the quasianalytic functions, the Gevrey classes are characterized by an exponential decay of their Fourier coefficients (see also [49]). In turn, this characterization has recently proved useful for showing that the solutions of various nonlinear partial differential equations are analytic. Foias and Temam [30] developed this technique in the context of the two-dimensional incompressible Navier-Stokes equations, thereby simplifying earlier work of Kahane [47] that had been more in the style of Gevrey. Doelman and Titi [22] subsequently applied the technique to the cubic CGL equation, and Ferrari and Titi [27] extended the results to a large class of analytic nonlinear parabolic equations. In particular, the authors of the last reference have exploited the fact that certain subclasses are normed algebras, thus making the estimation of polynomial nonlinearities a triviality. This is the strategy we will adopt because of its brevity and elegance.

A function $w \in C^\infty(\mathbb{T}^d)$ is said to be of *Gevrey class s* for some $s > 0$ if there exist constants $\rho > 0$ and $M < \infty$ such that for every $x \in \mathbb{T}^d$ and every $\alpha \in \mathbb{N}^d$ one has

$$(6.1) \qquad |\partial^\alpha w(x)| \leq M \left(\frac{\alpha!}{\rho^{|\alpha|}} \right)^s.$$

Here we employ the usual multi-index notation in which

$$(6.2) \qquad |\alpha| \equiv \sum_{j=1}^{2\sigma+1} \alpha_j \,, \qquad \alpha! \equiv \prod_{j=1}^{2\sigma+1} \alpha_j! \,, \qquad \partial^\alpha \equiv \prod_{j=1}^{d} \partial_{x_j}^{\alpha_j} \,.$$

The set of all functions of Gevrey class s is a vector space, denoted $G^s(\mathbb{T}^d)$. It is closed under multiplication and differentiation. Moreover, the composition of two functions of Gevrey class s is again of class s.

It is classical that $G^1(\mathbb{T}^d)$ is the space of real analytic functions $C^\omega(\mathbb{T}^d)$; a proof can be found, for example, in [45, page 65]. For $0 < s < 1$ the class $G^s(\mathbb{T}^d)$ is subclass of the analytic functions, while for $1 < s < \infty$ it contains the analytic functions. In fact, one has a hierarchy of spaces such that $0 < s_1 < s_2 < \infty$ implies the proper containments

$$(6.3) \qquad G^{s_1}(\mathbb{T}^d) \subset G^{s_2}(\mathbb{T}^d) \subset C^\infty(\mathbb{T}^d) \,.$$

Moreover, the union of the $G^s(\mathbb{T}^d)$ does not exhaust $C^\infty(\mathbb{T}^d)$ because there are quasianalytic functions that are not members of a Gevrey class [52].

In what follows it will be more convenient to characterize Gevrey classes in terms of the fractional Sobolev spaces $H^r(\mathbb{T}^d)$ (as defined on page 28 for every $r \in \mathbb{R}$) with $r \geq 0$, rather than uniform bounds such as (6.1).

LEMMA 6.1. *Given $s > 0$ and $r \geq 0$. Then $w \in G^s(\mathbb{T}^d)$ if and only if there are constants ρ, $M \in (0, \infty)$ that may depend on r, s and w such that for every $n \in \mathbb{N}$ one has*

$$(6.4) \qquad \|\nabla^n w\|_{H^r} = \left(\sum_{\xi \in 2\pi \mathbb{Z}^d} (1 + |\xi|^2)^r |\xi|^{2n} |\widehat{w}(\xi)|^2 \right)^{\frac{1}{2}} \leq M \left(\frac{n!}{\rho^n} \right)^s \,.$$

The proof is achieved by a direct application of the Sobolev embedding theorem [2] and shall be left to the reader.

This lemma enables us to characterize functions in G^s in terms of the decay of their Fourier coefficients, a result of the type first due to La Vallée Poussin [52, 59], although this attribution has been obscured in the recent literature. The construction here uses the operator $A = \sqrt{-\Delta}$ that, like $-\Delta$ itself, is nonnegative and self-adjoint so that arbitrary powers can be defined by spectral theory. For each $s \in (0, \infty)$ we define a family, parameterized by τ, of normed spaces

$$(6.5) \qquad \mathcal{D}(e^{\tau A^{1/s}} : H^r(\mathbb{T}^d)) \equiv \left\{ w \in H^r(\mathbb{T}^d) : \left\| e^{\tau A^{1/s}} w \right\|_{H^r} < \infty \right\} \,.$$

The functions in any such space have Fourier coefficients that decay faster than $\exp(-\tau |\xi|^{1/s})$. The next theorem recovers the Gevrey class $G^s(\mathbb{T}^d)$ as the union of all such classes. We believe the proof is new.

THEOREM 6.2. *For any $s > 0$ and $r \geq 0$,*

$$(6.6) \qquad G^s(\mathbb{T}^d) = \bigcup_{\tau > 0} \mathcal{D}(e^{\tau A^{1/s}} : H^r(\mathbb{T}^d)) \,.$$

REMARK 6.1. The use of the more general H^r rather than simply L^2 as the base space does not complicate the structure of the proof, but it is advantageous for the arguments that will follow.

PROOF. Let $w \in \mathcal{D}(e^{\tau A^{1/s}} : H^r(\mathbb{T}^d))$ for some $\tau > 0$ and let $\rho = \tau/s$. Then

$$
\begin{aligned}
\|\nabla^n w\|_{H^r}^2 &= \left(\frac{n!}{\rho^n}\right)^{2s} \sum_\xi (1 + |\xi|^2)^r \left(\frac{\rho^n |\xi|^{n/s}}{n!}\right)^{2s} |\widehat{w}(\xi)|^2 \\
&\leq \left(\frac{n!}{\rho^n}\right)^{2s} \sum_\xi (1 + |\xi|^2)^r\, e^{2s\rho|\xi|^{1/s}} |\widehat{w}(\xi)|^2 \\
&= \left(\frac{n!}{\rho^n}\right)^{2s} \left\|e^{\tau A^{1/s}} w\right\|_{H^r}^2 .
\end{aligned}
$$

(6.7)

By setting $M = \|e^{\tau A^{1/s}} w\|_{H^r}$ we obtain (6.4), whereby $w \in G^s(\mathbb{T}^d)$.

On the other hand, let $w \in G^s(\mathbb{T}^d)$. For an arbitrary $\tau \geq 0$ one has

$$
\begin{aligned}
\left\|e^{\tau A^{1/s}} w\right\|_{H^r}^2 &\equiv \sum_\xi (1 + |\xi|^2)^r\, e^{2\tau|\xi|^{1/s}} |\widehat{w}(\xi)|^2 \\
&= \sum_{m=0}^\infty \frac{(2\tau)^m}{m!} \sum_\xi (1 + |\xi|^2)^r\, |\xi|^{\frac{m}{s}}\, |\widehat{w}(\xi)|^2 .
\end{aligned}
$$

(6.8)

Now let ρ and M be such that (6.4) is satisfied. By interpolating (6.4) between $n = 0$ and any integer n such that $m/s \leq 2n$, the inner sum appearing in (6.8) can be bounded as

(6.9)
$$
\sum_\xi (1 + |\xi|^2)^r\, |\xi|^{\frac{m}{s}}\, |\widehat{w}(\xi)|^2 \leq M^2 \frac{(n!)^{\frac{m}{n}}}{\rho^m} .
$$

This bound is best if we choose $n = n_m \equiv [m/(2s)] + 1$, where $[\cdot]$ denotes the "greatest integer less than" function. Upon making this choice and applying the result in (6.8), one arrives at the bound

(6.10)
$$
\left\|e^{\tau A^{1/s}} w\right\|_{H^r}^2 \leq \sum_{m=0}^\infty \frac{(2\tau)^m}{m!} M^2 \frac{(n_m!)^{\frac{m}{n_m}}}{\rho^m} = M^2 \sum_{m=0}^\infty \left(\frac{2\tau}{\rho}\right)^m \frac{(n_m!)^{\frac{m}{n_m}}}{m!} .
$$

By making use of the Stirling formula [1] in the form

(6.11)
$$
\lim_{n\to\infty} \frac{e}{n}\, (n!)^{\frac{1}{n}} = 1 ,
$$

the limit of the mth root of the mth term in the last series of (6.10) can be evaluated as

(6.12)
$$
\lim_{m\to\infty} \frac{2\tau}{\rho} \frac{(n_m!)^{\frac{1}{n_m}}}{(m!)^{\frac{1}{m}}} = \frac{2\tau}{\rho} \lim_{m\to\infty} \frac{n_m}{m} = \frac{2\tau}{\rho} \frac{1}{2s} = \frac{\tau}{s\rho} .
$$

Hence, by the Hadamard root test, the series in (6.10) converges for every $\tau < \rho s$, whence $w \in \mathcal{D}(e^{\tau A^{1/s}} : H^r(\mathbb{T}^d))$. $\qquad\square$

REMARK 6.2. The above proof gives a sharp relationship between the τ of (6.5) and the ρ of (6.4). Had we been less careful in our choice of the n_m used in (6.10) then this relationship would have been missed.

REMARK 6.3. The theorem also holds over $\mathbb{R}^d$. The proof can proceed in the same way, using Fourier integrals in place of Fourier sums.

One reason Gevrey classes are useful in the context of nonlinear partial differential equations is that each G^s is closed under multiplication. It would be particularly useful if this property extended to each of the approximating (normed!) spaces $\mathcal{D}(e^{\tau A^{1/s}}: H^r(\mathbb{T}^d))$. The following theorem states that under certain conditions this is indeed the case.

THEOREM 6.3. *If $s \geq 1$, $\tau \geq 0$, and $r > d/2$ then $\mathcal{D}(e^{\tau A^{1/s}}: H^r(\mathbb{T}^d))$ is a Banach algebra. This means that it is closed under multiplication and that there exists a finite constant $C(r,d)$ such that any two functions v and w in $\mathcal{D}(e^{\tau A^{1/s}}: H^r(\mathbb{T}^d))$ satisfy the inequality*

$$(6.13) \qquad \left\| e^{\tau A^{1/s}}(vw) \right\|_{H^r} \leq C(r,d) \left\| e^{\tau A^{1/s}} v \right\|_{H^r} \left\| e^{\tau A^{1/s}} w \right\|_{H^r}.$$

The proof is a direct extension of the usual proof that $H^r(\mathbb{T}^d)$ is a Banach algebra when $r > d/2$ [2], a result that is recovered above by setting $\tau = 0$. A proof of this theorem for the case when $s = 1$ is given in [27]; this proof is easily generalized for any $s > 1$.

The spaces $\mathcal{D}(e^{\tau A^{1/s}}: H^r(\mathbb{T}^d))$ are naturally suited for the application to parabolic equations: τ will be identified with time so that $\mathcal{D}(e^{\tau A^{1/s}}: H^r(\mathbb{T}^d))$ evolves from being identical to H^r at $\tau = 0$ to being a subset of Gevrey class s in an arbitrarily short time. So if one can show that for some $T > 0$ the solutions to the equation with H^r initial data are in $\mathcal{D}(e^{t A^{1/s}}: H^r(\mathbb{T}^d))$ for every $t \in (0,T]$ then one has proved Gevrey regularity of class s over that interval.

Let us now proceed to show that the solution u of the CGL equation is of Gevrey class 1—in other words, is real analytic. To do this we employ the time dependent norm $\|w\|_t \equiv \|e^{tA} w\|_{L^2}$. As the L^2 norm of u is already controlled by (7.2), it is sufficient to derive a bound on the seminorm $\|A^r u\|_t$. A direct calculation gives

$$
\begin{aligned}
(6.14) \qquad \frac{1}{2}\frac{d}{dt}\|A^r u(t)\|_t^2 &= \mathrm{Re}\left(\int A^{r+1} e^{tA} u \, A^r e^{tA} u^* \, dx \right) \\
&\quad + \mathrm{Re}\left(\int A^r e^{tA} \partial_t u \, A^r e^{tA} u^* \, dx \right) \\
&= \|A^{r+1/2} u\|_t^2 + R\|A^r u\|_t^2 - \|A^{r+1} u\|_t^2 \\
&\quad - \mathrm{Re}\left((1+i\mu) \int A^r e^{tA}(|u|^{2\sigma} u) \, A^r e^{tA} u^* \, dx \right).
\end{aligned}
$$

The first term on the right is easily estimated by using

$$(6.15) \qquad \|A^{1/2} v\|_{L^2}^2 = (v, Av)_{L^2} \leq \|v\|_{L^2} \|Av\|_{L^2} \leq \tfrac{1}{4}\|v\|_{L^2}^2 + \|Av\|_{L^2}^2,$$

with $v = A^r e^{tA} u$. By using the Cauchy-Schwarz inequality on the integral in the last term of (6.14), one finds that it is bounded by

$$(6.16) \qquad \|A^r\left(u^{\sigma+1}(u^*)^\sigma\right)\|_t \|A^r u^*\|_t \leq c \|A^r u\|_t^{2\sigma+2}.$$

It is here that we use the fact that $\mathcal{D}(e^{tA}: H^r(\mathbb{T}^d))$ is an algebra. Then finally

$$(6.17) \qquad \frac{1}{2}\frac{d}{dt}\|A^r u(t)\|_t^2 \leq (R + \tfrac{1}{4}) \|A^r u(t)\|_t^2 + c \|A^r u(t)\|_t^{2\sigma+2}.$$

This differential inequality is easily integrated and shows that $\|A^r u(t)\|_t$ remains finite on some interval $[0,T]$. Given that σ is an integer, the local existence theorems of the last section guarantee that $u(t) \in H^r(\mathbb{T}^d)$ for any r and $t > 0$ as long as it exists. Hence by initializing (6.17) with $\|u(t)\|_{H^r} = \|A^r u(t)\|_0$ for some $r > d/2$ and every such t one can conclude the following.

THEOREM 6.4. *Provided that σ is a positive integer, a local classical solution is real analytic in x at any positive t as long as it exists.*

REMARK 6.4. In order to achieve global bounds with the method presented here, it is necessary to have *a priori* H^r control for some $r > d/2$. The method used originally by Doelman and Titi [22] requires only *a priori* H^1 control independent of d. However, that method requires a separate and fairly involved estimate on the nonlinear term because $\mathcal{D}(e^{tA} \colon H^1(\mathbb{T}^d))$ is not an algebra for $d > 1$. See Theorem 8.4 for the statement of the global result.

7. Global Classical Solutions

The task of this section is to show that there are regions of the $\mu\nu$-plane where we can elevate the local classical solutions found in Section 5 to global classical solutions. In these regions the CGL equation defines a dynamical system and we can investigate the properties of its global attractor. The values of μ and ν for which we establish this regularity will depend on the spatial dimension d and the degree of nonlinearity σ. The results of the last sections were very general—for example, we have the same local existence and regularity results for either sign of the real part of the coefficient of the nonlinear term—and assert substantially little more than the fact that the problem is well-posed. The estimates and results in this section, however, will depend much more delicately on the structure and dynamics of the CGL equation. They rely not only on the signs of the dissipative terms, but on the relative signs of the coefficients of the dispersive terms as well.

As was pointed out in the final remarks of Section 5, it suffices to obtain global control of any L^p norm where $p > \sigma d$. Such bounds can be obtained by computing the time evolution of the L^p norm, say, and then perform estimates on the various terms in order to obtain closed differential inequalities. This shall be first demonstrated for the L^2 norm which is simplest to control. A direct calculation and subsequent application of the Hölder inequality yields

$$\frac{1}{2}\frac{d}{dt}\int |u|^2\, dx = R\int |u|^2\, dx - \int |u|^{2\sigma+2}\, dx - \int |\nabla u|^2\, dx$$

(7.1)

$$\leq R\int |u|^2\, dx - \left(\int |u|^2\, dx\right)^{\sigma+1}.$$

In the appendix it is shown that for $\sigma > 0$ this differential inequality implies the L^2 norm of u satisfies

(7.2)
$$\|u(t)\|_{L^2} < \left(\frac{R}{1 - e^{-2\sigma Rt}}\right)^{\frac{1}{2\sigma}}.$$

This estimate is independent of the initial data u^{in} and decays to $R^{1/2\sigma}$ as $t \to \infty$. Hence for times bounded strictly away from 0, the L^2 norm of u is bounded uniformly in both time and initial condition. Whenever $\sigma d < 2$, Theorem 5.6 then implies that the CGL equation has global classical solutions for all initial data in $L^2(\mathbb{T}^d)$. For $\sigma d \geq 2$, one must control more than the L^2 norm.

Directly computing the evolution of the L^p norm gives

$$
\frac{1}{p}\frac{d}{dt}\int |u|^p \, dx = R\int |u|^p \, dx - \int |u|^{2\sigma+p} \, dx
$$

(7.3)

$$
+ \text{Re}\left(\int |u|^{p-2}u^*\Delta u \, dx\right) - \nu \, \text{Im}\left(\int |u|^{p-2}u^*\Delta u \, dx\right).
$$

The net contribution of the last two terms on the right side will be nonpositive for $p \geq 2$ provided

(7.4)
$$
|\nu| \leq \frac{2\sqrt{p-1}}{p-2}.
$$

This fact can be established many ways—either using Sobolev bounds as in [3], quadratic forms as in [25], or, as will be done here, as a consequence of Lemma 7.3. Upon supposing (7.4) is satisfied, the last two terms on the right of (7.3) can be neglected while the remaining terms can be treated through an application of a Hölder inequality analogous to the one for the L^2 estimate (7.1) thereby yielding the inequality

(7.5)
$$
\frac{1}{p}\frac{d}{dt}\int |u|^p \, dx \leq R\int |u|^p \, dx - \left(\int |u|^p \, dx\right)^{\frac{p+2\sigma}{p}}.
$$

The appendix again gives a uniform bound, identical to (7.2),

(7.6)
$$
\|u(t)\|_{L^p} < \left(\frac{R}{1-e^{-2\sigma Rt}}\right)^{\frac{1}{2\sigma}}.
$$

The existence of global classical solutions is then guaranteed by Theorem 5.6 in the subcritical case ($\sigma d < 2$)—as mentioned before—by taking $p = 2$. In the critical case ($\sigma d = 2$) one can still satisfy $\sigma d < p$ and (7.4) simultaneously by choosing p sufficiently close to 2. In the supercritical case ($\sigma d > 2$) however, the best we can do is to choose p sufficiently close to σd, making (7.4) a restriction on the possible values of ν for which the existence of global classical solutions can be established. This is summarized in the following.

THEOREM 7.1. *For $\sigma > 0$, the generalized CGL equation with C^2 initial data has unique global classical solutions provided either that $\sigma d \leq 2$ or that $\sigma d > 2$ and ν satisfies*

(7.7)
$$
|\nu| < \frac{2\sqrt{\sigma d - 1}}{\sigma d - 2}.
$$

Thus, in the supercritical case the CGL equation has global classical solutions only when ν lies in the strip around the μ-axis of the $\mu\nu$-plane given by (7.7) while μ may take on any parameter value.

REMARK 7.1. When global classical solutions exist, estimate (7.6) shows that the attractor is confined to the closed ball of radius $R^{1/(2\sigma)}$ in L^p provided p satisfies (7.4). Furthermore, estimate (7.6) is independent of the initial condition. This means that in an arbitrarily short time all solutions enter some ball of finite radius.

REMARK 7.2. The basic differential inequalities (7.1) on the L^2 norm of u and (7.5) on the L^p norm are sharp for the spatially homogeneous solutions (2.10) of the CGL equation. Moreover, their proofs use essentially the boundedness of the domain $\mathbb{T}^d$ and do not extend to $\mathbb{R}^d$. However, a bound on the L^p norm of u over

$\mathbb{R}^d$ that is exponentially growing in time can be obtained provided (7.4) is satisfied by neglecting the last three terms on the right of (7.3). Such a bound is enough to infer that Theorem 7.1, as stated, holds over $\mathbb{R}^d$.

An alternative to directly controlling the L^p norm is to rather first directly control the H^1 norm and then control L^p through Sobolev estimates. This is however subject to the constraint

$$(7.8) \qquad 1 \leq p < \begin{cases} \infty & \text{for } d = 1 \text{ or } 2\,, \\ \dfrac{2d}{d-2} & \text{for } d \geq 3\,. \end{cases}$$

Given that we are also constrained by $\sigma d < p$, this restricts us to those cases where

$$(7.9) \qquad d < 2 + \frac{2}{\sigma}\,,$$

which is incidentally the condition for local well-posedness of the underlying NLS problem, equation (3.7).

An estimate on the H^1 norm can be derived in a very similar way. As the L^2 norm is already controlled by (7.2), it suffices to control the L^2 norm of ∇u. To do so we assume $\sigma \geq \frac{1}{2}$ so that Theorem 5.4 gives enough regularity of the solution to take the gradient of the CGL equation. A direct calculation then shows

$$(7.10) \qquad \begin{aligned} \frac{1}{2}\frac{d}{dt}\int |\nabla u|^2\,dx &= R\int |\nabla u|^2\,dx - \int |\Delta u|^2\,dx \\ &\quad + \mathrm{Re}\left(\int |u|^{2\sigma} u^* \Delta u\,dx\right) + \mu\,\mathrm{Im}\left(\int |u|^{2\sigma} u^* \Delta u\,dx\right). \end{aligned}$$

The last two terms on the right have the same form as those on the right side of (7.3) upon identifying p with $2\sigma + 2$ and ν with $-\mu$. Upon treating them in the same way, their net contribution will be nonpositive provided (compare with (7.4))

$$(7.11) \qquad |\mu| \leq \frac{\sqrt{2\sigma + 1}}{\sigma}\,.$$

In that case, these terms can be neglected, and an elementary interpolation inequality applied to the second term (i.e. simply integrate by parts and use the Cauchy-Schwarz inequality) leads to

$$(7.12) \qquad \frac{1}{2}\frac{d}{dt}\int |\nabla u|^2\,dx \leq R\int |\nabla u|^2\,dx - \frac{\left(\int |\nabla u|^2\,dx\right)^2}{\int |u|^2\,dx}\,.$$

Upon using the L^2 bound for u provided by (7.2), this differential inequality gives uniform upper bounds on the L^2 norm of ∇u by following the procedure in the appendix. Indeed, even in this case the differential inequality (7.12) can be integrated exactly to obtain

$$(7.13) \qquad \|\nabla u(t)\|_{L^2}^2 < \Phi(\sigma, Rt)\left(\frac{R}{1 - e^{-2\sigma Rt}}\right)^{\frac{\sigma+1}{\sigma}}\,,$$

where Φ is given in terms of the hypergeometric function F (e.g. see [1]) by

$$(7.14) \qquad \Phi(\sigma, Rt) = \frac{2(1+\sigma)}{F(1, 1, 2 + \frac{1}{\sigma}; 1 - e^{-2\sigma Rt})}\,.$$

The function $\Phi(\sigma, Rt)$ is monotonically decreasing in t and one can easily find that $\Phi(\sigma, 0) = 2(1 + \sigma)$ and $\Phi(\sigma, \infty) = 2$. Thus (7.13) provides a uniform bound on $\|\nabla u\|_{L^2}$ analogous to the L^p estimate, equation (7.6). Then Theorem 5.6 shows the existence of global classical solutions in a strip around the ν-axis of the parameter plane (7.11) similar to the strip around the μ-axis (7.7) from Theorem 7.1. In the case when $p = 2 + 2\sigma$, these strips are completely symmetric, and both subject to the restriction $d < 2 + 2/\sigma$.

However, for no $p > 2$ does the union of these strips extend to all of the modulationally stable region of the underlying NLS equation where the energy functional is positive definite and hence no finite-time singularities can occur. As the CGL equation is a dissipative perturbation of the NLS equation (disregarding for the moment the linear growth proportional to R), we expect to have regularity at least in this region. And indeed, a "perturbation" of the NLS energy functional,

$$(7.15) \qquad F = \int |\nabla u|^2 + \frac{\beta^2}{\sigma + 1} |u|^{2\sigma+2} \, dx \,,$$

where $\beta > 0$ is to be chosen later, will help to considerably improve the region for which global existence of classical solutions can be shown. The technical reason why this particular combination yields better results than separate estimates on the H^1 and $L^{2+2\sigma}$ norms is that the problematic terms in the H^1 and $L^{2+2\sigma}$ estimates have the same functional dependence on the solution u and therefore taking the linear combination (7.15) and optimization in β provides for partial cancelation of these terms. This is reflected in the following result, the present form of which is due to Ginibre [**37**] and slightly improves upon the region of validity given in [**3, 25**].

THEOREM 7.2. *For $\sigma \geq \frac{1}{2}$ and $d < 2 + 2/\sigma$, the generalized CGL equation with C^2 initial data has unique global classical solutions if the dispersive parameters lie in the region of the $\mu\nu$-plane, bounded by hyperbolae,*

$$(7.16) \qquad -\frac{1 + \mu\nu}{|\mu - \nu|} \leq \frac{\sqrt{2\sigma + 1}}{\sigma} \,.$$

This region of the $\mu\nu$-plane completely contains the first and third quadrants where the underlying NLS equation is modulationally stable, the strips found from the previous separate H^1 and $L^{2+2\sigma}$ bounds (this by the very form of F). In addition, it contains the whole CGL modulationally stable region $1 + \mu\nu \geq 0$ and a bit more depending on the degree of nonlinearity. Indeed, for $d = 1, 2$ it reduces to the modulationally stable region in the limit of a high degree of nonlinearity $(\sigma \to \infty)$. However, part of the strip (7.7) obtained from a direct L^p bound lies outside this region when $p > 2 + 2\sigma$. This is summarized in Table 7.1 for integer values of σ.

Figure 7.1 illustrates the supercritical case $d = 3$, $\sigma = 1$; the corresponding picture for any case where $2 < \sigma d < 2 + 2\sigma$ is similar. The inner pair of hyperbolae bounds the modulationally stable region $1 + \mu\nu \geq 0$ while the outer pair limits the region where global classical solutions can be obtained from Theorem 7.2. The little unshaded strip outside the hyperbolic boundaries indicates the region where global classical solutions that can still be obtained by using the direct L^p result, Theorem 7.1.

REMARK 7.3. So far we have allowed for non-integer values of σ which is sufficient to obtain global classical solutions that are C^2 in the spatial variables. For

	$\sigma = 1$	$\sigma = 2$	$\sigma \geq 3$
$d = 1$	no restriction	no restriction	$\lvert \nu \rvert < \dfrac{2\sqrt{\sigma - 1}}{\sigma - 2}$ or $-\dfrac{1 + \mu\nu}{\lvert \mu - \nu \rvert} \leq \dfrac{\sqrt{2\sigma + 1}}{\sigma}$
$d = 2$	no restriction	$\lvert \nu \rvert < \sqrt{3}$ or $-\dfrac{1 + \mu\nu}{\lvert \mu - \nu \rvert} \leq \dfrac{\sqrt{5}}{2}$	$\lvert \nu \rvert < \dfrac{\sqrt{2\sigma - 1}}{\sigma - 1}$ or $-\dfrac{1 + \mu\nu}{\lvert \mu - \nu \rvert} \leq \dfrac{\sqrt{2\sigma + 1}}{\sigma}$
$d = 3$	$\lvert \nu \rvert < \sqrt{8}$ or $-\dfrac{1 + \mu\nu}{\lvert \mu - \nu \rvert} \leq \sqrt{3}$	$\lvert \nu \rvert < \dfrac{\sqrt{5}}{2}$	$\lvert \nu \rvert < \dfrac{2\sqrt{3\sigma - 1}}{3\sigma - 2}$
$d \geq 4$	$\lvert \nu \rvert < \dfrac{2\sqrt{d - 1}}{d - 2}$	$\lvert \nu \rvert < \dfrac{\sqrt{2d - 1}}{d - 1}$	$\lvert \nu \rvert < \dfrac{2\sqrt{\sigma d - 1}}{\sigma d - 2}$

TABLE 7.1. Sufficient restrictions for the existence of global classical solutions to the complex Ginzburg-Landau equation. The bounds on $\lvert \nu \rvert$ and the "no restriction" entries are a result of Theorem 7.1 while the hyperbolic boundaries arise from Theorem 7.2. The case where $d = 3$ and $\sigma = 1$ is illustrated in Figure 7.1.

$\sigma \geq n/2$ one can even get C^{n+2} solutions from Theorem 5.4, but for global smooth (i.e. C^∞) solutions σ needs to be integer. This case will the the focus of Section 8 where we derive explicit global bounds for all spatial derivatives.

In order to prove the results of this section, it is useful to have the following bounds on nonlinear functionals that occur throughout the proofs. They were first used by Ginibre [**37**] to obtain the improved hyperbolic boundaries (7.16).

LEMMA 7.3. *For $p \geq 2$ and a function $u \in H^2(\mathbb{T}^d) \cap L^{2p-2}(\mathbb{T}^d)$ the following inequalities hold:*

$$(7.17) \qquad \mathrm{Re}\left(\int \lvert u \rvert^{p-2} u^* \Delta u \, dx \right) \leq -\frac{2\sqrt{p - 1}}{p} \left\lvert \int \lvert u \rvert^{p-2} u^* \Delta u \, dx \right\rvert ,$$

$$(7.18) \qquad \left\lvert \mathrm{Im}\left(\int \lvert u \rvert^{p-2} u^* \Delta u \, dx \right) \right\rvert \leq \frac{p - 2}{p} \left\lvert \int \lvert u \rvert^{p-2} u^* \Delta u \, dx \right\rvert .$$

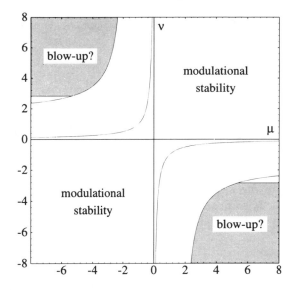

FIGURE 7.1. Parameter plane for the supercritical complex Ginz-burg-Landau equation (shown here is the case $d = 3$, $\sigma = 1$). Global existence of smooth solutions is established in the unshaded region only. The inner hyperbolae are the boundaries of the mod-ulationally stable region $1 + \mu\nu \geq 0$. The region within the outer pair of hyperbolae is a result of Theorem 7.1 while the small un-shaded strip outside this region is a result of Theorem 7.2. In the subcritical and critical case ($\sigma d \leq 2$) this strip extends to the whole parameter plane.

PROOF. Examine the identity, obtained through integration by parts,

$$2 \int |u|^{p-2} u^* \Delta u \, dx = -p \int |u|^{p-2} |\nabla u|^2 \, dx$$
$$- (p-2) \int |u|^{p-4} u^{*2} \nabla u \cdot \nabla u \, dx \, .$$

(7.19)

For any nonconstant u the first term on the right side is negative while the second term is smaller in magnitude by a factor no greater than $(p-2)/p$, whereby it follows that

$$\int |u|^{p-2} u^* \Delta u \, dx = - \left| \int |u|^{p-2} u^* \Delta u \, dx \right| \exp(i\theta) \, ,$$

(7.20)

for some θ in $|\theta| < \pi/2$ such that $|\sin\theta| \leq (p-2)/p$, or equivalently, such that $\cos\theta \geq 2\sqrt{p-1}/p$. By (7.20), these trigonometric bounds on θ translate into the bounds (7.17) and (7.18).   □

This result is more than enough to provide the missing links in the deriva-tions of the differential inequalities (7.5) and (7.12) and it also enables us to prove Theorem 7.2.

PROOF OF THEOREM 7.2. We compute the time evolution of F by taking the appropriate linear combination of (7.3) with $p = 2\sigma + 2$ and (7.10):

(7.21)
$$\frac{1}{2}\frac{dF}{dt} = R\int |\nabla u|^2 + \beta^2 |u|^{2\sigma+2}\,dx - \int |\Delta u|^2 + \beta^2 |u|^{4\sigma+2}\,dx$$
$$+ (1+\beta^2)\,\mathrm{Re}\left(\int |u|^{2\sigma} u^* \Delta u\,dx\right) + (\mu - \beta^2\nu)\,\mathrm{Im}\left(\int |u|^{2\sigma} u^* \Delta u\,dx\right).$$

The benefits of this linear combination are immediately obvious, noting that when μ and ν have the same sign, one can always find a β that makes the coefficient of the last term zero. In order to obtain an optimal result, we bound the last two terms by Lemma 7.3. In addition to that, one can use a fraction κ of the negative definite term to help control the last two terms. This is best done through the estimate

(7.22)
$$-\int |\Delta u|^2 + \beta^2 |u|^{4\sigma+2}\,dx \le -2\beta\left|\int |u|^{2\sigma} u^* \Delta u\,dx\right|.$$

Altogether, one obtains the differential inequality

(7.23)
$$\frac{1}{2}\frac{dF}{dt} = R\int |\nabla u|^2 + \beta^2 |u|^{2\sigma+2}\,dx - (1-\kappa)\int |\Delta u|^2 + \beta^2 |u|^{4\sigma+2}\,dx$$
$$- \left(2\kappa\beta + (1+\beta^2)\frac{\sqrt{2\sigma+1}}{\sigma+1} - |\mu - \beta^2\nu|\frac{\sigma}{\sigma+1}\right)\left|\int |u|^{2\sigma} u^* \Delta u\,dx\right|.$$

The last term can be neglected provided it is possible to choose β such that its coefficient is nonnegative. This is always the case if $\mu\nu \ge 0$. On the other hand, when $\mu\nu < 0$ we can without loss of generality consider the case where $\mu > 0$ and $\nu < 0$. Taking $\kappa = 1$, this results in the condition

(7.24)
$$2\beta + (1+\beta^2)\frac{\sqrt{2\sigma+1}}{\sigma+1} - (\mu - \beta^2\nu)\frac{\sigma}{\sigma+1} \ge 0.$$

Optimization in β then results in the restriction to the hyperbolic parameter region (7.16). As we have used up all of the negative definite term, one obtains a bound on F that is exponentially growing in time, sufficient for Theorem 5.6 to assert the existence of global classical solutions. $\qquad\square$

REMARK 7.4. For those μ and ν for which (7.16) holds as a strict inequality, we can keep $\kappa < 1$ and obtain uniform bounds on the H^1 and $L^{2\sigma+2}$ norms by closing the differential inequality (7.23) as follows:

(7.25)
$$F^2 \le \left(1 + \frac{\beta^2}{(\sigma+1)^2}\right)\left(\int |\nabla u|^2\,dx\right)^2 + \left(\beta^2 + \frac{\beta^4}{(\sigma+1)^2}\right)\left(\int |u|^{2\sigma+2}\,dx\right)^2$$
$$\le \left(1 + \frac{\beta^2}{(\sigma+1)^2}\right)\int |u|^2\,dx\int |\Delta u|^2 + \beta^2 |u|^{4\sigma+2}\,dx.$$

In the second step we have used integration by parts and Cauchy-Schwarz on the first and the Cauchy-Schwarz inequality on the second term. Hence, provided β can be chosen so that (7.24) holds as a strict inequality, F will satisfy the differential inequality

$$(7.26) \qquad \frac{1}{2}\frac{dF}{dt} \le (\sigma+1)RF - (1-\kappa)\frac{(\sigma+1)^2}{(\sigma+1)^2+\beta^2}\frac{F^2}{\int |u|^2\, dx}.$$

Upon again using the L^2 bound provided by (7.2), this differential inequality may be integrated as in the appendix to obtain the uniform bound

$$(7.27) \qquad F < \frac{(\sigma+1)^2+\beta^2}{(1-\kappa)(\sigma+1)}\left(\frac{R}{1-e^{-2\sigma Rt}}\right)^{\frac{\sigma+1}{\sigma}}.$$

This bound has the same scaling in R as $t \to \infty$ as the H^1 bound (7.13). Note that β and κ are subject to the constraint that the coefficient of the last term of (7.23) be nonnegative and, hence, are not independent. The task of optimizing the choice of β and κ in the prefactor of (7.27) is fairly complicated and gives little additional information. However, we conclude by observing that this prefactor blows up as κ approaches unity, which will be the case as (μ,ν) approaches the hyperbolic boundaries of the region (7.16).

8. Global Sobolev Bounds

This section is devoted to the derivation of global bounds on all spatial derivatives of the global smooth solutions found in the last section. As a consequence we will infer that the CGL equation possesses a compact global attractor when considered as a dynamical system over $H^1(\mathbb{T}^d)$. It will be assumed throughout that σ is a positive integer, so that the local existence of C^∞ solutions is ensured by Theorem 5.3, i.e. that the global classical solutions are in fact $C^\infty((0,\infty) \times \mathbb{T}^d)$. This does not yet give us uniform global bounds on norms of the derivatives (although some bounds might be derived by a subtle analysis of the bootstrapping arguments of Section 5). Here we explain how to directly control the norms of the Sobolev spaces $H^n(\mathbb{T}^d)$ and how to use those bounds to obtain global estimates on the analytic regularity of the solution.

Bartuccelli *et al.* [3] first estimated the H^n norms of solutions to the cubic CGL equation by constructing a "ladder"—a sequence of nonlinear functionals, each of which controls successively higher derivatives of u, and that are connected by differential inequalities. The method was subsequently applied to the generalized CGL equation [25] and the two-dimensional incompressible Navier-Stokes equation [23]. Recent work has shown that it is possible to find a differential inequality that yields global H^n control directly from any of the global L^p estimates of Section 7, i.e. for any n without the need for intermediate rungs and functionals [4]. This form of the estimate is presented below.

As the L^2 norm of u is already controlled by (7.2), it is sufficient to estimate the L^2 norm of $\nabla^n u$. Here $\nabla^n u$ is considered to be a symmetric n-tensor and

$$(8.1) \qquad |\nabla^n u|^2 \equiv \nabla^n u^* \cdot \nabla^n u,$$

where the dot denotes the usual tensor inner product which acts on a symmetric k-tensor and a symmetric l-tensor by contracting over $\min\{k,l\}$ indices to give a symmetric $|k-l|$-tensor.

A direct calculation shows that

$$\frac{1}{2}\frac{d}{dt}\|\nabla^n u\|_{L^2}^2 = R \|\nabla^n u\|_{L^2}^2 - \|\nabla^{n+1} u\|_{L^2}^2$$

(8.2)

$$- \operatorname{Re}\left((1+i\mu)\int \nabla^n(|u|^{2\sigma} u)\cdot \nabla^n u^* \, dx\right).$$

The second term on the right is easily estimated using the Sobolev inequality

(8.3)
$$\|\nabla^n u\|_{L^2} \le \|\nabla^{n+1} u\|_{L^2}^{\frac{n}{n+1}} \|u\|_{L^2}^{\frac{1}{n+1}}.$$

More work is required to achieve a bound on the nonlinear term. The estimate we will use is

(8.4)
$$\left|\int \nabla^n(|u|^{2\sigma} u)\cdot \nabla^n u^* \, dx\right| \le \delta \|\nabla^{n+1} u\|_{L^2}^2 + c(\delta) \|\nabla^n u\|_{L^2}^2 \|u\|_{L^p}^{\frac{2\sigma p}{p-\sigma d}}.$$

In this formula (which will be proved later as Lemma 8.3) we can choose $\delta > 0$ and $p > \sigma d$ to suit our purpose. In particular, by setting $\delta = |1+i\mu|^{-1}$ we can cancel the first term with half of the dissipative term of equation (8.2) to obtain

(8.5)
$$\frac{d}{dt}\|\nabla^n u\|_{L^2}^2 \le \left(2R + c \|u\|_{L^p}^{\frac{2\sigma p}{p-\sigma d}}\right) \|\nabla^n u\|_{L^2}^2 - \frac{\|\nabla^n u\|_{L^2}^{2(n+1)/n}}{\|u\|_{L^2}^{2/n}}.$$

One can use the *a priori* L^2 bound (7.2) to control the denominator of the last term. It is then clear that global L^p control with $p > \sigma d$—note that this is precisely the condition for local regularity from L^p data—is sufficient to achieve global H^n control for any n. Explicit bounds can be obtained by following the procedure in the appendix, although in general it will not be possible to integrate the differential inequality exactly. Hence we state the result in its simplest form.

THEOREM 8.1. *For all positive integers d, σ and n and for every $p > \sigma d$, when μ and ν lie in the region of the parameter plane where uniform L^p bounds on solutions of the CGL equation with L^p initial data exist, these solutions are global and smooth for all positive times with their H^n norm bounded uniformly for all initial data and $t \to \infty$, satisfying the differential inequality (8.5).*

REMARK 8.1. Estimate (8.5) does not give the best scaling in all cases, although known improvements are slight. For a more detailed discussion see [**4**].

REMARK 8.2. Nonlinear dissipation enters estimate (8.5) only implicitly via the L^2 and possibly the L^p bound. The explicit dependence on the sign of the nonlinear term has been lost in step (8.4). This in particular shows that by the same method global H^n bounds, although not necessarily uniform in time, can be obtained for equations of the type

(8.6)
$$\partial_t u = Ru + (1+i\nu)\,\Delta u - i\mu\,|u|^{2\sigma} u.$$

REMARK 8.3. Theorem 8.1 shows that the CGL equation possesses a compact global attractor when considered as a dynamical system over any $H^n(\mathbb{T}^d)$. (For purposes of comparison, it is natural to consider both the NLS and CGL equations as dynamical systems over $H^1(\mathbb{T}^d)$.) Indeed, because $H^{n+1}(\mathbb{T}^d)$ is compactly embedded in $H^n(\mathbb{T}^d)$, the existence of an absorbing ball in $H^{n+1}(\mathbb{T}^d)$, which is inferred from (8.5) and necessarily contains the attractor, implies the compactness of this ball in $H^n(\mathbb{T}^d)$. Furthermore, by the general theory developed by Mallet-Paret [**58**] and Mañé [**60**] the attractor has finite Hausdorff dimension and can be locally embedded in finite dimensional manifolds. This theory is reviewed in [**42**]. Explicit upper bounds for the Hausdorff dimension of the attractor were derived for $d = 1$ by Doering, Gibbon, Holm and Nicolaenko [**24**], for $d = 1, 2$ by Ghidaglia and Héron [**35**] and for $d = 3$ by Bartuccelli *et al.* [**3**]. The techniques are based on the estimation of global Lyapunov exponents which are discussed in [**13, 14, 78**].

REMARK 8.4. In some cases the attractor can be globally embedded in an *inertial manifold*, a notion introduced by Foias, Sell and Temam [**29**]. An inertial manifold is a finite dimensional, exponentially attracting, positively invariant manifold that contains the global attractor—for a general discussion see [**29, 15, 16**] and references therein. For $d = 1, 2$ the CGL equation is known to possess an inertial manifold while for $d \geq 3$ this is still an open question. Moreover, its dimension is bounded by the attractor dimension from below, while upper bounds can be proved following [**16, 24**]. This procedure relies on finding a large enough gap in the spectrum of the Laplacian, leading to a much larger estimate on the dimension of the manifold when $d = 2$ and failing for $d = 3$ as there are no arbitrarily large gaps in the spectrum. By contrast, it is not clear whether the Navier-Stokes equations in two spatial dimensions possess an inertial manifold. This question remains an active area of research.

To prove estimate (8.4), we first derive a preliminary estimate in which we impose that the Gagliardo-Nirenberg inequalities be used at their margin of validity, thus rendering the estimate independent of the dimension d. This result will then be used in two different ways to complete the proof.

LEMMA 8.2. *For all positive integers d, σ and n the following inequality holds:*

$$(8.7) \qquad \left| \int \nabla^n(|u|^{2\sigma}u) \cdot \nabla^n u^* \, dx \right| \leq c(n,q) \, \|\nabla^n u\|_{L^{2q}}^2 \, \|u\|_{L^{2r\sigma}}^{2\sigma} \,,$$

where

$$(8.8) \qquad \frac{1}{q} + \frac{1}{r} = 1, \qquad 1 \leq q, r \leq \infty.$$

REMARK 8.5. The advantage of this result, derived in [**69, 4**], in comparison with earlier estimates [**3, 25**] is that it permits a free choice of r while keeping the estimate independent of d.

PROOF. By moving the absolute value inside the integral, performing a Leibniz expansion on $\nabla^n(|u|^{2\sigma}u)$, and then using the triangle inequality, one obtains

$$(8.9) \qquad \left| \int \nabla^n(|u|^{2\sigma}u) \cdot \nabla^n u^* \, dx \right| \leq \sum_{\substack{\alpha \in \mathbb{N}^{2\sigma+1} \\ |\alpha|=n}} \frac{n!}{\alpha!} \int |\nabla^n u| \prod_{j=1}^{2\sigma+1} |\nabla^{\alpha_j} u| \, dx \,.$$

Here we have again employed the multi-index notation in which $\alpha!$ and $|\alpha|$ are defined as in (6.2). For a given multi-index α with $|\alpha| = n$ define θ_j and q_j by

$$(8.10) \qquad \theta_j = \frac{\alpha_j}{n}, \qquad \frac{1}{q_j} = \theta_j \frac{1}{2q} + (1 - \theta_j)\frac{1}{2\sigma r} \,.$$

These satisfy

$$(8.11) \qquad \sum_{j=1}^{2\sigma+1} \theta_j = \frac{|\alpha|}{n} = 1, \qquad \frac{1}{2q} + \sum_{j=1}^{2\sigma+1} \frac{1}{q_j} = \frac{1}{q} + \frac{1}{r} = 1.$$

Then the corresponding term on the right side of (8.9) can be estimated with a Hölder inequality followed by a Gagliardo-Nirenberg interpolation on each of the factors:

(8.12)
$$\int |\nabla^n u| \prod_{j=1}^{2\sigma+1} |\nabla^{\alpha_j} u| \, dx \le \|\nabla^n u\|_{L^{2q}} \prod_{j=1}^{2\sigma+1} \|\nabla^{\alpha_j} u\|_{L^{q_j}}$$

$$\le \|\nabla^n u\|_{L^{2q}} \prod_{j=1}^{2\sigma+1} c_j \|\nabla^n u\|_{L^{2q}}^{\theta_j} \|u\|_{L^{2\sigma r}}^{1-\theta_j}$$

$$= c \|\nabla^n u\|_{L^{2q}}^2 \|u\|_{L^{2\sigma r}}^{2\sigma} .$$

Finally, by setting this into (8.9), we obtain the stated result. $\qquad\square$

This result will now be used to establish estimate (8.4), thereby completing the proof of Theorem 8.1.

LEMMA 8.3. *For all positive integers d, σ and n and for every $p > \sigma d$ and every $\delta > 0$, one has the estimate*

(8.13)
$$\left| \int \nabla^n (|u|^{2\sigma} u) \cdot \nabla^n u^* \, dx \right| \le \delta \|\nabla^{n+1} u\|_{L^2}^2 + c(\delta) \|\nabla^n u\|_{L^2}^2 \|u\|_{L^p}^{\frac{2\sigma p}{p-\sigma d}} .$$

PROOF. Estimate the right side of (8.7) with a Gagliardo-Nirenberg inequality, followed by Young's inequality:

(8.14)
$$\left| \int \nabla^n (|u|^{2\sigma} u) \cdot \nabla^n u^* \, dx \right| \le c_1 \|\nabla^{n+1} u\|_{L^2}^{d/r} \|\nabla^n u\|_{L^2}^{2-d/r} \|u\|_{L^{2\sigma r}}^{2\sigma}$$

$$\le \delta \|\nabla^{n+1} u\|_{L^2}^2 + c_2(\delta) \|\nabla^n u\|_{L^2}^2 \|u\|_{L^{2\sigma r}}^{\frac{4\sigma}{2-d/r}} .$$

For the inequalities to be valid, we must have $2 - d/r > 0$. By setting $p = 2\sigma r$, we obtain the statement of Lemma 8.3.

If $d = 1$ however, the condition $r \ge 1$ from Lemma 8.2 would restrict the possible choices of p even further, so that for $\frac{1}{2} < r < 1$ we must complement the proof with a different interpolation in order to obtain the result in its full generality. Thus we estimate

(8.15)
$$\left| \int \partial^n (|u|^{2\sigma} u) \partial^n u^* \, dx \right| \le c_3 \|\partial^n u\|_{L^\infty}^{2\lambda} \|\partial^n u\|_{L^\infty}^{2(1-\lambda)} \|u\|_{L^{2\sigma}}^{2\sigma}$$

$$\le c_4 \left(\|\partial^{n+1} u\|_{L^2}^a \|u\|_{L^{2\sigma r}}^{1-a} \right)^{2\lambda}$$

$$\times \left(\|\partial^{n+1} u\|_{L^2}^{1/2} \|\partial^n u\|_{L^2}^{1/2} \right)^{2(1-\lambda)}$$

$$\times \left(\|\partial^{n+1} u\|_{L^2}^b \|u\|_{L^{2\sigma r}}^{1-b} \right)^{2\sigma} ,$$

where a and b are Gagliardo-Nirenberg exponents computed in the usual way. By the choice of $\lambda = 1/r - 1$, this equation reduces exactly to (8.14) so that the proof can proceed as before. $\qquad\square$

REMARK 8.6. The suggestive grouping of terms in equation (8.15) does not show a lower order term that has to be included in the estimate of $\|u\|_{L^{2\sigma}}$ because u is not a mean-zero function. However, in this case it is possible to absorb the extra term into the constant c_4. For details see [4].

REMARK 8.7. The ladder inequalities introduced in [3] and generalized in [25] use functionals of the form

$$(8.16) \qquad F_{n,m} = \int_{\mathbb{T}^d} |\nabla^n u|^{2m} + \alpha_{n,m} |u|^{2mn(\sigma+1)} \, dx \, ,$$

the choice of which is motivated by the observation that $|u|^{2\sigma}$ scales dimensionally like Δ in the CGL equation. Terms like $\nabla^n u$ and $|u|^{n\sigma} u$ have the same weight under this dimensional scaling. This however does not give any advantage over the direct approach as presented here.

REMARK 8.8. We remark that although the implications of Theorem 8.1 add nothing so far as the existence of global smooth solutions is concerned, the explicit estimates that have been derived are useful. Some of these estimates are discussed in detail in references [3, 4].

Finally we will use the global Sobolev bounds of this section to initialize the local Gevrey class estimate of Section 6. Recall that the differential inequality (6.17) shows that for every t for which $u(t) \in H^r(\mathbb{T}^d)$ there exists a positive finite time $T_{\max}$ such that for every $T \in [0, T_{\max}]$

$$(8.17) \qquad \|A^r u(t+T)\|_T \le 2 \, \|A^r u(t)\|_0 \le 2 \, \|u(t)\|_{H^r} \, .$$

We know from Theorem 8.1 that $\|u\|_{H^r}$ is bounded uniformly over all initial data. One therefore has the following.

THEOREM 8.4. *For positive integers d, σ and for every $r > d/2$ and $p > \sigma d$, when μ and ν lie in the region of the parameter plane where uniform L^p bounds on solutions of the CGL equation with L^p initial data exist, these solutions are global and real analytic for all positive times. Furthermore, there is a $T_{\max}$ such that the $\mathcal{D}(e^{TA} \colon H^r(\mathbb{T}^d))$ norm of $u(t)$ is bounded uniformly over all initial data, for every $T \in [0, T_{\max}]$ and for every time $t \ge T$.*

Theorem 8.4 implies that the Fourier modes of a solution to the CGL equation decay exponentially with the wavenumber ξ with a rate of decay which is uniform over large times. Doelman [21] first observed that the Fourier coefficients decay faster than any algebraic order. The exponential decay rate was established later in [22] by using methods similar to the ones presented in Section 6. This is consistent with numerical observations as well as with theoretical results concerning notions of finite-dimensionality which indicate that CGL dynamics is relatively low dimensional.

From the computational viewpoint the exponential decay of the Fourier coefficients means that simple Fourier-Galerkin truncation works well. The CGL dynamics is captured by a manageable number of Fourier modes [22, 26]. Furthermore, more sophisticated nonlinear Galerkin methods offer no advantage over Fourier-Galerkin truncation [46], a fact that is intrinsically linked to analytic regularity and the exponential decay of the Fourier coefficients.

A general question of current interest is whether the solution of a dynamical system can be determined by the time history of a finite set of data. The notion of *determining modes* was introduced by Foias and Prodi [28] in the context of the two dimensional incompressible Navier-Stokes equations. A set of Fourier modes is called determining if the knowledge of the time history of the corresponding Fourier coefficients completely specifies the solution. In other words, if the determining modes of any two solutions agree then the solutions are identical. This idea was subsequently extended to more general *determining degrees of freedom*—for example, the solution evaluated at nodes or averaged over cells [**31, 11**]. It is important to note that the dynamics of the full equation is not necessarily represented by the restricted dynamics on a set of determining modes. Hence, a low number of determining degrees of freedom has no direct implication for computational considerations. However, when an inertial manifold exists (as for CGL when $d = 1, 2$), a finite number of Fourier modes usually serve as its coordinates. The infinite dimensional dynamics of the remaining Fourier modes is therefore slaved by the dynamics of a finite dimensional set. In such cases one can generally establish bounds on the number of determining degrees of freedom in terms of the dimension of the inertial manifold [**11**]. Such bounds can be far from optimal. Indeed, the CGL equation can exhibit a very low number of determining degrees of freedom. Kukavica [**50**] showed that for $d = \sigma = 1$ the periodic CGL has sets of determining nodes consisting of only two points. His proof readily extends to the case where σ is any positive integer. Analytic regularity is a key ingredient in the argument. There is now a considerable effort to extend and sharpen these results.

Appendix. On the Integration of Differential Inequalities

Throughout Sections 7 and 8 we infer uniform bounds for any nonnegative quantity $F(t)$ that satisfies a differential inequality of the general form

$$(A.1) \qquad \frac{dF}{dt} \leq a(t)F - b(t)F^{1+s},$$

where $s > 0$ and $b(t) > 0$. Following Bernoulli, introduce the variable $Y = F^{-s}$ into (A.1) and obtain the linear differential inequality

$$(A.2) \qquad \frac{dY}{dt} \geq -s\,a(t)\,Y + s\,b(t).$$

Now multiply this equation by $\exp\!\left(\int_{t'}^{t} s\,a(t'')\,dt''\right)$ to obtain

$$(A.3) \qquad \frac{d}{dt}\left(Y \exp\!\left(\int_{t'}^{t} s\,a(t'')\,dt''\right)\right) \geq s\,b(t)\,\exp\!\left(\int_{t'}^{t} s\,a(t'')\,dt''\right).$$

Integration of (A.3) between 0 and t results in

$$(A.4) \quad Y(t) \geq Y(0)\,\exp\!\left(-\int_{0}^{t} s\,a(t')\,dt'\right) + \int_{0}^{t} s\,b(t')\,\exp\!\left(-\int_{t'}^{t} s\,a(t'')\,dt''\right)dt'.$$

The uniform upper bounds on $F(t)$ follow directly by neglecting the $Y(0)$ term, which effectively means taking the supremum over all initial conditions $F(0)$:

$$(A.5) \qquad F(t) < \left(\int_{0}^{t} s\,b(t')\,\exp\!\left(-\int_{t'}^{t} s\,a(t'')\,dt''\right)dt'\right)^{-1/s}.$$

This bound is satisfied by any nonnegative quantity $F(t)$ that satisfies the differential inequality (A.1), no matter what its initial value happens to be.

Acknowledgments

We are grateful to C.R. Doering and J.D. Gibbon for their helpful comments and for being kind enough to let us draw so heavily on the material in [**25**], and also to M. Bartuccelli, T. Cazenave, J. Ginibre, J.K. Hale, B.P. Luce, E.S. Titi, and F.B. Weissler for their helpful comments.

References

1. M. Abramowitz and I.A. Stegun, *Handbook of Mathematical Functions*, United States National Bureau of Standards Applied Mathematics Series **55**, 1964.
2. R.S. Adams, *Sobolev Spaces*, Academic Press, New York, 1975.
3. M. Bartuccelli, P. Constantin, C.R. Doering, J.D. Gibbon, and M. Gisselfält, *On the Possibility of Soft and Hard Turbulence in the Complex Ginzburg-Landau Equation*, Physica D **44** (1990), 421–444.
4. M. Bartuccelli, J.D. Gibbon, and M. Oliver, *Length Scales in Solutions of the Complex Ginzburg-Landau Equation*, Physica D (submitted 1995).
5. G.K. Batchelor, *The Theory of Homogeneous Turbulence*, Cambridge University Press, Cambridge, 1953.
6. P. Bollerman, A. Doelman, A. van Harten, P. Taháč, and E.S. Titi, *Analyticity of Essentially Bounded Solutions to Semilinear Parabolic Systems: A Complex Analytic Approach*, SIAM J. Math. Anal. (accepted 1995).
7. N. Bourbaki, *Eléments de mathématique III, Topologie générale, Ch. 10*, Herman, Paris, 1966.
8. J. Bourgain, *Fourier Transform Restriction Phenomena for Certain Lattice Subsets and Applications to Nonlinear Evolution Equations, Part I: Schrödinger Equations*, Geometric and Func. Anal. **3** (1993), 107–156.
9. J. Bourgain, *Exponential Sums and Nonlinear Schrödinger Equations*, Geometric and Func. Anal. **3** (1993), 157–178.
10. T. Cazenave and F. Weissler, *The Cauchy Problem for the Critical Nonlinear Schrödinger Equations in H^s*, Nonlinear Anal.: Theo., Meth. and Appl. **14** (1990), 807–836.
11. B. Cockburn, D.A. Jones, and E.S. Titi, *Determining Degrees of Freedom for Nonlinear Dissipative Equations*, C. R. Acad. Sci. (submitted 1995).
12. P. Constantin, *Navier-Stokes Equations and Incompressible Fluid Turbulence*, this volume.
13. P. Constantin and C. Foias, *Global Lyapunov Exponents, Kaplan-Yorke Formulas and the Dimension of the Attractor for 2D Navier-Stokes Equations*, Comm. Pure Appl. Math. **38** (1985), 1–27.
14. P. Constantin and C. Foias, *The Navier-Stokes Equations*, Chicago University Press, Chicago, 1988.
15. P. Constantin, C. Foias, B. Nicolaenko, and R. Temam, *Integral and Inertial Manifolds for Dissipative Partial Differential Equations*, Appl. Math. Sci. **70**, Springer-Verlag, New York, 1988.
16. P. Constantin, C. Foias, B. Nicolaenko, and R. Temam, *Spectral Barriers and Inertial Manifolds for Dissipative Partial Differential Equations*, J. Dynamics and Diff. Eq. **1** (1989), 45–73.
17. M.C. Cross and P.C. Hohenberg, *Pattern Formation Outside of Equilibrium*, Reviews of Modern Physics **65** (1993), 851–1089.
18. G. Cruz-Pacheco, C.D. Levermore, and B.P. Luce, *Complex Ginzburg-Landau Equations as Perturbations of Nonlinear Schrödinger Equations*, Physica D (submitted 1995).
19. R.J. DiPerna and P.-L. Lions, *On the Cauchy Problem for the Boltzmann Equation: Global Existence and Weak Stability Results*, Annals of Math. **130** (1990), 321–366.
20. R.C. DiPrima, W. Eckhaus, and L.A. Segel, *Non-Linear Wave-Number Interaction in Near-Critical Two-Dimensional Flows*, J. Fluid Mech. **49** (1971), 705–744.
21. A. Doelman, *Finite-dimensional Models of the Ginzburg-Landau Equation*, Nonlinearity **4** (1991), 231–250.
22. A. Doelman and E.S. Titi, *Regularity of Solutions and the Convergence of the Galerkin Method in the Ginzburg-Landau Equation*, Num. Func. Anal. and Opt. **14** (1993), 299–321.
23. C.R. Doering and J.D. Gibbon, *Note on the Constantin-Foias-Temam Attractor Dimension Estimate for Two Dimensional Turbulence*, Physica D **48** (1991), 471–480.
24. C.R. Doering, J.D. Gibbon, D.D. Holm and B. Nicolaenko, *Low Dimensional Behavior in the Complex Ginzburg-Landau Equation*, Nonlinearity **1** (1988), 279–309.
25. C.R. Doering, J.D. Gibbon, and C.D. Levermore, *Weak and Strong Solutions of the Complex Ginzburg-Landau Equation*, Physica D **71** (1994), 285–318.
26. J. Duan, E.S. Titi, and P. Holmes, *Regularity, Approximation and Asymptotic Dynamics for a Generalized Ginzburg-Landau Equation*, Nonlinearity **6** (1993), 915–933.
27. A.B. Ferrari and E.S. Titi, *Gevrey Regularity of Solutions of a Class of Analytic Nonlinear Parabolic Equations*, Comm. in P.D.E. (submitted 1994).

28. C. Foias and G. Prodi, *Sur le comportement global des solutions non stationaires des équations de Navier-Stokes en dimension deux*, Rend. Sem. Mat. Univ. Padova **39** (1967), 1–38.

29. C. Foias, G. Sell, and R. Temam, *Inertial Manifolds for Nonlinear Evolution Equations*, J. Diff. Eq. **73** (1988), 309–353.

30. C. Foias and R. Temam, *Gevrey Class Regularity for the Solutions of the Navier-Stokes Equations*, J. Func. Anal. **87** (1989), 359–369.

31. C. Foias and E.S. Titi, *Determining Nodes, Finite Difference Schemes and Inertial Manifolds*, Nonlinearity **4** (1991), 135–153.

32. A. Friedman, *Partial Differential Equations*, Holt, Rinehart and Winston, New York, 1969.

33. J.M. Gambaudo, P.A. Glendinning, and C. Tresser, *Stable Cycles with Complicated Structure*, J. Physique Lett. **46** (1985), 653–657.

34. M. Gevrey, *Sur la nature analytique des solutions des equations aux dérivées partielles*, Ann. Scient. Ec. Norm. Sup., 3ème série **35** (1918), 129–190.

35. J.M. Ghidaglia and B. Héron, *Dimension of the Attractors Associated to the Ginzburg-Landau Equation*, Physica D **28** (1987), 282–304.

36. J. Ginibre and G. Velo, *On a Class of Nonlinear Schrödinger Equations I: The Cauchy Problem, General Case*, Journal of Functional Analysis **32** (1979), 1–32.

37. J. Ginibre, *private communication*, 1993.

38. R. Glassey, *On the Blowing-up of Solutions to the Cauchy Problem for the Nonlinear Schrödinger Equation*, J. Math. Phys. **18** (1977), 1794–1797.

39. D. Goldman and L. Sirovich, *The One Dimensional Complex Ginzburg Landau Equation in the Low Dissipation Limit*, Nonlinearity **7** (1994) 417–439.

40. M. Goldman, *Plasma Turbulence*, Rev. Mod. Phys. **56** (1984), 709–735.

41. J. Hadamard, *Sur la généralisation de la notion de fonction analytique*, C.R. Séances Soc. Math. **40** (1912), 28.

42. J.K. Hale, L.T. Magalhães and W.M. Oliva, *An Introduction to Infinite Dimensional Dynamical Systems– Geometric Theory* Applied Mathematical Sciences **47**, Springer-Verlag, New York, 1984.

43. D. Henry, *Geometric Theory of Semilinear Parabolic Equations*, Lecture Notes in Mathematics **840**, Springer-Verlag, Berlin, 1981.

44. K. Horsch and C.D. Levermore, *Attractors for Lyapunov Cases of the Complex Ginzburg-Landau Equation*, Physica D (submitted 1995).

45. F. John, *Partial Differential Equations*, Springer, New York, 1982.

46. D. Jones, L. Margolin and E.S. Titi, *On the Effectiveness of the Approximate Inertial Manifolds—Computational Study*, Theoretical and Computational Fluid Dynamics (1995), to appear.

47. C. Kahane, *On the Spatial Analyticity of Solutions of the Navier-Stokes Equations*, Arch. Rat. Mech. Anal. **33** (1969), 386–405.

48. O. Kavian, *A Remark on the Blowing-up of Solutions to the Cauchy Problem for Nonlinear Schrödinger Equations*, Trans. Am. Math. Soc. **299** (1987), 193–203.

49. J. Kopeć and J. Musielak, *On Quasianalytic Classes of Functions, Expansible in Series*, Anal. Polon. Math. **7** (1960), 285–292.

50. I. Kukavica, *On the Number of Determining Nodes for the Ginzburg-Landau Equation*, Nonlinearity **5** (1992), 997–1006.

51. Y. Kuramoto, *Chemical Oscillations, Waves and Turbulence*, Series in Synergetics **19**, Springer, New York, 1984.

52. C. de La Vallée Poussin, *Quatre leçons sur les fonctions quasi-analytiques de variable réelle*, Bull. Soc. Math. France **52** (1924), 175–203.

53. M.J. Landman, G.C. Papanicolau, C. Sulem, and P.L. Sulem, *Singularities for the Nonlinear Schrödinger Equation*, Phys. Rev. A **38** (1988), 3837–3843.

54. B.J. LeMesurier, G.C. Papanicolau, C. Sulem, and P.L. Sulem, *Local Structure of the Self-Focusing Singularity of the Nonlinear Schrödinger Equation*, Physica D **32** (1988), 210–226.

55. J. Leray, *Sur le mouvement d'un fluide visqueux emplissant l'espace*, Acta Mathematica **63** (1934), 193–248.

56. C.D. Levermore and M. Oliver, *Distribution-Valued Initial Data for the Complex Ginzburg-Landau Equation*, in preparation, 1995.

57. B.P. Luce, *Homoclinic Explosions in the Complex Ginzburg-Landau Equation*, Physica D **83** (1995), 1–29.

58. J. Mallet-Paret, *Negatively Invariant Sets of Compact Maps and an Extension of a Theorem of Cartwright*, J. Diff. Eq. **22** (1976), 331–348.

59. S. Mandelbrojt, *Séries de Fourier et classes quasi-analytiques de fonctions*, Gauthier-Villars, Paris, 1935.

60. R. Mañé, *On the Dimension of the Compact Invariant Sets of Certain Nonlinear Maps*, Lecture Notes in Mathematics **898**, Springer-Verlag, New York, 1981, 230–242.

61. F. Merle, *Limit Behavior of Saturated Approximations of Nonlinear Schrödinger Equation*, Commun. Math. Phys. **146** (1992), 377–414.

62. F. Merle and Y. Tsutsumi, *L^2-Concentration of Blow-up Solutions for the Nonlinear Schrödinger Equation with Critical Power Nonlinearity*, J. Diff. Eq. **84** (1990), 205–214.

63. A. Mielke and G. Schneider, *Derivation and Justification of the Complex Ginzburg-Landau Equation as a Modulation Equation*, this volume.

64. A.C. Newell, T. Passot, and J. Lega, *Order Parameter Equations for Patterns*, Annual Rev. Fluid Mech. **25** (1993), 399–453.

65. A.C. Newell and J.A. Whitehead, *Finite Bandwidth, Finite Amplitude Convection*, J. Fluid Mech. **38** (1969), 279–303.

66. A.C. Newell and J.A. Whitehead, *Review of the Finite Bandwidth Concept*, in *Proceedings of the International Union of Theoretical and Applied Mechanics*, Symposium on Instability of Continuous Systems (1969), H. Leipholz ed., Springer-Verlag, Berlin 1971, 279–303.

67. P.K. Newton and L. Sirovich, *Instabilities of the Ginzburg-Landau Equation: Periodic Solutions*, Quart. Appl. Math. **44** (1986), 49–58.

68. L. Nirenberg, *On Elliptic Partial Differential Equations*, Annali della Scuola Norm. Sup. di Pisa **13** (1959), 115–162.

69. M. Oliver, Ph.D. dissertation, The University of Arizona, Tucson, 1996.

70. A. Pazy, *Semigroups of Linear Operators and Applications to Partial Differential Equations*, Appl. Math. Sci. **44**, Springer-Verlag, New York, 1983.

71. I. Rasmussen and K. Rypdal, *Blow-up in Nonlinear Schrödinger Equations I – A General Review*, Physica Scripta **33** (1987), 498–504.

72. L. Schwartz, *Functional Analysis*, Courant Institute Lecture Notes, New York University, New York, 1964.

73. L. Sirovich and J.D. Rodriguez, *Coherent Structures and Chaos: A Model Problem*, Phys. Lett. A **120** (1988), 211–214.

74. L.A. Segel, *Distant Side-Walls Cause Slow Amplitude Modulation of Cellular Convection*, J. Fluid Mech. **38** (1969), 203–224.

75. D. Stark, *Structure and Turbulence in the Complex Ginzburg-Landau Equation with a Nonlinearity of Arbitrary Order*, Ph.D. dissertation, The University of Arizona, Tucson, 1995.

76. K. Stewartson and J.T. Stuart, *A Nonlinear Instability Theory for a Wave System in Plane Poiseuille Flow*, J. Fluid Mech. **48** (1971), 529–545.

77. P. Takáč, *Invariant 2-Tori in the Time-Dependent Ginzburg-Landau Equation*, Nonlinearity **5** (1992), 289–321.

78. R. Temam, *Infinite-Dimensional Dynamical Systems in Mechanics and Physics*, Springer-Verlag, New York, 1988.

79. Y. Tsutsumi, *L^2-Solutions for NLS and Nonlinear Groups*, Funkcialaj Ekvacioj **30** (1987), 115–125.

80. F.B. Weissler, *Local Existence for Semilinear Parabolic Equations in L^p*, Indiana U. Math J. **29** (1980), 79–102.

81. V.E. Zakharov, *Collapse of Langmuir Waves*, Soviet J.E.T.P. **35** (1972), 908–914.

82. V.E. Zakharov and A.B. Shabat, *Exact Theory of Two-Dimensional Self-Focusing and One-Dimensional Self-Modulation of Waves in Nonlinear Media*, Sov. Phys. JETP **34(1)** (1973), 62–69.

83. V.E. Zakharov and A.B. Shabat, *Interaction Between Solitons in a Stable Medium*, Sov. Phys. JETP **37** (1973), 823–828.

84. V.E. Zakharov, V.V. Sobolev, and V.S. Synach, *Character of the Singularity and the Stochastic Phenomena of Self-Focusing*, Zh. Eksp. Teor. Fiz., Pis'ma Red **14** (1971), 390–393.

(C.D. Levermore) DEPARTMENT OF MATHEMATICS, UNIVERSITY OF ARIZONA, TUCSON, AZ 85721, USA

E-mail address: `lvrmr@math.arizona.edu`

(M. Oliver) PROGRAM IN APPLIED MATHEMATICS, UNIVERSITY OF ARIZONA, TUCSON, AZ 85721, USA

E-mail address: `oliver@math.arizona.edu`

Lectures in Applied Mathematics
Volume **31**, 1996

Derivation and Justification of the Complex Ginzburg–Landau Equation as a Modulation Equation

ALEXANDER MIELKE AND GUIDO SCHNEIDER

1 Introduction

The complex Ginzburg–Landau equation is a generic amplitude or modulation equation to describe many physical systems with dissipation near the onset of instability, see [NW69, DES71] and [NPL93] for a recent review. In typical situations, such as the Rayleigh–Bénard or the Taylor–Couette problem in hydrodynamics, one is concerned with a translationally invariant problem where the wavelength of the dominating pattern is much smaller than the size of the physical domain. In such situations it is advantageous to study the system on an infinite physical domain; this leads to new phenomena like sideband instabilities which have their counterpart in mathematical difficulties due to continuous spectra and noncompactness.

The main idea of the Ginzburg–Landau formalism is to describe slow spatial and temporal modulations of a basic periodic pattern by the solutions of a simple partial differential equation for the amplitude function A; this PDE for A is called the amplitude or modulation equation. It is our aim to present an introduction to this theory by considering the most simple example giving rise to this sort of pattern formation. Consider the following parabolic PDE, which is a mixture of the Swift–Hohenberg, the Korteweg–de Vries, and the Kuramoto–Shivashinsky equation:

$$u_t = -(1 + \partial_x^2)^2 u + (\alpha + \beta \partial_x^2) u_x + \varepsilon^2 u + f(u) + g(u) u_x \quad \text{for } t > 0, \ x \in I\!\!R. \quad (1.1)$$

AMS Subject Classification: Primary 35 K 55, 76 E 30; Secondary 35 B 35

Research partially supported by Deutsche Forschungsgemeinschaft (DFG) through Schwerpunktprogramm "Dynamische Systeme" under Mi 459/2–1.

Here $u = u(t, x) \in \mathbb{R}$ is the state variable, and f and g are smooth functions with $f(u) = f_2 u^2 + f_3 u^3 + \mathcal{O}(u^4)$ and $g(u) = g_1 u + g_2 u^2 + \mathcal{O}(|u|^3)$ for $u \to 0$. The parameter ε is small and ε^2 measures the distance from criticality. Linearizing at $u \equiv 0$ shows that the periodic patterns e^{ikx} are damped for $|1 - k^2| > \varepsilon$ and they are unstable in the other case with growth rates less than ε^2. The aim of the Ginzburg–Landau approximation is to describe the nonlinear evolution of these weakly unstable periodic patterns. For this purpose solutions are sought in the form

$$U_A^1(t, x) = \varepsilon A(T, X) e^{i(\omega t + x)} + \text{c.c.}, \quad \text{with } T = \varepsilon^2 t \text{ and } X = \varepsilon(x + c_g t),$$

where A is the slowly varying amplitude of the pattern, T is the slow time scale, and X the large spatial scale. Formal multiple scaling calculations, which are explained in Section 2, show that the associated modulation equation for A is the complex Ginzburg–Landau equation

$$\partial_T A = a \partial_X^2 A + bA - c|A|^2 A, \tag{1.2}$$

where the coefficients $a, b, c \in \mathbb{C}$ can be calculated explicitly from (1.1).

In Section 3 we discuss the question whether the function $\widehat{u}(t, x) = U_A^1(t)$ is really an approximation of a true solution u of (1.1). The theory involves two major steps. First, we show that a suitable approximate solution $\widetilde{u} \approx U_A^1$ has a small residual

$$\text{Res}(\widetilde{u}) = \partial_t \widetilde{u} - L(\partial_x)\widetilde{u} - \mathcal{N}(\varepsilon, \widetilde{u})$$

in a suitable function space. In the next step, a differential equation for the error $R = u - \widetilde{u}$ is established in the form

$$\partial_t R = L(\partial_x) R + D_u \mathcal{N}(\varepsilon, \widetilde{u}) R + \mathcal{M}(\varepsilon, \widetilde{u}, R) + \text{Res}(\widetilde{u}), \tag{1.3}$$

where $\mathcal{M} = \mathcal{O}(\|R\|^2)$. From (1.3) the error has to be controlled by using some kind of Gronwall inequality.

The second step goes through in a straight forward way for cases where the nonlinearity starts with cubic terms. However, the case of quadratic nonlinearities needs a more refined theory involving the structure of Fourier mode interaction. To this end we define mode projections to split u into $u_c + u_s$ where u_s contains stable modes and u_c the critical modes:

$$u_c = P_c u, \quad \text{where } \mathcal{F}(P_c u)(k) = \chi_c(k)(\mathcal{F}u)(k), \ k \in \mathbb{R},$$

where $\mathcal{F}$ is the Fourier transform and $\chi_c(k) = 1$ for $|1 - |k|| \leq 1/4$ and 0 elsewhere. In order to handle the case of quadratic nonlinearities it is essential to use the fact that the quadratic interactions of critical modes lead to stable modes, i.e. $P_c(P_c u \cdot P_c v) = 0$. The main result is contained in Theorem 3.6, and it states that for each solution $A = A(T)$ of (1.2) existing for $t \in [0, T_0]$ and each $d > 0$ there are $\varepsilon_0, D > 0$ such that for all $\varepsilon \in (0, \varepsilon_0)$ the following is true: the solution $u(t) = \mathcal{S}_t^\varepsilon(u_0)$ of (1.1) with $\|u_0 - U_A^1(0)\| \leq d\varepsilon^{5/4}$ exists for $t \in [0, T_0/\varepsilon^2]$ and satisfies

$$\|u(t) - U_A^1(t)\| \leq D\varepsilon^{5/4}$$

on this interval.

The whole mathematical theory of modulation problems depends very subtle on the choice of the underlying function spaces. In Section 3 of this paper we use $H^1(I\!R)$ as the basic phase space for (1.1) in order to present the theory as simple as possible. Other choices are spaces of bounded continuous functions or $L^\infty(I\!R)$, which were used in [CoE90, KSM92, Sch94a, Co94], and function spaces, which contain analytic functions having a bounded holomorphic extension into a complex strip around the real axis (see [vHa91]). As Fourier transform methods play an essential role it is also convenient to characterize the functions through its Fourier transform, e.g., in [Eck93] the functions u with $\mathcal{F}u \in L^1(I\!R) \cap L^\infty(I\!R)$ were used.

We prefer the spaces $H^n_{l,u}(I\!R)$ which were introduced in [Sch94b, Sch94c, MS94]; they consist of the uniformly local Sobolev spaces $H^n(I\!R) = W^{n,2}(I\!R)$ and contain all kind of bounded functions such as periodic or quasiperiodic spatial patterns as well as travelling fronts. Moreover, they are based on L^2 and, hence, the important tool of Fourier transformation is still available. Another nice property is that the global existence of solutions in these spaces can be shown for many classical dissipative system by the weighted energy estimates. This theory is recalled in Section 4.1. Moreover, we give a global existence proof for (1.2) with $\operatorname{Re} c > 0$ in Theorem 4.1 and derive explicit bounds for an absorbing ball in $H^1_{l,u}(I\!R)$.

Furthermore, Section 4 contains a brief account of further research which leads to a more detailed study of the system. In Section 4.2 we discuss the attractivity of the set of modulated periodic patterns. It was first observed in [Eck93], that the dynamics of the system is such that all small solutions develop after a certain time the structure of a modulated periodic pattern. This property is very similar to the attractivity of finite dimensional center manifolds; of course, in this case the existence of an invariant manifold is not known and seems very unlikely.

The approximation property derived in Section 3 and the attractivity property allow us to follow all small solutions u for all $t \in [0, \infty)$ by approximate solutions of Ginzburg–Landau type. However, every approximate solutions is only valid over a long but finite time interval; after that interval the error between the present approximation and the true solution may become large, but the attractivity tells us, that there is another modulated periodic pattern which is close to u. Using this new modulated pattern as a starting point for the Ginzburg–Landau approximation we again obtain an interval of good approximation. Thus, it is possible to construct a pseudo–orbit for (1.2) which shadows the true solution u for all $t > 0$. In [Sch94c] it was shown that the jumps in the pseudo–orbit can be controlled and are less than $C\varepsilon^{1/4}$ in the ε–independent Ginzburg–Landau equation.

A recent result in [MS94] concerns the existence of an attractor $\mathcal{A}^\varepsilon$ for (1.1) and of an attractor $\mathcal{A}_{\mathcal{G}}$ for (1.2) in the space $H^1_{l,u}(I\!R)$. Note that these attractors cannot be compact, as they are translationally invariant, nevertheless they are attractors in the norm topology of $H^1_{l,u}(I\!R)$. The existence theory is based on ideas of [Fei93]. Moreover, it is possible to compare the attractor $\mathcal{A}^\varepsilon$ with $\mathcal{A}_{\mathcal{G}}$ in a certain scaled way.

In this brief introduction many results in the field of modulation equations are not touched; here we mention a few of them in order to show the richness of the theory. Hyperbolic problems also show modulational effects and the most famous example is the nonlinear Schrödinger equation which describes the nonlinear evolution of the amplitude of a small wave packet in dispersive hyperbolic systems. See [CaE87] for formal derivations and [Kal89, KSM92, Sch94e] for a proof of the approximation property.

Generalizations of the above theory are available also for vector–valued problems and for systems on cylindrical domains one unbounded spatial direction. The stationary or time–periodic Navier–Stokes equation on an infinitely long cylinder was treated in [IMD89, IM91] and the stationary part of the Ginzburg–Landau equation was obtained. In [Sch94b, Bol94] the full parabolic system was treated and the Ginzburg–Landau equation was shown to produce valid approximations for the Navier–Stokes equations. Recall that the theory of modulation equations was originally invented to analyze exactly these problems in hydrodynamics, see [NW69, DES71].

Modulation equations can also be treated on domains with two unbounded spatial directions. In these systems one typically observes an competition between roll–like structures and hexagonal patterns; we refer to [NPL93] for a physical review on this topic. In [Sch94d] the Newell–Whitehead equation was studied as a modulation equation in the sense described above. However, it turns out, that it generally fails to describe the dynamics of the original problem on a suitably long time scale.

This negative example gives new motivation for the mathematical study of the connections between weakly nonlinear systems and their modulation equations. Not every equation derived formally is equally well suitable to describe the full system. The mathematical justification is one tool which helps us to decide which modulation equation is useful. In building up a catalogue of qualifying and disqualifying properties of modulation equations it will be possible to support the process of modelling the effects in pattern forming systems; thus, we can replace expensive physical or numerical experiments, which by now are the standard tools to test the validity of modulation equations.

2   The formal derivation

In this work we focus our attention to one of the simplest model problems which can be treated with the Ginzburg–Landau formalism:

$$u_t = -(1 + \partial_x^2)^2 u + (\alpha + \beta \partial_x^2) u_x + \varepsilon^2 u + f(u) + g(u) u_x \quad \text{for } t > 0, \ x \in \mathbb{R}. \quad (2.1)$$

Here $u = u(t, x) \in \mathbb{R}$ is the state variable, and f and g are smooth functions with $f(u) = f_2 u^2 + f_3 u^3 + \mathcal{O}(u^4)$ and $g(u) = g_1 u + g_2 u^2 + \mathcal{O}(|u|^3)$ for $u \to 0$. The parameter ε is considered to be small, such that ε^2 measures the distance from the onset of instability. Note that the parameter α can be changed easily by going into a moving frame: define $\widetilde{u}(t, x) = u(t, x - ct)$, then $\widetilde{u}$ satisfies (2.1) but with α replaced by $\alpha + c$. We will make use of this fact below.

The basic solution is $u \equiv 0$ and its stability can be studied by analyzing the linearization

$$v_t = -(1 + \partial_x^2)^2 v + (\alpha + \beta \partial_x^2) v_x + \varepsilon^2 v.$$

As this problem is independent of x and t it can be solved by exponential solutions in the form $v(t,x) = e^{\lambda t + ikx}$, where $k \in \mathbb{R}$ and

$$\lambda(k) = -(1 - k^2)^2 + (\alpha - \beta k^2)ik + \varepsilon^2.$$

The exponential growth rate is given by the real part of λ and we immediately see that growing solutions correspond to $|k| \in (\sqrt{1-\varepsilon}, \sqrt{1+\varepsilon})$. The most unstable wave number is where the maximum of $\operatorname{Re}\lambda$ is attained, in our case $k = \pm 1$.

The imaginary part $\omega(k) = \operatorname{Im}\lambda(k)$ gives the phase velocity of the spatially periodic pattern e^{ikx}. For $k = 1$ we find $\omega(1) = \alpha - \beta$. For modulated patterns, which can be understood as wave packets with slightly varying wave number, the group velocity c_g is important; it is defined as $c_g = \partial\omega/\partial k$. For the critical mode $k = 1$ we find $c_g = \alpha - 3\beta$. As the only stable wave numbers are in a ε-neighborhood of $k = \pm 1$ we let $k = 1 + \varepsilon K$. Now, the solutions can be expanded as follows:

$$v(t,x) = e^{\lambda t + ikx} = e^{i[(\alpha-\beta)t+x]+i\varepsilon[(\alpha-3\beta)Kt+Kx]+\varepsilon^2[1-(4+i3\beta)K^2]t+\mathcal{O}(\varepsilon^3)t}.$$

Hence, by choosing the frame speed such that the group velocity is 0 (i.e. $\alpha = 3\beta$), the solution is given as the basic periodic pattern $e^{i(2\beta t + x)}$ modulated on the slow time scale $T = \varepsilon^2 t$ and on the large spatial scale $X = \varepsilon x$:

$$v(t,x) = e^{[1-(4+i3\beta)K^2+\mathcal{O}(\varepsilon)]T}e^{iKX}\mathbf{E}, \quad \text{where } \mathbf{E} = e^{i(2\beta t + x)}.$$

For fixed t_0 we will use the special notation $\mathbf{E}_{t_0}$ to indicate the time–independent function $e^{i(2\beta t_0 + x)}$.

In a weakly unstable nonlinear theory it can be expected that the dynamics is dominated by the critical modes from linear theory. The interaction of these modes is driven by the nonlinear terms and occurs through modulations on the slow time scale $T = \varepsilon^2 t$ and the large spatial scale $X = \varepsilon x$. From now on we always assume that the group velocity of the critical pattern is zero, i.e. $\alpha = 3\beta$. Hence we are led to the ansatz

$$u(t,x) = U_A^m(t,x) + \mathcal{O}(\varepsilon^{m+1}), \quad \text{with} \\ U_A^m(t,x) = \varepsilon A(T,X)\mathbf{E} + \varepsilon\overline{A}(T,X)\overline{\mathbf{E}} + \sum_{n=2}^{m}\sum_{l=-n}^{n}\varepsilon^n A_{n,l}\mathbf{E}^l. \tag{2.2}$$

Here A is the amplitude function for the basic periodic pattern $\mathbf{E} = e^{i(x+2\beta t)}$. The amplitudes $A_{n,l} = \overline{A}_{n,-l}$ are the amplitudes of the powers of the basic pattern generated through the nonlinear interactions.

The Ginzburg–Landau formalism proceeds now as follows: we insert the ansatz (2.2) into the basic underlying equation (2.1) and equate the coefficients of the powers of $\varepsilon^n\mathbf{E}^m$ to zero. After finitely many steps we obtain a closed problem, which allows us to determine the amplitude A. Moreover, we may increase m and obtain a hierarchy of equations from which allows correction terms can be constructed.

In order to show how this procedure works we rewrite (2.1) in the form

$$\mathcal{L}u = \mathcal{N}(\varepsilon, u) = \varepsilon^2 u + \mathcal{N}_2(u,u) + \mathcal{N}_3(u,u,u) + \mathcal{O}(\|(u,u_x)\|^4), \\ \text{where } \mathcal{L}u = u_t - L(\partial_x)u = u_t + (1+\partial_x^2)^2 u + \beta(3+\partial_x^2)u_x, \\ \mathcal{N}_2(u,v) = f_2 uv + \tfrac{g_1}{2}(u_x v + uv_x), \\ \mathcal{N}_3(u,v,w) = f_3 uvw + \tfrac{g_2}{3}(uvw_x + uv_x w + u_x vw). \tag{2.3}$$

The nonlinear terms $\mathcal{N}_k$ are written as symmetric k–linear forms. By elementary calculations we obtain the following result.

Lemma 2.1
Let $m \in \mathbb{Z}$ and $B = B(T, X)$ with $T = \varepsilon^2$ and $X = \varepsilon x$. If B is sufficiently smooth, then

$$\mathcal{L}[B(\varepsilon^2 t, \varepsilon x)\mathbf{E}^m] = \mathbf{E}^m\Big\{[2\beta im - \lambda_0(m)]B + \varepsilon i \lambda_0'(m)B_X$$
$$+\varepsilon^2[B_T + \tfrac{1}{2}\lambda_0''(m)B_{XX}] - \varepsilon^3 \tfrac{i}{6}\lambda_0^{(3)}(m)\partial_X^3 B - \varepsilon^4 \tfrac{1}{24}\lambda^{(4)}(m)\partial_X^4 B\Big\}, \tag{2.4}$$

where $\lambda_0(k) = -(1 - k^2)^2 + \beta(3 - k^2)ik$, $k \in \mathbb{R}$.

Note that $2\beta im - \lambda_0(m) = 0$ if and only if $m = \pm 1$; this fact will be important for constructing the hierarchy in $A_{n,l}$.

Inserting (2.2) into the nonlinear terms we find

$$\mathcal{N}_2(U_A^m, U_A^m) = \varepsilon^2\Big(f_2[A^2\mathbf{E}^2 + 2|A|^2 + \overline{A}^2\mathbf{E}^{-2}] + g_1[iA^2\mathbf{E}^2 - i\overline{A}^2\mathbf{E}^{-2}]\Big)$$
$$+\varepsilon^3 \sum_{l=-3}^{3} n_A^{3,l}\mathbf{E}^l + \mathcal{O}(\varepsilon^4);$$

$$\mathcal{N}_3(U_A, U_A, U_A) = \varepsilon^3\Big((f_3 + ig_2)A^3\mathbf{E}^3 + (3f_3 + ig_2)|A|^2 A\mathbf{E} + \text{c.c.}\Big) + \mathcal{O}(\varepsilon^4),$$

where $n_A^{3,1} = (2f_2 + ig_1)(AA_{2,0} + \overline{A}A_{2,2})$. ¿From $\mathcal{L}(A\mathbf{E}) = \mathcal{O}(\varepsilon^2)$ it is obvious that all terms of order ε vanish. Comparing the terms corresponding to $\varepsilon^2\mathbf{E}^k$ yields $(2i\beta k - \lambda_0(k))A_{2,k} = n_A^{2,k}$ which results in

$$(0i - (-1))A_{2,0} = 2f_2|A|^2 + g_1 \cdot 0,$$
$$(2i\beta - 2i\beta)A_{2,1} = f_2 \cdot 0 + g_1 \cdot 0, \tag{2.5}$$
$$(4i\beta - (-9 - 2i\beta))A_{2,2} = (f_2 + ig_1)A^2.$$

($A_{n,l}$ with $l < 0$ is always given by the complex conjugate of $A_{n,-l}$.)

We are now ready to derive the associated Ginzburg–Landau equation by comparing the coefficients of $\varepsilon^3\mathbf{E}$:

$$A_T + \tfrac{\lambda_0''(1)}{2}A_{XX} + i\lambda_0'(1)A_{2,1} + (2i\beta - \lambda_0(1))A_{3,1}$$
$$= A + (2f_2 + ig_1)(AA_{2,0} + \overline{A}A_{2,2}) + (3f_3 + ig_2)|A|^2 A.$$

Using $\lambda_0(1) = 2i\beta$, $c_g = \lambda_0'(1) = 0$, and the previously obtained expressions for $A_{2,l}$ the result reads:

$$A_T = (4 + 3i\beta)A_{XX} + A + c|A|^2 A,$$
$$\text{where } c = (2f_2 + ig_1)\Big(2f_2 + \tfrac{f_2 + ig_1}{9 + 6i\beta}\Big) + 3f_3 + ig_2. \tag{2.6}$$

Of course, we can write out all the equations corresponding to $\varepsilon^n\mathbf{E}^l$. For $l \neq \pm 1$ we obtain an algebraic equation for $A_{n,l}$ in terms of A and all the lower order functions $A_{\tilde{n},\tilde{l}}$ with $\tilde{n} < n$. For $l = \pm 1$ and $n \geq 4$ we obtain a linear partial differential equation of the form

$$\partial_T A_{n-2,1} = (4 + 3i\beta)\partial_X^2 A_{n-2,1} + (1 + 2c|A|^2)A_{n-2,1} + cA^2 A_{n-2,-1} + \rho_A(T, X),$$

Here ρ_A depends only on A and $A_{\widetilde{n},\widetilde{l}}$ (and their derivatives) where either $\widetilde{l} \neq \pm 1$ and $\widetilde{n} \leq n - 1$ or $\widetilde{l} = \pm 1$ and $\widetilde{n} < n - 2$.

Therefore, we are able to calculate arbitrarily high correction terms $A_{n,k}$, as long as A is sufficiently smooth. Studying the exact interactions of derivatives and powers in ε we arrive at the following result.

Corollary 2.2
Assume that $A : [0, T_0] \times \mathbb{R} \to \mathbb{C}$ has bounded continuous partial derivatives $\partial_T^i \partial_X^j A$ for all $i, j \geq 0$ with $2i + j \leq m \geq 2$. Then, the above construction yields an approximate solution $\widetilde{u} = U_A^m$ such that

$$Res(\widetilde{u}) = \mathcal{L}\widetilde{u} - \mathcal{N}(\varepsilon, \widetilde{u}) = \mathcal{O}(\varepsilon^{m+1})$$

uniformly for $(t, x) \in [0, T_0/\varepsilon^2] \times \mathbb{R}$.

Thus, we have found approximate solutions in the sense that the residual for the original problem is as small as we like. It remains to show that small residuals also imply small errors, which is done using the boundedness of the associated linearized semigroup. Using the ansatz $u = U_A^m + \varepsilon^{\widetilde{m}} R$ we can derive a differential equation for the scaled error R. As U_A^m has size of order ε the error scaling $\varepsilon^{\widetilde{m}}$ will satisfy $1 < \widetilde{m} \leq m + 1$. If we are able to show that R remains of order 1 for $t \in [0, T_0/\varepsilon^2]$, the desired result is established. We ask for a time scale of order $1/\varepsilon^2$, which is absolutely necessary to make any sense out of the result. Recall that the modulations of A appear on an order 1 time scale in the slow time T. To see this effect in the approximate solution U_A^m it is necessary to follow the solution in the unscaled time t over an interval of length T_0/ε^2.

3   Error estimates

As mentioned above we can find approximate solutions U_A such that the residual is arbitrary small. The error r between this approximation and a suitable true solution u should be small. Letting $r = u - U_A = \varepsilon^n R$ we obtain for the error the equation

$$
\begin{aligned}
\mathcal{L}R &= \varepsilon^{-n}\Big(\mathcal{N}(\varepsilon, U_A + \varepsilon^n R) - \mathcal{N}(\varepsilon, U_A) - Res(U_A)\Big) \\
&= D_u\mathcal{N}(\varepsilon, U_A(t,x))R + \mathcal{M}(\varepsilon, U_A(t,x), R) - \varepsilon^{-n}Res(U_A),
\end{aligned}
\tag{3.1}
$$

where $\mathcal{M}(\varepsilon, \widetilde{u}, R) = \varepsilon^{-n}(\mathcal{N}(\varepsilon, \widetilde{u} + \varepsilon^n R) - \mathcal{N}(\varepsilon, \widetilde{u}) - D_u\mathcal{N}(\varepsilon, \widetilde{u})\varepsilon^n R) = \mathcal{O}(\varepsilon^n|R|^2)$ uniformly for $\varepsilon \in (0, 1]$. The main difficulty lies in the size of the linear part $D_u\mathcal{N}(\varepsilon, U_A)$, since it allows for an exponential growth rate of the error. As $U_A = \mathcal{O}(\varepsilon)$ we have $D_u\mathcal{N}(\varepsilon, U_A) = \mathcal{O}(\varepsilon^k)$, where $k = 1$ in the general case and $k = 2$ whenever $\mathcal{N}_2$ vanishes. The latter case is also called the cubic case, as the nonlinearity starts with cubic terms. We first treat this cubic case, which is much easier than the general quadratic case.

In order to build a mathematically basis for this theory we introduce a function space Z in which the solutions u of (2.1), the approximate solutions U_A, and the error in (3.1) will be considered. Associated to this is a function space Y for the solutions A of the Ginzburg–Landau equation (2.6). We consider the mapping $A \mapsto U_A$ as a mapping from the Ginzburg–Landau space Y into the phase space Z.

We choose the function spaces $Z = H^1(\mathbb{R})$ and $Y = H^1(\mathbb{R}) \oplus iH^1(\mathbb{R})$, where we continue to distinguish Z and Y, as Z contains functions depending on the unscaled spatial variable $x \in R$ whereas Y contains functions on the slow spatial variable $X \in \mathbb{R}$. For convenience we also use the abbreviations $Z^m = H^m(\mathbb{R})$ and $Y^m = H^m(\mathbb{R}) \oplus iH^m(\mathbb{R})$. We use the scaling mapping $W_\varepsilon : Y^n \to Z^n; A(\cdot) \mapsto A(\varepsilon \cdot)$ and the lift $Y \ni A \mapsto U_A^1 = \varepsilon((W_\varepsilon A)\mathbf{E} + (\overline{W_\varepsilon A})\mathbf{E}^{-1}) \in Z$, which satisfies the estimates

$$\|U_A^1\|_{\mathcal{C}^0} \le 2\varepsilon\|A\|_{\mathcal{C}^0}, \quad \|U_A^1\|_Z \le 2\varepsilon^{1/2}\|A\|_Y.$$

(We continue to use $\mathcal{C}^j$ for the space of functions on $\mathbb{R}$ which have j bounded continuous derivatives, where $\|\cdot\|_{\mathcal{C}^j}$ is the sum of the supremum norms of these derivatives.) The H^1–estimate does not contain the coefficient ε due to the fact that the scaling of the spatial variable $X = \varepsilon x$ diminishes the decay rate in U_A^1.

An important feature of the theory is to control the dominating linear operator $\mathcal{L} = \partial_t - L(\partial_x)$ with $L(\partial_x) = -(1 + \partial_x^2)^2 + \beta(3 + \partial_x^2)\partial_x$. In the chosen spaces the following result holds:

Lemma 3.1
The densely defined, closed linear operator $L(\partial_x) : Z^4 \subset Z^0 \to Z^0$ generates a holomorphic semigroup $S(t) = e^{L(\partial_x)t}$, $t \ge 0$, which satisfies, for all $m \ge n \ge 0$, the estimates

$$\|S(t)\|_{Z^n \to Z^n} \le 1, \quad \text{for } t \ge 0;$$
$$\|S(t)\|_{Z^n \to Z^m} \le C_{m-n}(1 + t^{-(m-n)/4}), \quad \text{for } t > 0.$$

Proof: The result follows easily by Fourier transform $\mathcal{F}$. Letting $\hat{u} = \mathcal{F}u$ the transformed equation reads $\hat{u}_t = \lambda_0(k)\hat{u}$ which has the solution $\hat{u}(t, k) = e^{\lambda_0(k)t}\hat{u}(0, k)$. Using the fact that the norm in Z^m is equivalent to $\|(1 + k^2)^{m/2}\hat{u}(k)\|_{Z^0}$, the result is easily established from $(1 + k^2)^{(m-n)/2}\left|e^{\lambda_0(k)t}\right| \le (1 + k^2)^{(m-n)/2}e^{-(1-k^2)^2t} \le 1$ for $m = n$ and $\le C_{m-n}(1 + t^{-(m-n)/4})$ for $m > n$. $\square$

The boundedness of the linearized flow at criticality ($\varepsilon = 0$) is one of the essential features in showing that approximate solutions U_A^m are close to true solutions; it allows us to turn small residuals onto small errors. We can now rewrite the nonlinear evolution problem (2.1) as integral equation

$$u(t) = S(t)u(0) + \int_0^t S(t - r)\mathcal{N}(\varepsilon, u(r))\, dr.$$

Here $\mathcal{N}(\varepsilon, \cdot)$ is a smooth mapping from Z into $Z^0 = L^2(\mathbb{R})$. By standard semilinear parabolic theory (cf. [Hen81]) we obtain a local, continuous semigroup $\mathcal{S}_t^\varepsilon$ on Z, that is, $u(t) = \mathcal{S}_t^\varepsilon(u(0))$ solves (2.1). Note that the longtime existence of the solutions is not clear, since standard theories involving Gronwall estimates guarantee existence of $\mathcal{S}_t^\varepsilon(u_0)$ only for $t \in [0, c\min\{1/\varepsilon^2, 1/\rho\}]$ for all u_0 with $\|u_0\|_Z \le \rho$.

Similarly, the Ginzburg–Landau equation can be treated in order to obtain the local semigroup $\mathcal{G}_T$ on Y, such that $A(T) = \mathcal{G}_T(A_0)$ solves (2.6). The question of justification of the Ginzburg–Landau formalism is now concerned with estimating the difference between the true solution $u(t) = \mathcal{S}_t^\varepsilon(U_{A_0}^1)$ and the approximate solution $\widetilde{u}(t) = U_A^1(t) = \varepsilon((W_\varepsilon \mathcal{G}_{\varepsilon^2 t}(A_0))\mathbf{E} + \text{c.c.})$.

3.1 The case of a cubic nonlinearity

Since the case with the cubic nonlinearity serves as an introductory part we make the further assumption that the nonlinearity in (2.1) is given solely by $f(u)$, that is $g \equiv 0$. This leads to the simplification that $\mathcal{N}$ is a smooth mapping from Z into itself rather than the larger space Z^0.

Theorem 3.2
(Cubic nonlinearity)
Let $f_2 = f''(0)/2 = 0$ and $g \equiv 0$. Let $A(0) \in H^4(\mathbb{R}) = Y^4 \subset Y$ be such that $A \in \mathcal{C}^0([0, T_0], Y^4)$ for some $T_0 > 0$. Then, for every $d > 0$ there exist $\varepsilon_0 > 0$ and $D > 0$ such that for all $\varepsilon \in (0, \varepsilon_0)$ the following statement holds: Let $u(t) = \mathcal{S}_t^\varepsilon(u(0))$ be a solution of (2.1) such that $\|u(0) - U_A^1(0)\|_Z \le d\varepsilon^{3/2}$, then u exists on the whole interval $[0, T_0/\varepsilon^2]$ and the estimate

$$\|u(t) - U_A^1(t)\|_Z \le D\varepsilon^{3/2}, \quad \text{for all } t \in [0, T_0/\varepsilon^2], \tag{3.2}$$

is satisfied.

Remark: Note that $\widetilde{u}(t) = U_A^1(t)$, and hence u, are of order ε in $L^\infty(\mathbb{R})$ and of order $\varepsilon^{1/2}$ in Z. Thus, the error of order $\varepsilon^{3/2}$ is relatively small.

Here we have assumed more smoothness on the initial condition $A(0)$ than is actually needed. We will improve on this when we consider the case of general nonlinearities.

Proof: We want to show that the error $u(t) - U_A^1(t)$ remains of order $\mathcal{O}(\varepsilon^{3/2})$ over the time interval $[0, T_0/\varepsilon^2]$. However, substituting U_A^1 into (2.1) leaves the residual terms of order ε^3, e.g. $\varepsilon^3 f_3 A^3 \mathbf{E}^3$. Integrating such a term over $[0, T_0/\varepsilon^2]$, leads to an error $\mathcal{O}(\varepsilon)$. To avoid this difficulty we follow the ideas of [KSM92] and use the improved approximation $\widetilde{u}(t) = \widetilde{U}_A^3$ which is given by

$$\widetilde{U}_A^3(t, x) = \varepsilon A \mathbf{E} + \varepsilon^3 (A_{3,3} \mathbf{E}^3 + \widetilde{A}_{3,1} \mathbf{E}) + \text{c.c.},$$

where $A_{3,3} = f_3/(64 + 24i\beta)A^3$ and $\widetilde{A}_{3,1} = 0$. As the coefficient $\widetilde{A}_{3,1}$ will not appear in the residual up to order ε^4 we may set it equal to zero rather than taking the correct value $A_{3,1}$.

From the construction and the considerations in Section 2 we find that all residual terms of order ε^j with $j = 1, 2, 3$ cancel. The remaining terms are at least of order ε^4 and are functions of $\partial_T^i \partial_X^k A$ with $4i + k \le 4$. From the assumption $A(0) \in Y^4$ we know that $A \in \mathcal{C}^0([0, T_0], Y^4) \cap \mathcal{C}^1([0, T_0], Y^2)$. Thus, the residual can be estimated by

$$\|\text{Res}(\widetilde{U}_A^3(t))\|_Z \le C_\rho \varepsilon^{7/2},$$

where again a factor $\varepsilon^{1/2}$ is lost by scaling.

The true solution will be written as $u(t) = \widetilde{u}(t) + \varepsilon^{3/2}R$, such that R satisfies

$$\|R(0)\|_Z \le d + \varepsilon^{-3/2}\|\widetilde{U}_A^3(0) - U_A^1(0)\|_Z = d + \mathcal{O}(\varepsilon) \le 2d$$

for $\varepsilon \in (0, \varepsilon_1]$ for some $\varepsilon_1 > 0$. Moreover, R satisfies the differential equation

$$\mathcal{L}R = D_u\mathcal{N}(\varepsilon, \widetilde{U}_A^3(t))R + \mathcal{M}(\varepsilon, \widetilde{U}_A^3(t), R(t)) + \rho_\varepsilon(t), \tag{3.3}$$

where $\rho_\varepsilon(t) = -\varepsilon^{-3/2}\text{Res}(\widetilde{U}_A^3(t))$ with $\|\rho_\varepsilon(t)\|_Z \le \varepsilon^2 C_\rho$ for $t \in [0, T_0/\varepsilon^2]$ and $\varepsilon \le \varepsilon_1$. The linear operator $D_u\mathcal{N} : Z \to Z$ is given by $R \mapsto (\varepsilon^2 + f'(\widetilde{U}_A^3))R$ and can be estimated by

$$\|D_u\mathcal{N}(\varepsilon, \widetilde{U}_A^3)\|_{Z \to Z} \le 2\|\varepsilon^2 + f'(\widetilde{U}_A^3)\|_{\mathcal{C}^1} \le C_1\varepsilon^2 \quad \text{for } \varepsilon \le \varepsilon_1.$$

The factor ε^2 in this estimate is essential for the following theory; it is due to the fact that the nonlinearity is cubic, that is $f'(u) = \mathcal{O}(u^2)$. In addition, the nonlinearity $\mathcal{M}$ allows for a good estimate, since the second derivative $D_u^2\mathcal{N}$ is still of order $\mathcal{O}(\|\widetilde{U}_A^3\|_{\mathcal{C}^0}) = \mathcal{O}(\varepsilon)$. In particular, for each $D > 0$ there is a constant $C_2(D)$ such that

$$\|\mathcal{M}(\varepsilon, \widetilde{U}_A^3(t), R)\|_Z \le \varepsilon^{5/2}C_2(D) \quad \text{for all } R \text{ with } \|R\|_Z \le D \text{ and all } \varepsilon \le \varepsilon_1.$$

The error satisfies the integral equation

$$R(t) = S(t)R(0) + \int_0^t S(t-\tau)\Big[D_u\mathcal{N}(\varepsilon, \widetilde{U}_A^3(\tau)R(\tau) + \mathcal{M}(\varepsilon, \widetilde{U}_A^3(\tau), R(\tau)) + \rho_\varepsilon(\tau)\Big]d\tau.$$

Using Lemma 3.1 and the estimates derived above and assuming that the solution R does not leave the ball of radius D in Z, we obtain

$$\|R(t)\|_Z \le 2d + \int_0^t \|S(t-\tau)\|_{Z \to Z}\Big[\|D_u\mathcal{N}\|_{Z \to Z}\|R(\tau)\|_Z + \|\mathcal{M} + \rho_\varepsilon\|_Z\Big]d\tau$$
$$\le 2d + \int_0^t C\Big[C_1\varepsilon^2\|R(\tau)\|_Z + \varepsilon^{5/2}C_2(D) + \varepsilon^2 C_\rho\Big]d\tau$$
$$\le 2d + CT_0\Big[\varepsilon^{1/2}C_2(D) + C_\rho\Big] + \int_0^t C\varepsilon^2 C_1\|R(\tau)\|_Z d\tau.$$

Gronwall's inequality yields

$$\|R(t)\|_Z \le \big(2d + CT_0[\varepsilon^{1/2}C_2(D) + C_\rho]\big)e^{CC_1\varepsilon^2 t} = \alpha(\varepsilon, D)e^{CC_1\varepsilon^2 t} \le \alpha(\varepsilon, D)e^{CC_1 T_0}.$$

We now choose the radius D of the ball in Z to be equal to $2\alpha(0,0)e^{CC_1 T_0}$, and then, ε_0 is chosen such that $\alpha(\varepsilon, D) \le 2\alpha(0,0)$. Now, $R(t)$ cannot leave the ball of radius D for $t \in [0, T_0/\varepsilon^2]$ and estimate (3.2) is established. Obviously, the solution R exists and is unique by standard local existence results. This proves that $u(t) = \widetilde{U}_A^3(t) + \varepsilon^{3/2}R(t)$ exists on the whole interval $[0, T_0/\varepsilon^2]$. $\qquad\square$

3.2   The case of quadratic nonlinearities

We now want to study the general case where $\mathcal{N}$ is not restricted to start with cubic terms. The problem with the previous approach is that the derivative $D_u\mathcal{N}(\varepsilon, U_A(t))$

is only of order ε; this would only allow for estimates on a time interval $[0, T_0/\varepsilon]$ which is too small to see interesting modulations.

The new idea for handling quadratic nonlinearities is to use the mode structures more explicitly. By quadratic interaction of the modes $\mathbf{E}^n$ and $\mathbf{E}^m$ we generate the modes $\mathbf{E}^l$ with $l \in \{n + m, n - m, -n + m, -n - m\}$. In particular, the critical modes with $n = \pm 1$ generate, through quadratic interactions, only modes $\mathbf{E}^l$ with $l \in \{-2, 0, 2\}$, but all these modes are uncritical, that is they are exponentially damped. Thus, we split the solutions and the error into two parts, one consisting of modes close to $k = \pm 1$ and the other contains all the remaining modes. This splitting is affected by mode projections defined in Fourier space. We let $\chi(k) = 1$ for $k \in I_c = [-5/4, -3/4] \cup [3/4, 5/4]$ and zero elsewhere, and we define the projections

$$P_c : Z \to Z; u \mapsto \mathcal{F}^{-1}[\chi\mathcal{F}(u)], \qquad P_s : Z \to Z; u \mapsto u - P_c u.$$

We shortly write $u = u_c + u_s$ with $u_c = P_c u$ the critical part and $u_s = P_s u$ the stable part. Obviously, these projections commute with the differential operator $L(\partial_x)$ and the following results hold.

Lemma 3.3
a) *For all $m \geq n \geq 0$ and all $t > 0$ we have*

$$\|P_s S(t)\|_{Z^n \to Z^m} \leq C_{m-n}(1 + t^{-(m-n)/4})e^{-t/6}, \quad \|P_c\|_{Z^n \to Z^m} \leq C_{m-n}.$$

b) *For all $u, v \in Z$ we have $P_c[(P_c u) \cdot (P_c v)] = 0$, where $\tilde{u} \cdot \tilde{v} \in Z$ means ordinary pointwise multiplication.*

Remarks:
1. The first estimate shows that the stable modes are strictly exponentially damped.
2. The second estimate in a) means that the critical mode projection is a smoothing operator: the functions in $P_c Z$ are in fact holomorphic functions.
3. The last assertion exactly is the manifestation of the fact that quadratic interactions of critical modes are not critical any more.

Proof: For part (a) we proceed as in Lemma 3.1 but now the supremum of the multiplier $(1+k^2)^{(m-n)/2}e^{-(1-k^2)^2 t}$ is only taken over the relevant intervals, namely $I_c = [-5/4, -3/4] \cup [3/4, 5/5]$ and $I_s = \mathbb{R} \setminus I_c$, respectively.

Part (b) follows immediately from the fact that the product in x–space is transformed to a convolution in Fourier space. Hence $\mathcal{F}(P_c u \cdot P_c v)$ has its support in $I_c + I_c = [-5/2, -3/2] \cup [-1/2, 1/2] \cup [3/2, 5/2]$, and $I_c \cap (I_c + I_c) = \emptyset$ implies the result. $\qquad\square$

Since we are now able to study the critical components separately from the stable ones we can use the damping in the stable components to compensate for lower powers in ε in the error equation. In order to estimate $u(t) - U_A^1(t)$ we introduce an intermediate approximation

$$\tilde{U}_A^2(t) = \varepsilon(P_0 W_\varepsilon A)\mathbf{E} + \varepsilon^2(P_0 W_\varepsilon A_{2,2})\mathbf{E}^2 + \frac{\varepsilon^2}{2}(P_0 W_\varepsilon A_{2,0}) + \text{c.c.},$$

where W_ε is the scaling operator $W_\varepsilon : Y^n \to Z^n; A \mapsto A(\varepsilon \cdot)$, and $P_0 : Z^0 \to Z^n; u \mapsto \mathcal{F}^{-1}[\chi_0 \mathcal{F} u]$ is a mode projection corresponding to the zero mode, namely $\chi_0(k) = 1$ for $|k| \le 1/12$ and 0 elsewhere. The amplitudes $A_{2,2}$ and $A_{2,0}$ are given in (2.5); note that $A_{2,1}$ is not included as it will not contribute to the relevant terms.

The estimation of the residual consists in giving rigorous bounds on the terms which where left over after equating the lowest order terms to zero. These remaining terms possess only formally the magnitude ε^n and have to be controlled in our function space Z^0. To this end it is necessary to use the smoothing properties of the solutions A of the Ginzburg–Landau equation: starting with $A(0) \in Y = Y^1 = H^1(\mathbb{R})$ implies

$$\|A(T)\|_{Y^{n+1}} + \|\partial_T A(T)\|_{Y^{n-1}} \le C_n(1 + T^{-n/2}), \quad \text{for } T \in (0, T_0]. \qquad (3.4)$$

Moreover, we have to control the scaling properties of $W_\varepsilon A$, which can only be measured by taking derivatives: $\partial_x(W_\varepsilon A) = \varepsilon W_\varepsilon(\partial_X A)$ is one order smaller than $W_\varepsilon A$ itself. Additionally it is important that the scaling concentrates the modes in Fourier space around the wave number zero, such that application of the projection P_0 does not change a scaled function too much.

Lemma 3.4
For each $n \in \mathbb{N}_0$ there exist constants C_n such that

$$\|W_\varepsilon A - P_0 W_\varepsilon A\|_{Z^n} \le C_n \varepsilon^{n-1/2} \|A\|_{Y^n}, \quad \|W_\varepsilon A\|_{Z^n} \le \varepsilon^{-1/2} \sum_{l=0}^{n} \varepsilon^l \|A\|_{Y^l}, \quad (3.5)$$

for all $A \in Y^n$.

Proof: The first result is obtained by Fourier transform:

$$\|W_\varepsilon A - P_0 W_\varepsilon A\|_{Z^n}^2 \le C \int_R \left|(1 + |k|^n)(1 - \chi_0(k)) \tfrac{1}{\varepsilon} (\mathcal{F}A)(\tfrac{k}{\varepsilon})\right|^2 dk$$
$$\le C \int_{\mathbb{R}} \left|2(12)^n \varepsilon^n (1 + \left|\tfrac{k}{\varepsilon}\right|^n)(\mathcal{F}A)(\tfrac{k}{\varepsilon})\right|^2 dk$$
$$= C(12)^{2n} \varepsilon^{2n-1} \int_{\mathbb{R}} \left|(1 + |K|^n)(\mathcal{F}A)(K)\right|^2 dK \le \tilde{C}(12)^{2n} \varepsilon^{2n-1} \|A\|_{Y^n}^2,$$

where $1 - \chi_0(k) = 0$ for $|k| \le 1/12$ was used essentially. The second estimate follows easily as each spatial derivative of the scaled function produces a factor ε. $\qquad \square$

The treatment of the quadratic nonlinearity follows the analysis in [Sch94a]; however, we use different function spaces, and we relax the smoothness assumption on $A(0)$. As seen above, it is necessary to use spatial derivatives in order to gain smallness which is expressed in powers of ε. But at the same time the finite smoothness of $A(T)$ for $T = 0$ will generate temporal singularities in the residual through the terms $\partial_X^n A(T)$. The residual estimate given below provides exact control on the temporal singularities and the associated powers of ε.

Theorem 3.5
Let $A = A(T) \in C^0([0, T_0], Y)$, U_A^1, and $\tilde{U}_A^2$ as above. Then, the following estimates

hold:

$$\|\widetilde{U}_A^2(t) - U_A^1(t)\|_{Z^n} \leq C_n \varepsilon^{3/2}(1 + t^{(1-n)/2}),$$
$$\varepsilon\|P_c\widetilde{U}_A^2(t)\|_{Z^n} + \|P_s\widetilde{U}_A^2(t)\|_{Z^n} \leq C_n\varepsilon^{3/2}, \tag{3.6}$$
$$\varepsilon\|P_c\widetilde{U}_A^2(t)\|_{C^n} + \|P_s\widetilde{U}_A^2(t)\|_{C^n} \leq C_n\varepsilon^2,$$

$$\|P_c Res(\widetilde{U}_A^2(t))\|_{Z^n} \leq C_n\varepsilon^{3/2}(1+t)^{-1},$$
$$\|P_s Res(\widetilde{U}_A^2(t))\|_{Z^n} \leq C_n\varepsilon^{5/2}(1+t^{-1/2}). \tag{3.7}$$

Proof: For the first estimate we consider U_A^2 which is defined as $\widetilde{U}_A^2$ but without the projections P_0. Then,

$$\|U_A^2 - U_A^1\|_{Z^n} \leq C\varepsilon^2(\|W_\varepsilon A_{2,2}\|_{Z^n} + \|W_\varepsilon A_{2,0}\|_{Z^n})$$
$$\leq C\varepsilon^{3/2}\left(1 + \sum_{l=1}^n \varepsilon^n\left[1 + (\varepsilon^2 t)^{-(l-1)/2}\right]\right) \leq C\varepsilon^{3/2}(1 + t^{(1-n)/2}),$$

where (3.4) and the second estimate in (3.5) was used. The difference $\widetilde{U}_A^2 - U_A^2$ can now be controlled by the first estimate in (3.5) and (3.4), and then our first estimate is established. The bounds on $\widetilde{U}_A^2$ follow immediately from

$$\|P_0 W_\varepsilon A\|_{Z^n} \leq \|P_0\|_{Z^0 \to Z^n}\|W_\varepsilon A\|_{Z^0} \leq C_n\varepsilon^{-1/2}\|A\|_{Z^0}.$$

and $\|W_\varepsilon A\|_{C^0} = \|A\|_{C^0}$. The first estimate in (3.6) is a consequence of (3.5), (3.4), and the explicit representations of $\widetilde{U}_A^2$ and U_A^1. This proves (3.6).

For the estimate of the residual we write $\widetilde{U}_A^2 = \varepsilon u_1 + \varepsilon^2 u_2$, where $\|u_j\|_{C^1} \leq C$, $\|u_j\|_Z \leq C\varepsilon^{-1/2}$, and $\mathcal{F}u_1$ has support in $[-13/12, -11/12] \cup [11/12, 13/12]$ and similarly $\mathcal{F}u_2$ has support in the three intervals of radius $1/12$ centered around $-2, 0,$ and 2. The residual has the expansion

$$\text{Res}(\widetilde{U}_A^2) = \partial_t\widetilde{U}_A^2 - L(\partial_x)\widetilde{U}_A^2 - \mathcal{N}(\varepsilon, \widetilde{U}_A^2) = \text{Res}_1 + \mathcal{O}(\varepsilon^{7/2})$$

$$\text{with } \text{Res}_1 = \varepsilon\partial_t u_1 + \varepsilon^2\partial_t u_2 - \varepsilon L u_1 - \varepsilon^2 L u_2 - \varepsilon^2\mathcal{N}_2(u_1, u_1)$$
$$-\varepsilon^3\big(u_1 + \mathcal{N}_2(u_1, u_2) + \mathcal{N}_3(u_1, u_1, u_1)\big) = \sum_{l=-3}^3 \delta_l,$$

where $\delta_{-l} = \overline{\delta}_l$ and $\text{supp}\mathcal{F}\delta_l \subset [l - 1/4, l + 1/4]$. The special splitting of the residual according to the modes in Fourier space allows for an easy distinction between the stable and the critical part, namely $P_c\text{Res}_1 = \delta_{-1} + \delta_1$. (This was the reason why the support of χ_0 was chosen as $[-1/12, 1/12]$.)

We immediately see $\|\delta_3\|_{Z^0} \leq C\varepsilon^{5/2}$. To estimate δ_2 we proceed as follows:

$$\delta_2 = \varepsilon^2\partial_t(P_0 W_\varepsilon A_{2,2})\mathbf{E}^2 - L(\partial_x)\big[(P_0 W_\varepsilon A_{2,2})\mathbf{E}^2\big] - \varepsilon^2\mathcal{N}_2(P_0 W_\varepsilon A\mathbf{E}, P_0 W_\varepsilon A\mathbf{E})$$
$$= \Big[\varepsilon^4 P_0 W_\varepsilon \partial_T A_{2,2} + \varepsilon^2\big((4i\beta - \lambda_0(2))P_0 W_\varepsilon A_{2,2} - M_2(\varepsilon)A_{2,2}\big)\Big]\mathbf{E}^2$$
$$-\Big[f_2(P_0 W_\varepsilon A)^2 + g_1 P_0 W_\varepsilon A\big(iP_0 W_\varepsilon A + \varepsilon P_0(W_\varepsilon\partial_X A)\big)\Big]\mathbf{E}^2,$$

where $M_2(\varepsilon) : Y^1 \to Z^0; B \mapsto \mathbf{E}^{-2}\big(\lambda_0(2) - L(\partial_x)\big)\big[(P_0 W_\varepsilon B)\mathbf{E}^2\big]$. Using the definition of $A_{2,2}$ from (2.5) and (3.5) we find

$$\|\delta_2\|_{Z^0} \leq C\varepsilon^{7/2}/\sqrt{\varepsilon^2 t} + \varepsilon^2\|M_2(\varepsilon)\|_{Y^1 \to Z^0}C$$
$$+\varepsilon^2|f_2 + ig_1|\|(P_0 W_\varepsilon A)^2 - P_0(W_\varepsilon A)^2\|_{Z^0} + \varepsilon^3|g_1|\|(P_0 W_\varepsilon A)(P_0 W_\varepsilon\partial_X A)\|_{Z^0}.$$

The fourth term is of order $\varepsilon^{5/2}$ uniformly for $t \in [0, T_0/\varepsilon^2]$. The third term can be estimated with (3.5) and

$$
\begin{aligned}
\|(P_0 B)^2 - P_0(B^2)\|_{Z^0} &\leq \|(P_0 B)^2 - B^2\|_{Z^0} + \|B^2 - P_0(B^2)\|_{Z^0} \\
&\leq \|P_0 B + B\|_{C^0} \|P_0 B - B\|_{Z^0} + \|P_0(B^2) - B^2\|_{Z^0};
\end{aligned}
$$

and we again obtain $\mathcal{O}(\varepsilon^{5/2})$. The linear operator $M_2(\varepsilon)$ is controlled via Fourier transform:

$$
\begin{aligned}
\|M_2(\varepsilon) B\|_{Z^0}^2 &= \int_{I\!R} \left| (\lambda_0(2) - \lambda_0(2+k)) \tfrac{\chi_0(k)}{\varepsilon} (\mathcal{F} B)(k/\varepsilon) \right|^2 dk \\
&\leq \int_{I\!R} |C|k|/\varepsilon (\mathcal{F} B)(k/\varepsilon)|^2 dk = C^2 \varepsilon \|\partial_X B\|_{Y^0}^2 \leq C^2 \varepsilon \|B\|_{Y^1}^2.
\end{aligned}
$$

Exactly the same procedure can be used to estimate δ_0 and the desired first estimate in (3.7) is shown.

The estimation of the critical part is a little more tiresome but follows the same scheme:

$$
\begin{aligned}
\delta_1 &= \varepsilon^3 (P_0 W_\varepsilon \partial_T A) \mathbf{E} - \varepsilon^3 (-\lambda_0''(1) P_0 W_\varepsilon \partial_X^2 A/2 + P_0 W_\varepsilon A) \mathbf{E} + \varepsilon [M_1(\varepsilon) A] \mathbf{E} \\
&\quad - \varepsilon^3 \left(2\mathcal{N}_2(P_0 W_\varepsilon A \mathbf{E}, P_0 W_\varepsilon A_{2,0} \mathbf{E}^0) + \mathcal{N}_2(P_0 W_\varepsilon A_{2,2} \mathbf{E}^2, \overline{P_0 W_\varepsilon A} \mathbf{E}^{-1}) \right. \\
&\qquad\qquad \left. + 3\mathcal{N}_3(P_0 W_\varepsilon A \mathbf{E}, P_0 W_\varepsilon A \mathbf{E}, \overline{P_0 W_\varepsilon A} \mathbf{E}^{-1}) \right),
\end{aligned}
$$

where $M_1(\varepsilon) B = \mathbf{E}^{-1}(\lambda_0(1) + \tfrac{1}{2}\lambda_0''(1)\partial_x^2 - L(\partial_x))[(P_0 W_\varepsilon B)\mathbf{E}]$. Inserting the Ginzburg–Landau equation to eliminate the leading linear part we obtain a nonlinear term which is close to the other nonlinear terms, such that the difference can be estimated as above, but now with the bound $C\varepsilon^{7/2}$. For the operator $M_1(\varepsilon) : Y^n \to Z^0$ we proceed as follows

$$
\begin{aligned}
\|M_1(\varepsilon) B\|_{Z^0}^2 &= \int_{I\!R} \left| (\lambda_0(1+k) - \lambda_0(1) - \lambda_0'(1)k - \lambda_0''(1)\tfrac{k^2}{2}) \tfrac{\chi_0(k)}{\varepsilon} (\mathcal{F} B)(k/\varepsilon) \right|^2 dk \\
&\leq \int_{I\!R} |C|k|^p/\varepsilon (\mathcal{F} B)(k/\varepsilon)|^2 dk = C^2 \varepsilon^{2p-1} \|\partial_X^p B\|_{Y^0}^2 \leq C^2 \varepsilon^{2p-1} \|B\|_{Y^p}^2,
\end{aligned}
$$

where $p = 1, 2,$ or 3. Thus, we find

$$
\varepsilon \|[M_1(\varepsilon) A(T)]\mathbf{E}\|_{Z^0} \leq C\varepsilon \min\{\varepsilon^{1/2}, \frac{\varepsilon^{3/2}}{\sqrt{\varepsilon^2 t}}, \frac{\varepsilon^{5/2}}{\varepsilon^2 t}\} \leq C \frac{\varepsilon^{3/2}}{1+t},
$$

and the second estimate in (3.7) is established by noting $\varepsilon^{7/2} \leq T_0 \varepsilon^{3/2}/(1+t)$ on the relevant time interval. $\qquad\square$

Theorem 3.6
(General nonlinearity)
Let $A_0 \in H^1(I\!R) = Y$ be such that $A(T) = \mathcal{G}_T(A_0)$ satisfies $A \in \mathcal{C}^0([0, T_0], Y)$ for some $T_0 > 0$. Then, for every $d > 0$ there exist $\varepsilon_0 > 0$ and $C > 0$ such that for all $\varepsilon \in (0, \varepsilon_0)$ the following statement holds:

Let $u(t) = \mathcal{S}_t^\varepsilon(u(0))$ be a solution of (2.1) such that $\|u(0) - U_A^1(t)\|_Z \leq d\varepsilon^{5/4}$, then $u(t) = S_t^\varepsilon(u(0))$ exists on the whole interval $[0, T_0/\varepsilon^2]$ and the estimate

$$
\|P_c[u(t) - \tilde{U}_A^2(t)]\|_Z \leq C\varepsilon^{5/4} \qquad \|P_s[u(t) - \tilde{U}_A^2(t)]\|_Z \leq C(\varepsilon^{5/2} + \varepsilon^{5/4} e^{-t/6}), \quad (3.8)
$$

for all $t \in [0, T_0/\varepsilon^2]$, is satisfied.

Remarks: Note that the initial condition $u(0)$ is only close to the first order approximation $U_A^1(0)$, but the result states closeness of the solution to the second order approximation such that the error in the stable part is even one order smaller. This is a first result in the direction of attractivity, which means errors are damped out.

It is clear from the proof below that the result of the theorem can be changed in such a way that $\varepsilon^{5/4}$ can be replaced by ε^β, with $\beta \in (1, 3/2)$, or $\varepsilon^{3/2} \log(1/\varepsilon)$.

Proof: We write the solution $u = u(t)$ of the full system in the form

$$u = \varepsilon \widetilde{u}_c + \varepsilon^2 \widetilde{u}_s + \varepsilon^{5/4} R_c + \varepsilon^{9/4} R_s, \tag{3.9}$$

where $\widetilde{u}_c = \varepsilon^{-1} P_c \widetilde{U}_A^2$, $\widetilde{u}_s = \varepsilon^{-2} P_s \widetilde{U}_A^2$, $R_c = P_c R_c$, and $R_s = P_s R_s$. By construction we know that $\widetilde{u}_c$ and $\widetilde{u}_s$ are uniformly bounded in $C^1(\mathbb{R})$ for $t \in [0, T_0/\varepsilon^2]$, thus we will estimate them henceforth by C.

Using Theorem 3.5 it is sufficient to show that R_s and R_c are bounded uniformly in $\varepsilon \in (0, \varepsilon_0]$ and $t \le T_0/\varepsilon^2$ in the appropriate norms. We define

$$\mathcal{K}(\varepsilon, \widetilde{u}_c, \widetilde{u}_s, R_c, R_s) = \mathcal{N}(\varepsilon, \widetilde{U}_A^2 + \varepsilon^{3/2} R_c + \varepsilon^{5/2} R_s) - \mathcal{N}(\varepsilon, \widetilde{U}_A^2) - \mathrm{Res}(\widetilde{U}_A^2),$$

and note the expansion

$$\|\mathcal{K}(\varepsilon, \widetilde{u}_c, \widetilde{u}_s, R_c, R_s) - \varepsilon^{13/4}(R_c + \varepsilon R_s) - 2\varepsilon^{9/4} \mathcal{N}_2(\widetilde{u}_c + \varepsilon \widetilde{u}_s, R_c + \varepsilon R_s)$$
$$- \varepsilon^{5/2} \mathcal{N}_2(R_c + \varepsilon R_s, R_c + \varepsilon R_s) + \mathrm{Res}(\widetilde{U}_A^2)\|_Z^0 \tag{3.10}$$
$$\le C_0 \varepsilon^{13/4} \big[\|\widetilde{u}_c + \varepsilon \widetilde{u}_s\|_{C^1}^2 + \varepsilon^{1/4} C_\mathcal{N}(\|R_c + \varepsilon R_s\|_Z)\big]\|R_c + \varepsilon R_s\|_Z,$$

where $C_\mathcal{N}$ is some continuous functions from $[0, \infty)$ into itself. Inserting (3.9) into (2.1) and projecting onto the critical and stable part we obtain the coupled system

$$\partial_t R_c = L(\partial_x) R_c + \mathcal{K}_c(\varepsilon, \widetilde{u}_c, \widetilde{u}_s, R_c, R_s),$$
$$\partial_t R_s = L(\partial_x) R_s + \mathcal{K}_s(\varepsilon, \widetilde{u}_c, \widetilde{u}_s, R_c, R_s), \tag{3.11}$$

where $\mathcal{K}_c = \varepsilon^{-5/4} P_c \mathcal{K}$ and $\mathcal{K}_s = \varepsilon^{-9/4} P_s \mathcal{K}$.

Note that $\mathcal{N}_2(u, c) = f_2 uv + g_1(u_x v + u v_x)/2$ can be estimated in two ways, namely

$$\|c \mathcal{N}_2(u, v)\|_{Z^0} \le C \min\{\|u\|_{C^1}\|v\|_Z, \|u\|_Z\|u\|_Z\}.$$

We will always use the C^1-norm for $\widetilde{u}_{c,s}$ but the Z-norm for the errors $R_{c,s}$.

Using (3.10), Lemma 3.3(b) and Theorem 3.5 we obtain the following estimates

$$\|\mathcal{K}_c(\ldots)\|_{Z^0} \le C \varepsilon^2 \Big(\|R_c\|_Z + \|R_s\|_Z + \varepsilon^{1/4} C_\mathcal{K}(\|R_c + \varepsilon R_s\|_Z)\Big) + \varepsilon^{1/4}/(1 + t),$$
$$\|\mathcal{K}_s(\ldots)\|_{Z^0} \le C \Big(\|R_c + \varepsilon R_s\|_Z + \varepsilon^{1/4} C_\mathcal{K}(\|R_c + \varepsilon R_s\|_Z) + \varepsilon^{1/4}(1 + t^{-1/2})\Big).$$

The first estimate relies heavily on the cancellations $P_c \mathcal{N}_2(\widetilde{u}_c, R_c) = P_c \mathcal{N}_2(R_c, R_c) = 0$.

We now rewrite (3.11) with the variations of constants formula using the semi-group $S(t)$ which satisfies the estimates in Lemma 3.3(a). By the assumption on the initial condition and the first estimate in (3.6) (with $n = 1$) we know $\|R_c(0) + \varepsilon R_s(0)\|_Z \leq C$. Assuming $\|R_c(t) + \varepsilon R_s(t)\| \leq D$ for $[0, t_1]$, we obtain

$$\|R_c(t)\|_Z \leq C + C\varepsilon^2 \int_{\tau=0}^t \left[\|R_c(\tau)\|_Z + \|R_s(\tau)\|_Z + \varepsilon^{1/4} C_\mathcal{K}(D) + \frac{\varepsilon^{1/4}}{1+\tau} \right] d\tau,$$
$$\|R_s(t)\|_Z \leq Ce^{-t/6}/\varepsilon + C \int_{\tau=0}^t (1 + (t-\tau)^{-1/4})e^{-(t-\tau)/6} \left[\|R_c(\tau) + \varepsilon R_s(\tau)\|_Z \right.$$
$$\left. + \varepsilon^{1/4} C_\mathcal{K}(D) + \varepsilon^{1/4}(1+\tau^{-1/2}) \right] d\tau.$$

The stable part R_s may have an initial condition of order $1/\varepsilon$.

We introduce the functions $r_{c,s}(t) = \max\{ \|R_{c,s}(\tau)\|_Z : \tau \in [0, t] \}$ and obtain from the second inequality $\|R_s(t)\| \leq$

$$\frac{C}{\varepsilon}e^{-t/6} + C \int_0^t (1 + (t-\tau)^{-1/4})e^{-(t-\tau)/6}d\tau \left[r_c(t) + \varepsilon r_s(t) + \varepsilon^{1/4} C_\mathcal{K}(D) \right] + C\varepsilon^{1/4}.$$

Thus, we obtain $r_s(t) \leq Ce^{-t/6}/\varepsilon + 6C[r_c(t) + \varepsilon r_s(t) + \varepsilon^{1/4} C_\mathcal{K}(D)] + C\varepsilon^{1/4}$ which, for $\varepsilon \leq 1/(12C)$ leads to

$$\|R_s(t)\|_Z \leq r_s(t) \leq 2Ce^{-t/6}/\varepsilon + 12Cr_c(t) + 12C\varepsilon^{1/4}(C_\mathcal{K}(D) + 1).$$

This result can be inserted into the first estimate to give

$$r_c(t) \leq C + C\varepsilon^2 \int_0^t \left[r_c(\tau) + e^{-\tau/6}/\varepsilon + \varepsilon^{1/4} C_\mathcal{K}(D) \right] d\tau + C\varepsilon^{1/4} \int_0^t 1/(1+\tau)\, d\tau$$
$$\leq C\varepsilon^2 \int_0^t r_c(\tau)\, d\tau + C^* \left[\varepsilon + \varepsilon^{1/4} \left(T_0 C_\mathcal{K}(D) + 1 + T_0 + \log(T_0/\varepsilon^2) \right) \right],$$

for all $t \in [0, T_0/\varepsilon^2]$. The classical Gronwall estimate yields on the same time interval
$$r_c(t) \leq \alpha_c(\varepsilon, D) \quad \text{and} \quad r_s(t) \leq C_s e^{-t/6}/\varepsilon + \alpha_s(\varepsilon, D),$$

where the continuous functions $\alpha_{c,s}$ do not depend on D when ε is set equal to 0.

Now, we choose $D = 2\alpha_c(0,0) + 2\alpha_s(0,0) + 2C_s$, then, there is a ε_0 such that for all $\varepsilon \in (0, \varepsilon_0]$ we have $\alpha_c(\varepsilon, D) + C_s + \varepsilon\alpha_s(\varepsilon, D) \leq D$. Hence, the solutions $R_c(t) + \varepsilon R_s(t)$, $t \in [0, T_0/\varepsilon^2]$ cannot leave the ball of radius D in Z, and the desired estimates follow.   □

We may refine the above by giving estimates in smoother spaces.

Corollary 3.7

The estimate (3.8) can be generalized to

$$\|u(t) - \widetilde{U}_A^2(t)\|_{Z^{n+1}} \leq C_n(1 + t^{-n/4})\varepsilon^{5/4},$$
$$\|\widetilde{U}_A^2(t) - U_A^1(t)\|_{Z^{n+1}} \leq C_n(1 + t^{-n/2})\varepsilon^{3/2} \tag{3.12}$$

where $n \in \mathbb{N}$ is arbitrary.

Proof: This result is obtained the same way as in the proof of Theorem 3.6, we only have to estimate R_c and R_s in Z^{n+1}. Having the boundedness of $R_{c,s}$ in Z, it is classical linear regularity theory for parabolic PDEs (cf. [Hen81]) to derive the estimates

$$\|R_c(t)\|_{Z^{n+1}} + \|R_s(t)\|_{Z^{n+1}} \leq C \left(1 + t^{-n/4} [\|R_c(0)\|_Z + \|R_s(0)\|_Z] \right).$$

Recall that the nonlinearity $\mathcal{N}$ is arbitrarily smooth and $\mathrm{Res}(\widetilde{U}_A^2(t))$ is bounded in each Z^n. Using the first estimate in (3.6) the assertion is established. $\qquad\square$

In general, it is possible to use $U_A^n(t)$, $n \geq 2$ as an approximate solution; then, the corresponding residual is smaller: $\|\mathrm{Res}(U_A^n(t))\|_{Z^0} = \mathcal{O}(\varepsilon^n)$. In such situations it is possible to use the ansatz

$$u = \varepsilon\widetilde{u}_c + \varepsilon^2\widetilde{u}_s + \varepsilon^{n-3/4}R_c + \varepsilon^{n+1/4}R_s,$$

with smaller coefficients in front of the errors $R_{c,s}$. Employing the above theory it is possible to derive rigorous error estimates of the form

$$\|u(t) - U_A^n(t)\|_Z \leq C\varepsilon^{n-3/4}, \quad \text{for } t \in [0, T_0/\varepsilon^2],$$

where $\|u(0) - U_A^n(0))\|_Z \leq d\varepsilon^{n-3/4}$ is assumed.

4 Further results

4.1 Absorbing balls in uniformly local function spaces

To develop the theory further it is necessary to switch to more appropriate function spaces which have nicer properties. First of all, we should mention that it is of practical interest to include bounded functions into the analysis rather than functions in $H^1(\mathbb{R})$, which have to decay at infinity. In particular, periodic (or quasiperiodic) functions and travelling fronts are the center of many research studies in the theory of pattern formation on unbounded domains. But there is also a major mathematical reason to use different function spaces, namely the fact that the Ginzburg–Landau equation

$$\partial_T A = a\partial_X^2 A + bA - c|A|^2A, \quad a,b,c \in \mathbb{C}, \ \mathrm{Re}\, a > 0, \tag{4.1}$$

does not have an absorbing ball in $H^1(\mathbb{R})$. This can be seen by considering the real case with $a,b,c > 0$ and starting with any real initial condition $A_0 \in H^1(\mathbb{R})$ and $A_0(X) \in (0, \sqrt{b/c})$. Then, from the maximum principle we conclude $0 < A(T,X) < \sqrt{b/c}$, and moreover, $A(T,X) \to \sqrt{b/c}$ for $T \to \infty$ and X fixed. It can be shown that $\|A(T,\cdot)\|_{Y^0} \geq C\sqrt{T}$ for large T.

Following [MS94] we introduce function spaces $\widehat{Z}^n$ and $\widehat{Y}^n$ which will replace the old spaces Z^n and Y^n. The new spaces contain all suitably smooth and bounded functions and are based on L^2 theory such that Fourier transform methods are still available. First we choose a positive weight function $\rho \in \mathcal{C}^2(\mathbb{R}, (0,\infty))$ which is bounded, has a finite integral $\int_{\mathbb{R}} \rho(x)\, dx$, and satisfies $|\rho'(x)| \leq \rho(x)$ for all x. As a consequence we obtain $\rho(x+y) \leq e^{|y|}\rho(x)$ for all x,y. (We may fix ρ once and for all to $\rho(x) = 1/\cosh(x)$ or $\rho(x) = 2/(2+x^2)$.) Next we let

$$\widetilde{L}_{l,u}^2(\mathbb{R}) = \{\, u \in L_{\mathrm{loc}}^2(\mathbb{R}) \ : \ \|u\|_{L_{l,u}^2} < \infty \,\}, \quad \text{with } \|u\|_{L_{l,u}^2}^2 = \sup_{y \in \mathbb{R}} \int_{\mathbb{R}} \rho(y+x)u(x)^2 dx,$$

and define the translation operator $T_y : \widetilde{L}^2_{l,u}(\mathbb{R}) \to \widetilde{L}^2_{l,u}(\mathbb{R}); u \mapsto u(\cdot + y)$. Our final space of uniformly local L^2 functions is given as

$$L^2_{l,u}(\mathbb{R}) = \{ u \in \widetilde{L}^2_{l,u}(\mathbb{R}) : \|T_y u - u\|_{L^2_{l,u}} \to 0 \text{ as } y \to 0 \}.$$

Note that different weight functions lead to the same uniform space with equivalent norms. For $n \in \mathbb{N}$ we define the associated Sobolev spaces $H^n_{l,u}$ by requiring that the first n distributional derivatives lie in $L^2_{l,u}(\mathbb{R})$. The condition that the translations T_y are continuous is needed to guarantee that the spaces $H^n_{l,u}(\mathbb{R})$, $n \in \mathbb{N}$, are dense in $L^2_{l,u}(\mathbb{R})$, see [MS94] Lemma 3.1.

We now define the corresponding phase spaces $\widehat{Z}^n = H^n_{l,u}(\mathbb{R})$ for the functions u and the phase space $\widehat{Y}^n = H^n_{l,u}(\mathbb{R}) \oplus iH^n_{l,u}(\mathbb{R})$ for the Ginzburg–Landau equation. As a first result we derive an upper bound for the radius of an absorbing set for (4.1). We use weighted energy norms similar to [CoE90, Co94], however we provide explicit bounds in terms of the coefficients a, b, c.

Theorem 4.1
Let $a_r = \operatorname{Re} a$, $b_r = \operatorname{Re} b$, and $c_r = \operatorname{Re} c$ be positive. Then, all solutions $A = A(T, \cdot)$ of (4.1) with $A(0, \cdot) \in \widehat{Y}^1$ exist for all $T > 0$ and satisfy

$$\|A(T)\|^2_{\widehat{Y}^0} \le e^{-\alpha T}\|A(0)\|^2_{\widehat{Y}^0} + (1 - e^{-\alpha T})\Delta_0,$$

$$\limsup_{T \to \infty} \|\partial_X A(T)\|^2_{\widehat{Y}^0} \le \tfrac{1+a_r}{a_r}\Big(2(1 + \alpha + \delta\Delta_0) + \tfrac{1+a_r}{a_r}\delta^2\Delta_0\Big),$$

where $\alpha = 2b_r + |a|^2/a_r$, $\Delta_0 = C_\rho\alpha/c_r$, $\delta = C_\rho \max\{0, 2|c| - 4c_r\}$, and C_ρ is a constant depending only on the weight ρ.

Proof: We proceed as for standard energy estimates but use weighted norms. Partial integration then involves a derivative of the weight function ρ which can be estimated by ρ due to our assumption.

$$\tfrac{d}{dT}\int_{\mathbb{R}} \rho|A|^2 dX = 2\operatorname{Re}\int_{\mathbb{R}} \rho\overline{A}A_T\, dX \;=\; 2\operatorname{Re}\int_{\mathbb{R}} \rho\overline{A}(aA_{XX} + bA - c|A|^2 A)\, dX$$

$$= 2\operatorname{Re}\Big\{-a\int_{\mathbb{R}}(\rho'\overline{A} + \rho\overline{A}_X)A_X\, dX\Big\} + 2\int_{\mathbb{R}} \rho(b_r|A|^2 - c_r|A|^4)\, dX.$$

$$\le -a_r\int_{\mathbb{R}} \rho|A_X|^2 dX + \int_{\mathbb{R}} \rho(\alpha|A|^2 - \gamma|A|^4)dX,$$

where $\alpha = 2b_r + |a|^2/a_r$ and $\gamma = 2c_r$. Here we have used $|\rho'(X)| \le \rho(X)$ for all $X \in \mathbb{R}$ and

$$2\operatorname{Re} a\int \rho'\overline{A}A_X\, dX \le 2|a|\int \rho|A||A_X|\, dX \le a_r\int \rho|A_X|^2 dX + \frac{|a|^2}{a_r}\int \rho|A|^2 dX.$$

With $\alpha|A|^2 - \gamma|A|^4 \le \alpha(\alpha - \gamma|A|^2)/\gamma$ we obtain $\tfrac{d}{dT}\int_{\mathbb{R}} \rho|A|^2 \le \int_{\mathbb{R}} \rho(\alpha^2/\gamma - \alpha|A|^2)\, dX$, and an application of Gronwall's inequality yields

$$\int_{\mathbb{R}} \rho(X)|A(T, X)|^2 dX \le e^{-\alpha T}\int_{\mathbb{R}} \rho(X)|A(0, X)|^2 dX + (1 - e^{-\alpha T})\tfrac{\alpha}{\gamma}\int_{\mathbb{R}} \rho(X)\, dX$$

$$\le e^{-\alpha T}\|A(0)\|_{\widehat{Y}^0} + (1 - e^{-\alpha T})\tfrac{\alpha}{\gamma}C_\rho.$$

Since this estimate is also true when ρ is replaced by the translated weight $T_y\rho$, the first estimate is proved.

For the first derivative we proceed similarly and obtain

$$\frac{d}{dT}\int_{\mathbb{R}}\rho|A_X|^2 dX \leq -a_r\int_{\mathbb{R}}\rho|A_{XX}|^2 dX + \int_{\mathbb{R}}\rho\Big(\alpha|A_X|^2 + 2(|c|-2c_r)|A|^2|A_X|^2\Big)\,dX.$$
(4.2)

In order to shorten the following formulae we introduce the abbreviations $e_j(T) = \|A(T,\cdot)\|_{\hat{\mathcal{G}}_j}$ and $r_j(T) = (\int_{\mathbb{R}}\rho|\partial_X^j A(T,X)|^2 dX)^{1/2}$ for $j = 0, 1$, and 2. From partial integration we find $r_1^2 \leq r_0(r_1 + r_2)$ and a trivial variant of Sobolev's embedding theorem yields $\|A\|_{\mathcal{C}^0}^2 \leq C_\rho e_0(e_0 + e_1)$. With these stipulations (4.2) takes the form

$$\begin{aligned}\tfrac{d}{dT}r_1^2 &\leq -a_r r_2^2 - (\beta+1)r_1^2 + (\beta+1+\alpha+\delta e_0(e_0+e_1))r_0(r_1+r_2)\\ &\leq -(\beta+1)r_1^2 + s(t)r_1 + s^2(t)/(4a_r) \leq -\beta r_1^2 + (1+1/a_r)s^2(t)/4,\end{aligned}$$

where $\beta > 0$ is arbitrary, $\delta = 2C_\rho\max\{0, |c| - 2c_r\}$, and $s(t) = (\beta+1+\alpha+\delta e_0(e_0+e_1))r_0$. Applying Gronwall's inequality and using the same estimate for all translated weights we obtain

$$\begin{aligned}e_1^2(T) &\leq e^{-\beta T}e_1^2(0) + \tfrac{1+a_r}{4a_r}\int_0^T e^{-\beta(T-\tau)}\Big(\beta+1+\alpha+\delta\big[e_0^2(\tau)+e_0(\tau)e_1(\tau)\big]\Big)^2 d\tau\\ &\leq e^{-\beta T}e_1^2(0) + \tfrac{1+a_r}{2a_r}\int_0^T e^{-\beta(T-\tau)}\Big(\big[\beta+1+\alpha+\delta e_0^2(\tau)\big]^2 + \delta^2 e_0^2(\tau)e_1^2(\tau)\Big)d\tau.\end{aligned}$$

Note that β is still arbitrary in the above estimate. We may now use Lemma 4.2 below and $\limsup_{T\to\infty} e_0^2(T) \leq \Delta_0$ in order to see that $e_1(T)$ is also bounded and satisfies

$$\limsup_{T\to\infty} e_1^2(T) \leq \frac{\frac{1+a_r}{2a_r}(\beta+1+\alpha+\delta\Delta_0)^2}{\beta - \frac{1+a_r}{2a_r}\delta^2\Delta_0}.$$

Setting $\beta = 1 + \alpha + \delta\Delta_0 + \frac{1+a_r}{a_r}\delta^2\Delta_0$ we obtain the desired result. $\qquad\square$

Lemma 4.2
Let $\mu, \nu \in \mathcal{C}^0(\mathbb{R}, \mathbb{R})$ be bounded functions with $\nu(t) \geq 0$ for all t. Assume that $\beta > \bar{\nu} = \limsup_{t\to\infty}\nu(t)$ and that the continuous function $\rho \in \mathcal{C}^0(\mathbb{R}, \mathbb{R})$ satisfies

$$\rho(t) \leq \rho(0)e^{-\beta t} + \int_0^t e^{-\beta(t-\tau)}[\mu(\tau) + \nu(\tau)\rho(\tau)]\,d\tau,$$

for all $t \geq 0$. Then, ρ is bounded on $[0,\infty)$ and satisfies

$$\limsup_{t\to\infty}\rho(t) \leq \frac{1}{\beta-\bar{\nu}}\limsup_{\tau\to\infty}\mu(\tau).$$

Proof: We let $w(t) = \rho(0)e^{-\beta t} + \int_0^t e^{-\beta(t-\tau)}[\mu + \nu\rho]\,d\tau$, then $\rho(t) \leq w(t)$, $w(0) = \rho(0)$, and

$$\frac{d}{dt}w = -\beta w + \mu + \nu\rho \leq (-\beta + \nu)w + \mu.$$

From Gronwall's inequality we find $w(t) \leq \rho(0)e^{a(t)} + \int_0^t e^{a(t)-a(\tau)}\mu(\tau)\,d\tau$, where $a(t) = \int_0^t (\nu(\tau) - \beta)\,d\tau$. Let $\overline{\mu} = \limsup_{\tau \to \infty} \mu(\tau)$ and take $\varepsilon > 0$ with $\overline{\nu} + \varepsilon < \beta$. Then, there is a time t_0 such that $\mu(t) \leq \overline{\mu} + \varepsilon$ and $\nu(t) \leq \overline{\nu} + \varepsilon$ for $t \geq t_0$. With $\mu(t), \nu(t) \leq M$ for $t \in [0, t_0]$ we find for $t \geq t_0$ the estimate $a(t) - a(t_0) \leq -(\beta - \overline{\nu} - \varepsilon)(t - t_0)$ and

$$w(t) = \left(\rho(0)e^{a(t_0)} + \int_0^{t_0} e^{a(t_0)-a(\tau)}\mu(\tau)\,d\tau \right) e^{a(t)-a(t_0)} + \int_{t_0}^t e^{a(t)-a(\tau)}\mu(\tau)\,d\tau$$
$$\leq (\rho(0) + t_0 M)e^{Mt_0}e^{-(\beta-\overline{\nu}-\varepsilon)(t-t_0)} + \int_{t_0}^t e^{-(\beta-\overline{\nu}-\varepsilon)(t-\tau)}\,d\tau(\overline{\mu} + \varepsilon).$$

Thus, we have $\limsup_{t \to \infty} w(t) \leq (\overline{\mu} + \varepsilon)/(\beta - \overline{\nu} - \varepsilon)$, and since $\rho(t) \leq w(t)$ and $\varepsilon > 0$ was arbitrary the result follows. $\square$

In the uniformly local functions spaces $H_{l,u}^n(\mathbb{R})$ the method of Fourier transform is still available, namely in the form of the so–called multiplier theory. An operator $M : H_{l,u}^q \to L_{l,u}^2$ is called a multiplier if it is defined by multiplying the Fourier transform $\widehat{u} = \mathcal{F}u$ by a function $\widehat{m} \in L^\infty(\mathbb{R}, \mathbb{C})$ and then doing an inverse Fourier transform. Using the following lemma, which is proved in [Sch94b], allows us to study the mapping properties of $M : u \mapsto \mathcal{F}^{-1}(\widehat{m}\mathcal{F}u)$. Natural applications are convolution operators $Mu(x) = \int_{\mathbb{R}} m(x - y)u(y)\,dy$ with m integrable, where $\widehat{m} = \mathcal{F}m$.

Lemma 4.3
Let $q, s \in \mathbb{N}_0$ and $w_{s-q}(k) = (1 + k^2)^{(s-q)/2}\widehat{m}(k) \in \mathcal{C}_b^2(\mathbb{R}, \mathbb{C})$. Then $M : H_{l,u}^q \to H_{l,u}^s; u \mapsto \mathcal{F}^{-1}(\widehat{m}\mathcal{F}u)$ is well defined with the estimate

$$\|Mu\|_{H_{l,u}^s} \leq C(q,s)\|w_{s-q}\|_{\mathcal{C}_b^2(\mathbb{R},\mathbb{C})}\|u\|_{H_{l,u}^q},$$

where $C(q,s)$ does not depend on $\widehat{m}$.

Using this result it is trivial to see that the operator $G(T) = e^{aT\partial_x^2}$ defines a holomorphic semigroup on each $\widehat{Y}^n$ and similarly $S(t) = e^{tL(\partial_x)}$ defines a holomorphic semigroup on each $\widehat{Z}^n$. Moreover, the analogues of the estimates in Lemma 3.1 hold:

$$\|G(T)\|_{\widehat{Y}^n \to \widehat{Y}^m} \leq C_{n-m}(1 + T^{-(n-m)/2}), \quad \|S(t)\|_{\widehat{Z}^n \to \widehat{Z}^m} \leq C_{n-m}(1 + t^{-(n-m)/4}),$$
$$\tag{4.3}$$

where $n \leq m$.

In fact, all the theory of Section 3 can be carried through in these spaces with only one modification, namely the definition of the mode projections P_s, P_c, and P_0. According to Lemma 4.3 we have to use smooth cut–off functions in order to define the mode separation: we choose $\widehat{\chi} \in \mathcal{C}^2(\mathbb{R}, [0,1])$ with $\widehat{\chi}(k) = 1$ for $|k| \leq 1/5$ and 0 for $|k| \geq 1/4$. The critical part is now separated using E_c, which is defined via $E_c u = \mathcal{F}^{-1}(\widehat{\chi}_c \mathcal{F}u)$, where $\widehat{\chi}_c(k) = \widehat{\chi}(k + 1) + \widehat{\chi}(-1 - k)$. This operator E_c is a bounded operator from $\widehat{Z}^0$ into $\widehat{Z}^n$ for each $n \in \mathbb{N}_0$, however, it is no longer a projection. Since E_c deletes all Fourier modes outside the interval $[-5/4, -3/4] \cup [3/4, 5/4]$ while other modes are kept or diminished, the operator E_c is called a mode filter. In addition, we define $E_s u = u - E_c u$ and the splitting

$u = u_c + u_s$ with $u_c = E_c u$. Moreover, we need the mode filter E_0 which is given via $E_0 u = \mathcal{F}^{-1}(\widehat{\chi}(3k)\mathcal{F}u)$ and extracts the Fourier modes concentrated around 0.

When all $P_{c,s,0}$ are replaced by $E_{c,s,0}$, respectively, all the results of Section 3 can be worked out completely similar as in the L^2–case. We refer to [Sch94c, MS94] for the details and give here an overview on some further results.

4.2 The attractivity of the set of modulated patterns

Here we treat the question why modulated waves are of such an importance. The Ginzburg–Landau equation is formally derived to describe the evolution of the linearly unstable modes. In weakly nonlinear theory one expects that these modes dominate the dynamics of the full problem under arbitrary initial conditions. In the last section error estimates are shown under the assumption that the initial conditions already possess the scaled mode structure. It is our aim to explain why after the time $t_1 = T_1/\varepsilon^2$ all solutions of the full problem develop this mode structure and, henceforth, can be described by the solutions of the associated Ginzburg–Landau equation.

The first result in this direction was obtained in [Eck93]. For our choice of function spaces the result is given in [Sch94c]. We note here that all the results given below are only established for the case that the Ginzburg–Landau equation (4.1) has real coefficients, $a, b, c \in \mathbb{R}$. However, it is clear that all statements and proofs can easily be translated into the present context of the complex Ginzburg–Landau equation.

Since multiple scaling analysis is only a local theory this attractivity can only be expected for initial conditions in a small neighborhood of the trivial solution.

Theorem 4.4
(Attractivity of the set of modulated patterns)
Fix $C_1 > 0$, then there exists $C_0 > 0$, $\varepsilon_0 > 0$, and $T_1 > 0$, such that the following is true. For all $\varepsilon \in (0, \varepsilon_0)$ and $u_0 \in Z$ with $\|u_0\|_{\widehat{Z}^1} \leq C_1 \varepsilon$ the solution $u(t) = \mathcal{S}_t^\varepsilon(u_0)$ exists for $t \in [0, T_1/\varepsilon^2]$ and there is an $A_0 \in \widehat{Y}^1$ with

$$\|A_0\|_{\widehat{Y}^1} \leq C_0 \quad \text{and} \quad \|\mathcal{S}_{T_1/\varepsilon^2}^\varepsilon(u_0) - \varepsilon(W_\varepsilon A_0 \mathbf{E}_{T_1/\varepsilon^2} + \overline{W_\varepsilon A_0} \mathbf{E}_{T_1/\varepsilon^2}^{-1})\|_{\widehat{Z}^1} \leq C_0 \varepsilon^{5/4}. \quad (4.4)$$

Note that $\mathbf{E}_{T_1/\varepsilon^2}$ can be replaced by the function $\mathbf{E}_0$ as the constant factor $e^{i2\beta T_1/\varepsilon^2}$ can be compensated into A_0. Thus, we may consider the set of modulated patterns,

$$\mathbf{MP} = \{\,\varepsilon(W_\varepsilon A_0 \mathbf{E}_0 + \overline{W_\varepsilon A_0}\mathbf{E}_0^{-1})\,:\,\|A_0\|_{\widehat{Y}^1} \leq C_0\,\},$$

as a small very flat ellipsoid in the phase space $\widehat{Z}^1$, and $\mathbf{MP}$ attracts all solution starting in the ball of radius $C_1 \varepsilon$ around $0 \in \widehat{Z}^1$, at least up to an error of order $\varepsilon^{3/2}$.

Using the result from above it is possible to apply the approximation theory of the previous section. In particular, the assumptions of Theorem 3.6 are fulfilled when $u(0) = \mathcal{S}_{T_1/\varepsilon^2}^\varepsilon(u_0)$ and $A(0) = A_0$ is chosen and $T_0 > 0$ is any time such that $A(T) = \mathcal{G}_T(A_0)$ exists for $T \in [0, T_0]$.

The development of the mode structure is essentially a linear effect, and for illustrative purposes we establish below the linear analogue of Theorem 4.4. We find that the time scale $\mathcal{O}(1/\varepsilon^2)$ is necessary for the solutions $u(t)$ to develop the scaled mode structure. For the nonlinear theory it is then necessary to show that the solutions of the nonlinear problem exist over the desired time interval $t \in [0, T_1/\varepsilon^2]$. For general problems with quadratic nonlinearity, solutions of order ε exists only on the time scale $1/\varepsilon$, but again the fact can be used that quadratic interactions of critical modes generate damped modes. The time T_1 has to be chosen sufficiently small to avoid blow-up of solutions, which may occur in the case when $\operatorname{Re} c < 0$ in (4.1) (see below for the longtime existence in the case $\operatorname{Re} c > 0$).

The key step for isolating the Ginzburg–Landau mode $A \in Y$ from a general function $u \in Z$ is to find an approximate inverse of the mapping $\widehat{Y}^n \ni A \mapsto U_A^1 \in \widehat{Z}^n$. We let

$$\Phi(\varepsilon, t)u = \frac{1}{\varepsilon} W_\varepsilon^{-1} E_0 \big[u \mathbf{E}_t^{-1} \big].$$

Using the counterpart of Lemma 3.4 in $H_{l,u}^n(\mathbb{R})$ (see [Sch94c]) we have

$$u = \varepsilon \big(W_\varepsilon(\Phi(\varepsilon, t)u)\mathbf{E}_t + \overline{W_\varepsilon(\Phi(\varepsilon, t)u)}\mathbf{E}_t^{-1} \big) + \mathcal{O}(\varepsilon^2).$$

The operator $\Phi(\varepsilon, t) : \widehat{Z}^n \to \widehat{Y}^n$ has a norm of order ε^{-n-1}, and a function u may be called to have mode structure when $\varepsilon \|\Phi(\varepsilon, t)u\|_{\widehat{Y}^n} \approx \|u\|_{\widehat{Z}^n}$. The following lemma shows that the linear part develops the desired mode structure for times $t \geq T_1/\varepsilon^2$ with $T_1 > 0$.

Lemma 4.5
Let $S(t) = e^{L(\partial_x)t}$, $t \geq 0$, then for each $n \in \mathbb{N}$ there is a constant such that

$$\|\Phi(\varepsilon, t)S(t)\|_{\widehat{Z}^0 \to \widehat{Y}^n} \leq \frac{C_n}{\varepsilon}\Big(1 + \frac{1}{\varepsilon^n(t^{n/4} + t^{n/2})}\Big).$$

Hence for each $T_1 > 0$ the operator $\varepsilon\Phi(\varepsilon, t)S(t) : \widehat{Z}^0 \to \widehat{Y}^n$ is bounded independently of ε for all $t \in [T_1/\varepsilon^2, \infty)$.

Proof: Letting $A = \Phi(\varepsilon, t)S(t)u$ we find that the Fourier transform of A is given by $\mathcal{F}A(K) = \widehat{m}(\varepsilon, t, K)(\mathcal{F}u)(1 + \varepsilon K)$, where

$$\widehat{m}(\varepsilon, t, K) = \widehat{\chi}(3(1 + \varepsilon K))e^{(\lambda_0(1+\varepsilon K) - \lambda_0(1))t}.$$

Here the old wave number k is replaced by $1 + \varepsilon K$ due to the shift from $\mathbf{E}_t$ and the spatial scaling. The factor $1/\varepsilon$ is compensated by transforming the Fourier integral from k to K. Hence, The operator $\Phi(\varepsilon, t)S(t)$ can be written as the composition of $M(\varepsilon, t)N(\varepsilon, t)$ where $N(\varepsilon, t)u = \frac{1}{\varepsilon}W_\varepsilon^{-1}(u\mathbf{E}_t^{-1})$ and M is defined by the multiplier associated to the function $\widehat{m}(\varepsilon, t, K)$. Using Lemma 4.3 we find

$$\|M(\varepsilon, t)\|_{\widehat{Y}^0 \to \widehat{Y}^n} \leq \|\widehat{m}(\varepsilon, t, \cdot)\|_{C^2} \leq C_n\big(1 + 1/[\varepsilon^n(t^{n/4} + t^{n/2})]\big).$$

Moreover $\|N(\varepsilon, t)\|_{\widehat{Z}^0 \to \widehat{Y}^0} \leq C/\varepsilon$, and the result is established. $\square$

From the estimate in the lemma we see that for small t the smoothing of $\Phi(\varepsilon, t)S(t)$ occurs through the fourth order operator $L(\partial_x)$ which leads to the weak singularity $t^{-n/4}$. However, this smoothing does not generate the properly scaled modes. The mode scaling is only enforced by the quadratic part of $\lambda_0(1 + \varepsilon K)$ and thus is associated to the stronger singularity $t^{-n/2}$.

4.3 Shadowing by pseudo–orbits

Finally we consider the case where the Ginzburg–Landau equation is stable, i.e., $c_r = \operatorname{Re} c > 0$ in (4.1). From Theorem 4.1 we know that the solutions of the Ginzburg–Landau equation stay inside an absorbing ball or converge exponentially to it. Hence, we may use the approximation property as well as the attractivity to control any solution of the original problem (2.1) for arbitrarily long time intervals. The idea is to make a first step to find the mode structure $u(t_1) = U^1_{A_1}(t_1) + \mathcal{O}(\varepsilon^2)$ where $A_1(\varepsilon^2 t_1) = \Phi(\varepsilon, t_1)u(t_1)$, then the solution can be described over a time interval $[t_1, t_1 + T_1/\varepsilon^2]$ by the Ginzburg–Landau approximation $U^1_{A_1}(t)$, where A_1 solves the Ginzburg–Landau equation with the above initial condition. At the end of this interval the error between the true solution and the approximation might be large; however, from the attractivity we know that the true solution $u(t_2)$ is close to a different modulated pattern $U^1_{A_2}$ from which the approximation theory may start on the next interval. The exponential attractivity of the absorbing ball controls the size of the solution A and guarantees that A stays bounded for all $T > 0$.

In this way we can shadow the true orbit $u(t) = \mathcal{S}^\varepsilon_t(u_0)$ for all $t > 0$ by a sequence of solutions of the Ginzburg–Landau equation.

Definition 4.6
Let $T_1 > 0$ and $\kappa > 0$. A function $A \in L^\infty((0, \infty), \widehat{Y}^1)$ is called a (T_1, κ)–pseudo–orbit in $\widehat{Y}^1$ for (4.1) if for all $n \in I\!N$ the relations

$$A((n-1)T_1 + \tau) = \mathcal{G}_\tau(A((n-1)T)) \text{ for all } \tau \in [0, T_1),$$
$$\|A(nT_1 - 0) - \mathcal{G}_{T_1}(A((n-1)T_1))\|_{\widehat{Y}^1} \leq \kappa$$

hold, where $\mathcal{G}_T$ is the semigroup associated with (4.1) and $A(T-0) = \lim_{\tau \nearrow T} A(\tau)$.

The following result is Theorem 3 of [Sch94c].

Theorem 4.7
(Shadowing by pseudo–orbits)
Let $\operatorname{Re} c > 0$ in (4.1); then for all $T_1 > 0$ there exist positive constants ε_0, C, and T_0 such that for all $\varepsilon \in (0, \varepsilon_0]$ the following is true:

For all initial conditions u_0 with $\|u_0\|_{\widehat{Z}^1} \leq \varepsilon$ the solution $u(t) = \mathcal{S}_t(u_0)$ exists for all time and there is a $(T_1, C\varepsilon^{1/4})$–pseudo–orbit A for (4.1) which satisfies $\|A(0)\|_Y \leq C$ and approximates $u(t)$ as follows:

$$\|u(t + T_0/\varepsilon^2) - U^1_A(t)\|_{\widehat{Z}^1} \leq C\varepsilon^{5/4} \quad \text{for all } t \geq 0.$$

This shadowing technique can be used to show that in fact all solutions starting in a small but ε–independent neighborhood $\mathcal{U}$ of 0 in $\widehat{Z}^1$ exist for all time and are finally absorbed into a ball of radius $C\varepsilon$. This result is interesting in so far as it devices a new way to construct absorbing sets, namely by studying the dynamics rather than applying energy estimates which would result in bounds of order $\varepsilon^{1/2}$, see the discussion in [MS94].

Theorem 4.8
(Locally absorbing set)
Let (4.1) be the modulation equations associated to (2.1) such that $c_r > 0$. Then there exist positive constants ε_0, δ_0, and C such that for all $\varepsilon \in (0, \varepsilon_0]$ all solutions $u(t) = S_t(u_0)$ of (2.1) with $\|u_0\|_{\widehat{Z}^1} \leq \delta_0$ exists for all $t > 0$ and satisfy $\limsup_{t \to \infty} \|u(t)\|_{\widehat{Z}^1} \leq C\varepsilon$.

4.4   Comparison of attractors

Using the weighted energy estimates which are introduced in the proof of Theorem 4.1 it is not only possible to construct absorbing balls for the semigroups $\mathcal{G}_T$ on $\widehat{Y}^1$ and $\mathcal{S}_t^\varepsilon$ on $\widehat{Z}^1$, but we are also able to find attractors in the following sense. We refer to [MS94] for the following two results.

Theorem 4.9
(Existence of an attractor)
Let $\mathcal{G}_T$ be the nonlinear semigroup of the Ginzburg–Landau equation (4.1) with $c_r > 0$ posed in $\widehat{Y}^1$. Then, there is a non–empty closed bounded set $\mathcal{A}_\mathcal{G} \subset \widehat{Y}^1$ with the following properties:

(i) $\mathcal{A}_\mathcal{G}$ attracts bounded sets in $\widehat{Y}^1$, i.e., for any bounded $B \subset \widehat{Y}^1$, we have

$$\mathrm{dist}_{\widehat{Y}^1}(\mathcal{G}_T(B), \mathcal{A}_\mathcal{G}) = \sup_{b \in B} \inf_{a \in \mathcal{A}_\mathcal{G}} \|\mathcal{G}_T(b) - a\|_{\widehat{Y}^1} \to 0 \text{ for } T \to \infty,$$

(ii) $\mathcal{A}_\mathcal{G}$ is time and translation invariant, i.e., $\mathcal{G}_T(\mathcal{A}_\mathcal{G}) = T_y \mathcal{A}_\mathcal{G} = \mathcal{A}_\mathcal{G}$ for all $T \geq 0$, $y \in \mathbb{R}$.

(iii) $\mathcal{A}$ is localized compact in the sense that $\mathcal{A}_\mathcal{G}$ is bounded in $\widehat{Y}^1$ and compact when the weighted norm $\|A\|_\rho = \left(\int_\mathbb{R} \rho(|A|^2 + |A'|^2) dX \right)^{1/2}$ is used in $\widehat{Y}^1$.

The proof of this theorem is based on ideas in [Fei93]. Note that the attractor $\mathcal{A}_\mathcal{G}$ is nontrivial as it contains all the attractors $\mathcal{A}_\mathcal{G}^\ell$ which are obtained by restricting (4.1) ti the space of functions with period ℓ.

Similarly there exists attractors $\mathcal{A}^\varepsilon$ for the semigroups $\mathcal{S}_t^\varepsilon$. The question is now as to how well the attractors $\mathcal{A}^\varepsilon$ can be described by the attractor of the limit problem which is the Ginzburg–Landau equation. A first result is given in the following theorem.

Theorem 4.10
(Comparison of attractors)
For every $\sigma > 0$ there exist $C, \varepsilon_0 > 0$ such that for all $\varepsilon \in (0, \varepsilon_0]$ the estimates

$$\mathrm{dist}_{\widehat{Y}^1}(\Phi(\varepsilon, 0)\mathcal{A}^\varepsilon, \mathcal{A}_\mathcal{G}) \leq \sigma \quad \text{and} \quad \mathrm{dist}_{\widehat{Z}^1}(E_s \mathcal{A}^\varepsilon, \{0\}) \leq C\varepsilon^{5/4},$$

hold.

This result means that $\mathcal{A}^\varepsilon$ is upper semicontinuous towards the attractor $\mathcal{A}_\mathcal{G}$ of the Ginzburg–Landau equation in the sense that

$$\mathrm{dist}_Y(\Phi(\varepsilon, 0)\mathcal{A}^\varepsilon, \mathcal{A}_\mathcal{G}) + \frac{1}{\varepsilon}\mathrm{dist}_Z(E_s \mathcal{A}^\varepsilon, \{0\}) \to 0 \quad \text{for } \varepsilon \to 0.$$

Thus, the solutions u in the attractor $\mathcal{A}^\varepsilon$ have relatively small stable parts $u_s = E_s u$ and the critical part u_c is given approximately by a Ginzburg–Landau mode $U^1_A(0)$ where A is in the limit attractor $\mathcal{A}_\mathcal{G}$. In this way we have obtained an upper bound on the richness of the attractor $\mathcal{A}^\varepsilon$. It is an unsolved problem to show the opposite direction, namely that $\mathcal{A}^\varepsilon$ is also as rich as the attractor $\mathcal{A}_\mathcal{G}$, which means a lower bound on the complexity of the attractor. In mathematical terms this means lower semicontinuity in the sense $\mathrm{dist}_Y(\mathcal{A}_\mathcal{G}, \Phi(\varepsilon, 0)\mathcal{A}^\varepsilon) \to 0$ for $\varepsilon \to 0$.

References

[Bol94] **P. Bollerman.** Validity of the Ginzburg–Landau approximation in 2–dimensional Poiseuille flow. Manuscript University of Utrecht, 1994.

[CaE87] **F. Calogero and W. Eckhaus.** Nonlinear evolution equations, rescalings, model PDEs and their integrability: I + II. *Inverse Problems,* **3** (1987) 229–262; and **4**, (1988) 11–33.

[Co94] **P. Collet.** Thermodynamic limit of the Ginzburg–Landau equations. *Nonlinearity* **7** (1994) 1175–1190.

[CoE90] **P. Collet, J.-P. Eckmann.** The time dependent amplitude equation for the Swift–Hohenberg problem. *Comm. Math. Phys.,* **132** (1990) 139-153.

[DES71] **R.C. DiPrima, W. Eckhaus, and L.A. Segel.** Non–linear wave–number interaction in near–critical two–dimensional flows. *J. Fluid Mechanics,* **49** (1971) 705–744.

[Eck93] **W. Eckhaus.** The Ginzburg–Landau equation is an attractor. *J. Nonlinear Science,* **3** (1993) 329-348.

[Fei93] **E. Feireisl.** Locally compact attractors for semilinear damped wave equations on R^n. Manuscript, Institute of Math. CSAV, Prague, 1993.

[Hen81] **D. Henry.** *Geometric Theory of Semilinear Parabolic Equations.* Lecture Notes in Mathematics Vol. **840**, Springer–Verlag 1981.

[IM91] **G. Iooss, A. Mielke.** Bifurcating time–periodic solutions of Navier–Stokes equations in infinite cylinders. *J. Nonlinear Science,* **1** (1991) 107–146.

[IMD89] **G. Iooss, A. Mielke, Y. Demay.** Theory of steady Ginzburg–Landau equation, in hydrodynamic stability problems. *Europ. J. Mech., B/Fluids,* **8**(3) (1989) 229-268.

[Kal89] **L.A. Kalyakin.** Long wave asymptotics, integrable equations as asymptotic limits of non–linear systems. *Russian Math. Surveys,* **44**(1), 3–42, 1989.

[KSM92] **P. Kirrmann, G. Schneider, A. Mielke.** The validity of modulation equations for extended systems with cubic nonlinearities. *Proc. Royal Society Edinburgh,* **122A** (1992) 85–91.

[MS94] **A. Mielke, G. Schneider.** Attractors for modulation equations on un-bounded domains — existence and comparison—. IfAM–Preprint Universität Hannover 1994. Submitted to *Nonlinearity*.

[NPL93] **A.C. Newell, T. Passot, J. Lega.** Order parameter equations for patterns. *Annual Reviews Fluid Mech.*, **25** (1993) 399–453.

[NW69] **A. Newell, J. Whitehead.** Finite bandwidth, finite amplitude convection. *J. Fluid Mech.*, **38** (1969) 279–303.

[Sch94a] **G. Schneider.** A new estimate for the Ginzburg–Landau approximation on the real axis. *J. Nonlinear Science,* **4** (1994) 23–34.

[Sch94b] **G. Schneider.** Error estimates for the Ginzburg–Landau approximation. *J. Appl. Math. Physics (ZAMP),* **45** (1994) 433–457.

[Sch94c] **G. Schneider.** Global existence via Ginzburg–Landau formalism and pseudo–orbits of Ginzburg–Landau approximations. *Comm. Math. Phys.,* **164** (1994) 157–179.

[Sch94d] **G. Schneider.** Validity and limitation of the Newell–Whitehead equation. IfAM–Preprint Universität Hannover 1994. Submitted to *Math. Nachrichten.*

[Sch94e] **G. Schneider.** Justification of modulation equations for hyperbolic systems via normal forms. IfAM–Preprint Universität Hannover 1994. Submitted to *Mathematische Zeitschrift.*

[vHa91] **A. van Harten.** On the validity of Ginzburg–Landau's equation. *J. Nonlinear Science,* **1** (1991) 397–422.

Address:
Alexander Mielke, Guido Schneider
Institut für Angewandte Mathematik
Universität Hannover
Welfengarten 1
30167 Hannover
GERMANY
E–Mail: mielke@ifam.uni–hannover.de

Section IV

Fluid Mechanics and Turbulence

Lectures in Applied Mathematics
Volume **31**, 1996

Navier-Stokes Equations and Incompressible Fluid Turbulence

PETER CONSTANTIN

1. Introduction

In this talk I will discuss informally mathematical questions related to turbulence. The equations of motion of incompressible fluids are described as follows: Let $x \in \mathbf{R}^n$ denote a point in space, $n = 2$ or 3 and let $t \geq 0$ denote time. One associates to the velocity $u(x, t) \in \mathbf{R}^n$ a first order differential operator, D_t:

$$D_t = \partial_t + u(x, t) \cdot \nabla.$$

D_t is the so-called material derivative; its characteristics – solutions of the ODE

$$\frac{dX}{dt} = u(X, t)$$

are called particle trajectories. The Navier-Stokes equations are

$$D_t u + \nabla p = \nu \Delta u + f,$$

$$\nabla \cdot u = 0.$$

The number $\nu > 0$ is the kinematic viscosity of the fluid. If one sets $\nu = 0$ in the equation above one obtains the Euler equations of ideal fluids. The scalar function $p(x, t)$ represents pressure; its mathematical role is to maintain the constraint of incompressibility $\nabla \cdot u = 0$. The functions f represent body forces. The fluid occupies a region $G \subset \mathbf{R}^n$ and appropriate boundary conditions are prescribed. A great variety of physical situations are described by the nature of body forces and boundary conditions. Two examples, Taylor-Couette and Rayleigh-Benard turbulence can serve as a guide to posing questions regarding turbulence. In the Taylor – Couette setting fluid is placed in the space between two concentric vertical cylinders and one of the cylinders is rotated, entraining the fluid. In the Rayleigh – Benard setting fluid is placed in a closed container

---

1991 *Mathematics Subject Classification*. Primary 76D05, 76F10.

partially supported by NSF/DMS 9207080 and DOE/DE-FG02-92ER-25119

and heated from below. The external conditions are encoded in non-dimensional parameters: Reynolds number, Rayleigh number. They represent a measure of the strength of the externally supplied energy (determined for instance by how fast the cylinders are rotated or how much the fluid is heated). The term non-dimensional refers to scale invariance. The experimenters prepare an experiment at given values of the control parameters. The experiment ([1], [2]) is allowed to run for a sufficiently long time in order to make sure that one registers the time asymptotic regime. The information (one point measurements or rarely two point measurements) is recorded and processed, taking time averages. New values of the control parameters are selected and the process is repeated. As the non-dimensional control parameters are increased the behavior of the fluid changes, very roughly speaking from simple to complex. At the high end of the parameter scale we find incompressible turbulence. The appropriate mathematical turbulence problem is thus: study the long time behavior of solutions to Navier-Stokes equations at a fixed Reynolds number (fixed $\nu > 0$, boundary conditions and body forces f). Record it using appropriate measures or averages. Then increase the Reynolds number (for instance, decrease viscosity, keeping all other conditions the same) and study the large Reynolds number limit.

2. Dynamical Systems Paradigm

The finite dimensional dynamical system paradigm ([3]) has predictions ([4]) regarding the transition to turbulence that have been verified experimentally. Consider the case of two dimensional, spatially periodic Navier-Stokes equations with time independent forces. At a fixed Reynolds number the Navier-Stokes equations are solved by a nonlinear semigroup $S(t)$ in a Hilbert space H, the space of divergence-free square integrable velocities. The norm $|\cdot|_H$ is the square root of the total kinetic energy and $S(t)u_0$ represents the velocity at time t. The semigroup is dissipative: there exists a compact set in H such that all trajectories $S(t)u_0$ belong to it for large enough t. The semigroup has a global attractor $\mathcal{A}$: a compact invariant $(S(t)(\mathcal{A}) = \mathcal{A})$ set that contains all omega limit sets. All Borel measures which are invariant under $S(t)$ are supported in $\mathcal{A}$. This set has finite fractal dimension ([5]). The dimension is related to the Reynolds number ([6]) as predicted by the argument of Landau ([7]) concerning the number of degrees of freedom of turbulence. The existence of a finite dimensional attractor and the time analyticity of solutions are used in results that are valid beyond the transition to turbulence. Here are two examples.

Consider the set of global solutions

$$\mathcal{G} = \{u_0 \in H;\; S(t)u_0 \text{ extends to } t \in \mathbf{R}\} \,.$$

The global attractor can be described by

$$\mathcal{A} = \{u_0 \in \mathcal{G};\; S(t)u_0 \text{ is bounded for } t \in \mathbf{R}\} \,.$$

Consider now the set of those global solutions that grow at most exponentially backward in time

$$\mathcal{M} = \left\{ u_0 \in \mathcal{G}; \lim_{t \to -\infty} \sup \frac{\log |S(t)u_0|_H}{|t|} < \infty \right\}.$$

One can prove ([8]) that $\mathcal{M}$ is weakly dense in H. The dynamics on $\mathcal{M} \backslash \mathcal{A}$ can be related to the inviscid Eulerian dynamics: rescaling appropriately velocity and time, one obtains the Euler equations as the infinite negative time limit. Thus, for velocities selected from the rich invariant set $\mathcal{M} \backslash \mathcal{A}$, quasi-Eulerian dynamics describe the past and finite dimensional dynamics on the Navier-Stokes attractor $\mathcal{A}$, the future.

One of the most common ways to analyze temporal data, such as the temperature $\theta(t)$ recorded at a given location in the Rayleigh-Benard convection experiment is to compute the power spectrum,

$$P(\omega) = \lim_{T \to \infty} \frac{1}{T} \left| \int_0^T e^{-i\omega t} \theta(t) dt \right|^2.$$

The mathematical justification of this operation is based on classical work on stochastic signals, going back to Wiener, Khintchin and Kolmogorov. At finite Rayleigh and Reynolds numbers though, the signal is not a random process, rather it is the output of a finite dimensional dynamical system. Using the properties of solutions of Navier-Stokes equations one can still make mathematical sense of the procedure. Moreover one can prove that the power spectrum must decay at least exponentially at high frequencies ([9]). Specifically, the power spectrum of the temporal data obtained by evaluating a solution at some fixed location is a well defined positive Borel measure $P(d\omega)$ and satisfies

$$\int_{-\infty}^{\infty} e^{\tau|\omega|} P(d\omega) < \infty.$$

The positive constant $\tau = \tau(\mathcal{A})$ depends only on the attractor, in other words only on the control parameters.

The dynamical systems paradigm is applicable, under the assumption of regularity, to the three dimensional Navier-Stokes equations in a bounded domain, at fixed Reynolds number. Even if one is willing to accept this assumption (a reasonable position, because finite time singularities in the Navier-Stokes equations are not physical), the finite dimensional dynamical system picture is incomplete: there are no (known) inertial manifolds. A finite dimensional Lipschitz manifold $\mathcal{I}$ in the phase space H is an inertial manifold if it is forward invariant ($S(t)\mathcal{I} \subset \mathcal{I}$) and attracts exponentially all trajectories, $dist_H(S(t)u_0, \mathcal{I}) \leq C \exp(-k_\mathcal{I} t)$. Such manifolds are known to exist for dissipative systems other than the Navier-Stokes one ([10], [11]), but not for Navier-Stokes.

3. Dissipation

One of the most common, imporant and accessible effects of turbulence is bulk dissipation. Torque in Taylor-Couette flow, drag in flow past an obstacle, and heat transfer in Rayleigh-Benard convection are examples of bulk dissipation quantities. These are one-point quantities (that is, numbers, not functions) that depend on the key parameters (Reynolds number, Rayleigh number) in a reproducible manner: they obey empirical laws obtained in physical experiments ([1], [2]).

I will describe a method ([12], [13], [14]) to estimate rigorously such quantities directly from the equations of motion. I will use a simplified Taylor-Couette setting to illustrate this method. Let us consider the incompressible Navier-Stokes equations

$$(1) \qquad\qquad \partial_t u + u \cdot \nabla u + \nabla p = \Delta u$$

$$\nabla \cdot u = 0$$

in a 2-D strip, with boundary conditions

$$u(x + \ell, y, t) = u(x, y, t)$$

$$u(x, 0, t) = 0; \qquad u(x, 1, t) = \mathbf{Re}\,\hat{x}$$

The Reynolds number $\mathbf{Re}$ is a large positive number, $\hat{x}$ is the direction of the x axis and $\ell > 0$ represents the aspect ratio. Consider

$$< \|u\|^2 >= \mathrm{limsup}_{T\to\infty} \frac{1}{T} \int_0^T \|u(t)\|^2 dt$$

where $\|u\|^2 = D(u)$ is the bulk dissipation:

$$\|u\|^2 = \int_0^\ell \int_0^1 |\nabla u|^2 \, dx dy.$$

The problem is to estimate $< \|u\|^2 >$ on solutions as a function of $\mathbf{Re}$ as $\mathbf{Re} \to \infty$.

The inhomogeneous boundary conditions together with the finiteness of the kinetic energy and of the dissipation determine an affine set $\mathcal{U}$ in function space. We consider any time independent function $b \in \mathcal{U}$ ("the background") and associate to it a corresponding linear operator $\mathcal{L}_b$. In the Taylor-Couette case this operator is computed as follows:

$$\mathcal{L}_b = A + 2\mathcal{S}_b$$

where

$$\mathcal{S}_b v = \mathbf{P}\left(\frac{((\nabla b) + (\nabla b)^*)}{2} v \right),$$

$\mathbf{P}$ is the Leray-Hodge projector on divergence-free vectors in L^2 and A the Stokes operator $\mathbf{P}(-\Delta)$. Let $\lambda(b)$ denote the bottom of the spectrum of $\mathcal{L}_b$ in L^2.

THEOREM 1. *Assume that $b \in \mathcal{U}$ satisfies $\mathbf{P}(b \cdot \nabla b) = 0$ and $\lambda(b) \geq 0$. Then, every solution $u(t)$ in $\mathcal{U}$ satisfies*

$$< \|u(t)\|^2 > \leq \|b\|^2.$$

The interpretation of this result is the following. Take any steady solution of the inviscid (Euler) equation $\mathbf{P}(b \cdot \nabla b) = 0$ with the correct boundary conditions ($b \in \mathcal{U}$) and compute its quadratic form stability as if it were a steady solution of the Navier-Stokes equation with half the given viscosity ($\frac{1}{2}$ in our non-dimensional setting; this is related to the fact that the nonlinearity is quadratic). If the solution is stable ($\lambda(b) \geq 0$) then its bulk dissipation is an upper bound for the long time average bulk dissipation of *any* solution of the viscous problem.

Examples of such functions are obtained by choosing flat shear flow backgrounds with sharp boundary layers. Choosing the size of the boundary layer of the order of $\mathbf{Re}^{-1}$ ensures that $\lambda(b) > 0$. The requirement that the background must satisfy the steady Euler equations can be dropped; then the bulk dissipation is bounded by a more complicated expression ([13]).

Using Theorem 1 one can prove that

$$< \|u(t)\|^2 > \leq C\mathbf{Re}^3.$$

Moreover, C can be estimated explicitly and the results agree with the physical experiment ([2], [12], [13], [14]). In order to improve the estimate one is naturally lead to variational problems with spectral side conditions. A set of backgrounds $\mathcal{X}$ is chosen, such that if $b \in \mathcal{X}$ then $\mathbf{P}(b \cdot \nabla b) = 0$. The set of shear backgrounds

$$\mathcal{X} = \left\{ b \in \mathcal{U}; b = \begin{pmatrix} \mathbf{Re}\psi(y) \\ 0 \end{pmatrix}, \psi(0) = 0, \psi(1) = 1 \right\}$$

is a natural and simple example. In $\mathcal{X}$ one considers the subset

$$\mathcal{C}_\mathcal{X} = \{b \in \mathcal{X}; \lambda(b) > 0\}.$$

The minimization problem is to compute

$$\inf_{b \in \mathcal{C}_\mathcal{X}} \|b\|^2.$$

This problem leads to new, nonlinear Orr-Sommerfeld-like equations ([13]). Bifurcations in the boundary of the sets $\mathcal{C}_\mathcal{X}$ as the control parameters are varied can be responsible for sudden transitions in the scaling law of the bulk quantity. This can happen as a consequence of the finite size (discrete spectrum) of the domain. If the condition of divergence-free is dropped, the problem simplifies considerably. Then there are no sudden transitions and the constrained Euler-Lagrange equations lead to a nonlinear eigenvalue problem for a cubic steady one dimensional Schrodinger equation. This can be solved explicitly, at each Reynolds number and the asymptotic behavior, as the Reynolds number is increased can be computed explicitly. The outcome is an improvement in the prefactor C but the power law remains the same. This law is consistent with the Kolmogorov dissipation law (see next section) which, however was not envisioned to apply in such boundary-dominated situations. The fundamental question remains: does the presence of boundaries alter the Kolmogorov dissipation law? In particular, does the logarithmic wall friction law rigorously represent the infinite Reynolds number asymptotics? All our preliminary results, including in addition to Taylor-Couette turbulence, Rayleigh-Benard convection and channel flow are consistent with the asymptotic predictions of the Kolmogorov dissipation law.

4. Stochastic PDE Paradigm

The idea to describe turbulence in terms of statistical rather than deteministic solutions is not new [15]. Contemporary theories ([16], [17]) use tools of statistical physics, but the contact with the Navier-Stokes equations is not complete. The mathematical existence theory for statistical solutions of the Navier-Stokes equations ([18], [19], [20]) is developed to a similar extent as is the existence theory for deterministic solutions. The temporal power spectrum of wind tunnel data provides strong experimental verification of one of the beacons in the subject of turbulence, the Kolmogorov theory ([21]). Grounded in dimensional analysis, this theory proposes the existence of universal scaling behavior in fully developed turbulence. According to the Kolmogorov dissipation law, the energy dissipation rate

$$\epsilon = \nu \left\langle |\nabla u(x,t)|^2 \right\rangle$$

is a positive constant that is bounded independently of viscosity. The braces $< \cdots >$ represent ensemble average (functional integration). Kolmogorov assumed homogeneity and isotropy (invariance with respect to translations and rotations of the underlying probability distributions). In addition, the Kolmogorov theory assumes that there exists an interval, the so-called inertial range, which extends from a small length (the Kolmogorov dissipation scale η), to a large one (the integral scale ρ) such that, for $r \in [\eta, \rho]$, the variation

$$s(r) = \left\langle |u(x+y) - u(x)|^2 \right\rangle^{\frac{1}{2}}$$

of velocity across a distance $r = |y|$ can be determined from dimensional analysis:

$$s(r) \sim (\epsilon r)^{\frac{1}{3}}.$$

This is the Kolmogorov two-thirds law (two-thirds because one has $s^2 \sim (\epsilon r)^{\frac{2}{3}}$). It is one of the most important predictions of this theory. It implies that the energy spectrum behaves approximately like a $-\frac{5}{3}$ power in a range of wave numbers. There exists experimental numerical and theoretical evidence that seems to indicate the possibility of small departures from simple scaling. More precisely, the equal time generalized structure functions (see below) scale with distance like powers

$$< |u(x+y,t) - u(x,t)|^m >^{\frac{1}{m}} \sim U \left(\frac{|y|}{L} \right)^{\zeta_m}$$

where the value of the exponents ζ_m is close to $\frac{1}{3}$ but depends on m. The ζ_m must be nonincreasing in m because of the Hölder inequality.

I will describe here the results of ([**14**], [**22**]) regarding the scaling of velocity structure functions for the Navier-Stokes equations. We consider ensembles of solutions of the Navier-Stokes equations in the whole 3-D space. We assume that there exist uniform bounds for the velocities in the ensembles

$$\sup_{x,t} |u(x,t)| \leq U \quad (*)$$

This assumption implies regularity of the solutions. We will consider driving body forces f which are bounded uniformly.

$$\sup_{x,t} |f(x,t)| \leq B \quad (**)$$

The forces are deterministic. We assume also:

$$\int_{|y| \leq \rho} s_1^{(\rho)}(y) \frac{dy}{|y|^3} \leq cU \quad (***)$$

where $s_1^{(\rho)}$ is defined below.

These are the standing assumptions. There are no assumptions of homogeneity or isotropy.

We define the averaging procedure M_ρ by

$$M_\rho(h(x,t)) = \text{AV} \sup_{x_0 \in R^3} \lim_{T \to \infty} \sup \frac{1}{T} \frac{3}{(4\pi\rho^3)} \int_0^T \int_{B_\rho(x_0)} h(x,t) dx dt$$

AV means ensemble average and $\sup_{x_0}$ is a supremum over all Euclidean balls B_ρ with center x_0 and radius ρ.

We set

$$\epsilon_{(\rho)} = \nu M_{\frac{\rho}{2}} \left(|\nabla u(x,t)|^2 \right),$$

and

$$s_m^{(\rho)}(y) = [M_\rho \left(|u(x+y,t) - u(x,t)|^m \right)]^{\frac{1}{m}}.$$

We denote

$$\mathbf{Re} = \frac{U\rho}{\nu}.$$

If

$$cU \left(\frac{|y|}{\rho} \right)^{\zeta_m} \leq s_m^{(\rho)}(y) \leq CU \left(\frac{|y|}{\rho} \right)^{\zeta_m}$$

holds on an interval

$$\mathbf{Re}^{-\frac{1}{1+\zeta m}} \leq \frac{|y|}{\rho} \leq 1$$

then we say that we have m-scaling. This definition implies that the local Reynolds number $\frac{|y| s_m^{(\rho)}(y)}{\nu}$ ranges between c and $C\mathbf{Re}$ in the scaling region.

Here are the main results.

THEOREM 2. *Assume* $(*), (**), (* * *)$. *Then*

$$\epsilon_{(\rho)} \leq C \left(\frac{U^3}{\rho} + BU + \nu \frac{U^2}{\rho^2} \right)$$

Note that the dissipation is bounded uniformly as $\nu \to 0$. Thus, the Kolmogorov dissipation law is true as an upper bound. A lower bound is much more difficult to obtain; as far as I know it is still an open problem.

We show that structure functions for the pressure are bounded in terms of structure functions for the velocity. Once the pressure is controlled then the result follows via local energy inequalities. We relate the second structure function to the first:

THEOREM 3. *Assume* $(*), (**), (* * *)$. *Then*

$$s_2^{(\rho)}(y) \leq cU \frac{|y|}{\rho} Re^{\frac{1}{2}}$$

In particular, if 1 and 2-scaling occur then

$$\zeta_1 \geq \zeta_2 \geq \frac{1}{3}.$$

This is seen as a result deduced in the neighborhood of the dissipation scale. Without the postulate that scaling persists down to lengths where the corresponding local Reynolds number is order one we can not assert a direct relation between ζ_1 and ζ_2.

It is widely believed that $\zeta_3 = \frac{1}{3}$. The evidence is both numerical and theoretical. The traditional theoretical arguments are based on the assumption of homogeneity and isotropy ([15]). A conjecture of Onsager corresponds to the statement that $\zeta_3 > \frac{1}{3}$ implies $\epsilon = 0$; a mathematical formulation and a proof of this conjecture ([23]) offer additional arguments in support of $\zeta_3 = \frac{1}{3}$. Here is a brief description of the result in ([23]). Denote first

$$u_\epsilon = \int \phi_\epsilon(x - y) u(y) dy$$

a standard mollification of u. Then denote by

$$r_\epsilon(u, u)(x) = \int \phi_\epsilon(y) \left(\delta_y u(x) \otimes \delta_y u(x) \right) dy$$

where

$$\delta_y u(x) = u(x - y) - u(x).$$

The following identity holds pointwise:

$$(u \otimes u)_\epsilon = u_\epsilon \otimes u_\epsilon + r_\epsilon(u, u) - (u - u_\epsilon) \otimes (u - u_\epsilon).$$

THEOREM 4. *If u is a weak solution of Euler's equation which belongs to the space $L^3((0, T); \mathcal{B}_3^{\alpha, \infty})$ for some $\alpha > \frac{1}{3}$ then*

$$\|u(\cdot, t)\|_{L^2} = \|u(\cdot, 0)\|_{L^2}$$

for $0 \leq t \leq T$.

The proof uses the well-known facts about Besov spaces:

$$\|u - u_\epsilon\|_{L^3} \leq C\epsilon^\alpha \|u\|_{\mathcal{B}_3^{\alpha, \infty}},$$

$$\|\delta_y u\|_{L^3} \leq C|y|^\alpha \|u\|_{\mathcal{B}_3^{\alpha, \infty}},$$

$$\|\frac{d}{d\epsilon} u_\epsilon\|_{L^3} \leq C\epsilon^{\alpha-1} \|u\|_{\mathcal{B}_3^{\alpha, \infty}},$$

and

$$\|\nabla u_\epsilon\|_{L^3} \leq C\epsilon^{\alpha-1} \|u\|_{\mathcal{B}_3^{\alpha, \infty}}.$$

It follows that

$$\|r_\epsilon(u, u)\|_{L^{\frac{3}{2}}} \leq C\epsilon^{2\alpha} \|u\|_{\mathcal{B}_3^{\alpha, \infty}}^2.$$

The proof follows from the identity and from

$$\frac{1}{2} \frac{d}{dt} \|u_\epsilon\|_{L^2}^2 = \int \mathrm{Tr}\left((u \otimes u)_\epsilon (\nabla u_\epsilon)\right) dx$$

by the ususal incompressible cancellation and by straightforward Holder inequalities. One has to mollify in time but that is not a problem.

Corrections to the Kolmogorov-Obukhov exponent of $\zeta = \frac{1}{3}$ are referred to by the name of intermittency. They are believed to be connected to the existence of statistically significant large variations in the velocity gradients, over small regions in physical space. We have some mathematical evidence for this connection:

THEOREM 5. *Assume $(*), (**), (* * *)$, and 4-scaling. If*

$$\nu M_{\frac{\rho}{4}} \left(|\nabla u(x + y, t) - \nabla u(x, t)|^2\right) \geq c\epsilon_{(\frac{\rho}{2})} > 0$$

for some y satisfying

$$\frac{|y|}{\rho} \leq C(Re)^{-\beta}$$

then

$$\zeta_4 \leq \frac{1}{4\beta}.$$

Thus, if the gradient can be decorrelated by translations over very small distances then intermittency sets in. Note that $\beta = \frac{3}{4}$ is the value of the exponent of the Kolmogorov length. If $\zeta_3 = \frac{1}{3}$ and if ζ_4 makes sense, then it must be equal or less than $\frac{1}{3}$. If $\beta = \frac{3}{4}$ then our result is consistent with $\zeta_4 = \frac{1}{3}$, i.e, with the K41 theory. If $\beta > \frac{3}{4}$ however, then the preceding result gives sufficient and testable conditions for intermittency corrections.

5. The inviscid limit

Here we consider the inviscid limit, that is we fix time and let the control parameter tend to infinity. Let us focus on the Navier-Stokes equation. The first question is: what are the limiting equations? In realistic closed systems where the boundary effects are important, unstable boundary layers drive the system: the limit is not well understood. In the case of no boundaries (periodic solutions or solutions decaying at infinity) the issue becomes one of smoothness and rates of convergence. Indeed in $n = 2$, if the initial data are very smooth then the limit is the Euler equation and the difference between Navier-Stokes solutions and corresponding Euler solutions is optimally small ($O(\nu)$). However, if the initial data are not that smooth, for instance in the case of vortex patches, then the situation changes. Vortex patches are solutions whose vorticity (antisymmetric part of the gradient) is a step function. They are the building blocks for the phase space of an important statistical theory ([24], [25]). When one leaves the realm of smooth initial data the inviscid limit becomes more complicated: internal transition layers form because the smoothing effect present in the Navier-Stokes solution is absent in the Eulerian solution. In the case of vortex patches with smooth boundaries, the inviscid limit is still the Euler equations, but there is a definite price to pay for rougher data: the difference between solutions (in L^2) is only $O(\sqrt{\nu})$ ([26]). This drop in rate of convergence actually occurs – there exist exact solutions providing lower bounds. The question of the inviscid limit for the whole phase space of the statistical theory of [24] and [25] is open. If the initial data are more singular then even the classical notion of weak solutions for the Euler equations might need revision ([27]) except when the vorticity is of one sign ([28]).

In the case of three spatial dimensions and smooth initial data, the inviscid limit is the Euler equation as long as the corresponding solution to the Euler equation is smooth ([29],[30]) This might be a true limitation, because of the possibility of finite time blow-up. The blow-up problem for the Euler ($\nu = 0$) equations is the following: do smooth data (for instance $f = 0$, and smooth, rapidly decaying initial velocity) guarantee smooth solutions for all time? The answer is known to be yes only for $n = 2$, not known for $n = 3$.

The three dimensional incompressible Euler equations are equivalent to the requirement that two vector fields Ω and D_t commute:

$$[D_t, \Omega] = 0.$$

Both vector fields are associated to divergence free vectors. The first one, D_t is associated to the velocity field $u = u(x, t)$:

$$D_t = \partial_t + u(x, t) \cdot \nabla$$

The second vector field, Ω, is associated to the vorticity $\omega = \omega(x, t)$:

$$\Omega = \omega(x, t) \cdot \nabla.$$

The integral curves of Ω are vortex lines. The commutation relation is equivalent to the stretching equation

$$D_t \omega = \omega \cdot \nabla u.$$

Many incompressible inviscid hydrodynamical models are obtained from the commutation relation by specifying a coupling relation between the coefficients u and ω of D_t and Ω. In many cases this coupling has the form

$$u = \mathcal{K} * \omega.$$

The 3D Euler coupling kernel $\mathcal{K}$ is given by the Biot-Savart law

$$\mathcal{K}_{ij}^{(\text{Euler})}(y) = \epsilon_{ijk} \frac{\hat{y}_k}{|y|^2}$$

where $\hat{y} = \frac{y}{|y|}$, ϵ_{ijk} is the signature of the permutation $(1, 2, 3) \mapsto (i, j, k)$ and repeated indices are summed. Note that the singularity at the origin is of the order $2 = n - 1$. A natural analogue of the three dimensional Euler coupling in two dimensions is ([14], [33], [32])

$$\mathcal{K}_{ij}^{(\text{QGASE})}(y) = \delta_{ij} \frac{1}{|y|}$$

This defines the quasi-geostrophic active scalar equation (QGASE). Its coupling kernel has the analogous singularity strength $1 = n-1$. The QGASE is physically significant in its own right: it is a model for temperature in a quasi-geostrophic (Coriolis forces balance pressure gradients) approximation of atmospheric flow. We will discuss simultaneously the 3D Euler equations and QGASE. The notation is purposely ambivalent. Because the QGASE systems are 2D they can be described by the time evolution of scalars, functions $\theta(x, t)$ that obey

$$(\partial_t + u \cdot \nabla) \theta = 0$$

where u is obtained from ω as above and θ is related to ω by

$$\omega = \nabla^\perp \theta.$$

($\nabla^\perp$ denotes the gradient rotated counter-clockwise by 90 degrees). The integral lines of Ω are material i.e., they are carried by the flow. The QGASE analogues of vortex lines are iso-θ lines. Their length element is $|\omega|$. They stretch according to

$$D_t|\omega| = \alpha|\omega|.$$

It is well-known ([31] for the 3D incompressible Euler equations, but the result holds for QGASE as well) that

$$\int_0^T \|\omega(\cdot, t)\|_{L^\infty(dx)} dt = \infty$$

is a necessary and sufficient condition for blow-up.

It is also easy to prove that the singularities cannot occur without small scales developing ([14]) in ω. By that I mean that that large spatial gradients of its magnitude must develop, at a fast enough rate. This is related, perhaps, to intermittency (see previous section).

The stretching rate alpha is given by

$$\alpha(x) = (\nabla u(x)) \xi(x) \cdot \xi(x)$$

and the direction field ξ by

$$\xi(x) = \frac{\omega(x)}{|\omega(x)|}.$$

The region $\{x : |\omega(x)| > 0\}$ is material. Both α and ξ are defined in it. The stretching rate α has a remarkable integral representation

$$\alpha(x) = \text{P. V.} \int D\left(\hat{y}, \xi(x+y), \xi(x)\right) |\omega(x+y)| \frac{dy}{|y|^n}.$$

Here n is the dimension of space (3 for Euler equations, 2 for QGASE) and

$$D^{(\text{Euler})}\left(\hat{y}, \xi(x+y), \xi(x)\right) = ((\hat{y} \cdot \xi(x)) \text{Det}\left(\hat{y}, \xi(x+y), \xi(x)\right))$$

and

$$D^{(\text{QGASE})}\left(\hat{y}, \xi(x+y), \xi(x)\right) = ((\hat{y} \cdot \xi^\perp(x))(\xi(x+y) \cdot \xi^\perp(x))).$$

Note that the geometric factors D vanish not only in the spherical average, but also if the vectors $\xi(x+y)$ and $\xi(x)$ are parallel or anti-parallel. More precisely, if $\cos\phi = \xi(x) \cdot \xi(x+y)$ then $|D| \leq |\sin\phi|$. The geometric factors vanish also if $\hat{y}$ is parallel to $\xi(x)$.

Based on these properties we prove ([32], [36]) that if the direction field ξ is smooth in regions of high $|\omega|$ then blow up does not occur.

Let u be the velocity of either a three dimensional incompressible Euler solution with smooth and localized initial data or of a two dimensional QGASE solution with smooth and localized initial data. Assume that the corresponding solution is smooth on a time interval $0 \leq t < T$. The velocity field $u = u(x, t)$ is used to define particle trajectories $X(q, t)$, solutions of

$$\frac{dX}{dt} = u(X, t),$$

$$X(q, 0) = q.$$

If W_0 is a set we denote by W_t its image at time t

$$W_t = X(t, W_0)$$

$B_r(W)$ denotes the neighborhood of W formed with points situated at Euclidean distance not larger than r from W. We say that a set W_0 is *smoothly directed* if there exists $\rho > 0$ and r, $0 < r \le \frac{\rho}{2}$ such that the following three conditions are satisfied. First, for every $q \in W_0^*$,

$$W_0^* = \{q \in W_0; |\omega_0(q)| \ne 0\}.$$

and all $t \in [0, T)$, the function $\xi(\cdot, t)$ has a Lipschitz extension (denoted by the same letter) to the Euclidean ball of radius 4ρ centered at $X(q, t)$ and

$$M = \lim_{t \to T} \sup_{q \in W_0^*} \int_0^t \|\nabla \xi(\cdot, t)\|^2_{L^\infty(B_{4\rho}(X(q, t)))} dt < \infty.$$

Secondly,

$$\sup_{B_{3r}(W_t)} |\omega(x, t)| \le m \sup_{B_r(W_t)} |\omega(x, t)|$$

holds for all $t \in [0, T)$ with $m \ge 0$ constant. And finally,

$$\sup_{B_{4\rho}(W_t)} |u(x, t)| \le U$$

holds for all $t \in [0, T)$.

The first assumption means that the direction field ξ is well behaved in a fixed neighborhood of a bunch of trajectories. The second assumption states that this neighborhood is large enough to capture the local intensification of $|\omega|$. In other words, the competing and significantly stronger blow up trajectories, if they exist, do not come near this bunch. Our regularity result is

THEOREM 6. *Assume W_0 is smoothly directed. Then there exists $\tau > 0$ and Γ such that*

$$\sup_{B_r(W_t)} |\omega(x, t)| \le \Gamma \sup_{B_\rho(W_{t_0})} |\omega(x, t_0)|$$

holds for any $0 \le t_0 < T$ and $0 \le t - t_0 \le \tau$.

The numerical evidence for QGASE ([32]) supports strongly two statements: first that sharp fronts do form in finite time (sharp means large gradients of θ, i.e. large $|\omega|$; based on the present computations, blow up cannot be predicted). Secondly that there is a marked difference in the rate of development of these sharp fronts caused by the nature of ξ. There exist initial data for which ξ develops only antiparallel, regularly directed singularities. The formation of fronts is then depleted. For other initial data, a saddle point in θ provides a Lipschitz singularity in the ξ direction field. This is the source of much more intense gradient growth. Qualitatively similar numerical results were obtained also for the Euler equations ([37]).

Now I will address briefly the role of viscosity. In the three dimensional incompressible Navier-Stokes equations there exist suitable weak solutions (this is a technical term) that satisfy

$$< |\omega(x,t)||\nabla\xi(x,t)|^2 > \leq \frac{\Gamma}{\nu}$$

where $\nu > 0$ is the viscosity, Γ is given in terms of the initial data and $< \cdots >$ is an appropriate space and time average ([35], [34]). Consequently, typical regions of high vorticity have Lipschitz ξ.

Moreover, if one assumes

Assumption (A)

There exist constants $\Omega > 0$ and $\rho > 0$ such that

$$|P^{\perp}_{\xi(x,t)}(\xi(x+y,t))| \leq \frac{|y|}{\rho}$$

holds if both $|\omega(x,t)| > \Omega$ and $|\omega(x+y,t)| > \Omega$, and $0 \leq t \leq T$.

$(P^{\perp}_{\xi(x)}\xi(x+y)$ is the projection of $\xi(x+y)$ orthogonal to $\xi(x))$, then ([34]):

THEOREM 7. *Under the assumption (A) the solution of the initial value problem for the Navier–Stokes equation is smooth (C^{∞}) on the time interval $[0, T]$.*

Based on these theoretical considerations and on the numerical evidence on active scalars ([32]) and on Euler equations ([37]) we speculate that blow up occurs in the Euler equations and that this event occurs in the neighborhood of a singular object in physical space (such as a vortex tube pair with a conical singularity). We continue to speculate that for the Navier-Stokes equation these events are not blow-up events but trigger topological change and large dissipation events; their presence is felt at the dissipation scales and is perhaps the source of small scale intermittency.

REFERENCES

1. B. Castaing, G. Gunaratne, F. Heslot, L.P. Kadanoff, A. Libchaber, S. Thomae, X.-Z. Wu, S. Zaleski, G. Zanetti, Scaling of hard thermal turbulence in Rayleigh-Benard convection, J. Fluid Mech. **204**, 1-23 (1989).
2. D. P. Lathrop, J. Fineberg, H. L. Swinney, Turbulent flow between concentric rotating cylinders at large Reynolds number, Phys. Rev. Lett **68**, 1515-1518 (1992).
3. D. Ruelle, F. Takens, On the nature of turbulence, Commun. Math. Phys. **20**, 167-192 (1971).
4. M.J. Feigenbaum, Quantitative universality for a class of nonlinear transformations, J. Stat. Phys **19**, 25-52 (1978).
5. C. Foias, R. Temam, Some analytic and geometric properties of the solutions of the evolution Navier-Stokes equations, J. Math. Pures Appl. **58**, 339-368 (1979).
6. P. Constantin, C. Foias, R. Temam, Attractors representing turbulent flows, Memoir AMS **53**, 314 (1985).
7. L.D. Landau, E.M. Lifschitz, Fluid Mechanics, Pergamon Press (1959).
8. P. Constantin, C. Foias, I. Kukavica, A. Majda, Dirichlet quotients and periodic 2D Navier-Stokes equations, J. Math. Pures Appl., to appear.
9. H. Bercovici, P. Constantin, C. Foias, O. Manley, Exponential decay of the power spectrum of turbulence, J. Stat. Phys., to appear.

10. C. Foias, G. Sell, R. Temam, Inertial manifolds for nonlinear evolutionary equations, J. Diff. Eq. **73**, 309-353 (1988).

11. P. Constantin, C. Foias, B. Nicolaenko, R. Temam, Integral manifolds and inertial manifolds for dissipative partial differential equations, Appl. Math. Sciences, Springer Verlag (1989).

12. Ch. R. Doering and P. Constantin, Energy dissipation in shear driven turbulence, Phys. Rev. Lett **69**, 1648 - 1651, (1992).

13. P. Constantin, C.R. Doering, Variational Bounds on Energy Dissipation in Incompressible Flows, I Shear Flow, Physical Review E **49**, 4087-4099 (1994).

14. P. Constantin, Geometric Statistics in Turbulence, SIAM Review, **36**, 73-98 (1994).

15. A.S. Monin, A.M. Yaglom, Statistical Fluid Mechanics, M.I.T. Press, ISBN 0262130629 (1987).

16. R.H. Kraichnan, Some modern developments in the statistical theory of turbulence. In: Statistical mechanics, S. Rice, K. Freed, J. Light, (eds), Chicago University Press, 1971.

17. A. Chorin, Scaling laws in the lattice vortex model of turbulence, Commun. Math. Phys., **114**, 167 - 176 (1988); Spectrum, dimension and polymer analogies in fluid turbulence, Phys. Rev. Lett. **60**, 1947 - 1949, (1988).

18. E. Hopf, Statistical hydrodynamics and functional calculus, J. Rat. Mech. Anal.,**1**, 87 - 123, (1952).

19. C. Foias, Statistical study of Navier-Stokes equations I and II, Rend. Sem. Mat. Univ. Padova, **48**, 219 - 348, (1972) and **49**, 9 - 123, (1973).

20. M.J. Vishik, A.V. Fursikov, Solutions statistiques homogenes des systemes differentiels paraboliques et du systeme de Navier-Stokes, Ann. Sc. Norm. Sup. Pisa Sci **4**, 531 - 576, (1977); and Homogeneous in x space-time statistical solutions of the Navier-Stokes equations and individual solutions with infinite energy, Dokl. Akad. Nauk. SSR **239**, 1025 - 1032 (1978).

21. A. N. Kolmogorov, Local structure of turbulence in an incompressible fluid at very high Reynolds numbers, Dokl. Akad. Nauk. SSSR **30**, 299-303 (1941).

22. P. Constantin, Ch. Fefferman, Scaling exponents in fluid turbulence: some analytic results, Nonlinearity, **7**, 41-57 (1994).

23. P. Constantin, W. E, E. Titi, Onsager's conjecture on the energy conservation for solutions of Euler's equation, Commun. Math. Phys., **165**, 207-209 (1994).

24. J. Miller, Statistical mechanics of Euler equations in two dimensions, Phys. Rev. Lett. **65**, 2137-2140 (1990).

25. R. Robert, A maximum entropy principle for two dimensional perfect fluid dynamics, J. Stat. Phys **65**, 531-543 (1991).

26. P. Constantin, J. Wu, Inviscid limit for vortex patches, Institut Mittag Leffler **7** ISSN 1103-467X (1994).

27. R. Diperna, A. Majda, Oscillations and concentrations in weak solutions of the incompressible fluid equations, Commun. Math. Phys. **108** 667-689 (1987)

28. J.M. Delort, Existence de nappes de tourbillon en dimension deux, J. Amer. Math. Soc. **4**, 553-586 (1991).

29. T. Kato Nonstationary flows of viscous and ideal fluids in R3, J. Funct. Anal. **9**, 296-305 (1972).

30. P. Constantin, Note on loss of regularity for solutions of the 3D incompressible Euler and related equations, Commun. Math. Phys. **104**, 311-326 (1986).

31. T. Kato , J.T. Beale, A. Majda, Remarks on the breakdown of smooth solutions for the 3-D Euler equations, Commun. Math. Phys. **94**, 61 (1989).

32. P. Constantin, A. Majda, E. Tabak, Formation of strong fronts in the 2D quasi-geostrophic thermal active scalar, Nonlinearity, **7**, 1495-1533 (1994).

33. P. Constantin, Geometric and analytic studies in turbulence, in Trends and Perspectives in Appl. Math., L. Sirovich ed., Appl. Math. Sciences **100**, Springer-Verlag, (1994).

34. P. Constantin, Ch. Fefferman, Direction of vorticity and the problem of global regularity for the Navier-Stokes equations, Indiana Univ. Math. Journal, **42** 775-789 (1993).

35. P. Constantin, Navier-Stokes equations and area of interfaces, Commun. Math. Phys. **129**,

    241 - 266 (1990).
36. P. Constantin, C. Fefferman, A. Majda, manuscript.
37. R. M. Kerr, Evidence for a singularity of the three dimensional incompressible Euler equa-
    tions, Phys. Fluids A, **6** 1725-1739 (1993).

DEPARTMENT OF MATHEMATICS, THE UNIVERSITY OF CHICAGO

Lectures in Applied Mathematics
Volume **31**, 1996

Turbulence as a Near-Equilibrium Process

Alexandre J. Chorin[1]

Department of Mathematics
University of California
Berkeley, California 94720-3840

Abstract. The small-scale structure of fully developed homogeneous turbulence in an incompressible flow is described as a perturbation of an ensemble of vortices in thermal equilibrium. A cartoon model that illustrates how an equilibrium spectrum can coexist with an energy cascade is displayed. Analytical and numerical results that explain why the vortex model is reasonable are summarized; an important role is played by a generalization of the Kosterlitz-Thouless analysis of vortex phase transitions.

1991 *Mathematics Subject Classification.* Primary 65C99, 76C05, 76D05, 76F99, 82B35, 82C05.

[1]This work was supported in part by Applied Mathematical Sciences subprogram of the Office of Energy Research, U.S. Department of Energy, under contract no. DE–AC03–76SF00098, and in part by the National Science Foundation under grants DMS94–14631 and DMS89–19074.

Introduction

The inertial scales of three-dimensional fully developed turbulence, which mediate between the energy-containing range and the dissipation range, are a main focus of turbulence theory. They exhibit the well-known Kolmogorov spectrum, $E(k) = \text{constant} \cdot \varepsilon^{\frac{2}{3}} k^{-\frac{5}{3}}$, where $E(k)$ is the energy spectrum (energy as a function of the wave number k) and ε is the rate of energy dissipation per unit mass. Most derivations of this spectrum rely on the idea of an energy cascade: energy is fed into the system at large scales, it is dissipated at small scales, and in the inertial range the energy flow is not only irreversible but also far from a Gibbsian thermal equilibrium. Thermal equilibrium for the equation of fluid mechanics is, on the other hand, identified with an "equipartition" ensemble, in which the mean energy per Fourier mode is constant, and the resulting spectrum is proportional to k^2 — a very different spectrum indeed. This general picture has survived in the literature (see e.g. [28]) even though the idea of a unidirectional cascade contradicts experiment [29], closely related fields (e.g. plasma turbulence [19], superfluid turbulence [17]) do not exhibit such a chasm between equilibrium and non-equilibrium phenomena, recent analyses of two-dimensional turbulence make ample use of equilibrium ideas, and the "equipartition" ensemble is of doubtful relevance to fluid mechanics. Indeed, turbulence, as it is usually described, is the only major example of an irreversible system with many degrees of freedom which cannot be connected in a significant way to an appropriate equilibrium process (e.g., via a fluctuation-dissipation theorem). It has already been argued in detail, e.g. in [10], that the "equipartition" (or "Hopf") equilibrium is an artifact produced by divergent spectral approximations and has no relevance to fluid mechanics, and this argument will not be repeated here. We shall emphasize instead the contrary point of view, according to which turbulence can indeed be usefully studied as a small-to-moderate perturbation of a Gibbsian thermal equilibrium in vortex (or "magnet") variables. The idea is sensible because it is plausible that the small scales of turbulence have intrinsic time scales much smaller than the time scale of overall decay, and there is enough time for equilibrium to be reached. The emphasis here is on the motivation and on the plausibility of the framework presented rather than on detailed calculations, especially when the latter are already available in print. Note that except possibly in the next section, "equilibrium" means "Gibbsian equilibrium" and not "statistical steady state".

Example of a cascade through an equilibrium spectrum

In subsequent sections we shall present models whose spectrum is determined by an equilbrium ensemble and which also exhibit an energy transfer between large and small scales. The usual belief is that equilibrium and turbulent energy transfer are mutually exclusive, unless "equilibrium" means "statistical steady state". In the present section we present a simple example of a system where an irreversible energy cascade proceeds across a spectrum whose form is determined by what can be reasonably viewed as "thermal equilibrium" considerations, and where in addition there is a relation between the amplitude of the energy spectrum and a rate of energy dissipation. More elaborate examples of systems with similar behavior are available in [12]. In no way do we imply that there is never a difference between statistical steady states and Gibbsian thermal equilibrium- only that it is possible for the spectral distribution of the energy to be similar in both cases.

Consider the unit square $D = \{x, y \mid 0 \leq x \leq 1, \ 0 \leq y \leq 1\}$; we are going to implement on D a discrete version of the baker's transformation (see e.g. [30]): $x' = 2x - [2x]$, $y' = \frac{1}{2}(x + [2x])$, where $[x]$ denotes the largest integer $\leq x$. This transformation is well known to be ergodic and mixing, and thus not an absurd example of the growth in disorder typical of turbulence.

To discretize the transformation, divide D into boxes $D_{ij} = \{x, y \mid ih \leq x \leq (i+1)h, \ jh \leq y \leq (j+1)h\}$, where $h = \frac{1}{n}$, $n = 2^m$ for some integer m, and i, j takes the values $0, 1, \ldots, (n-1)$. Furthermore, consider the discrete function $\phi = \{\phi_{ij}\}$, $\phi_{ij} = 0$ or 1 on each D_{ij}.

The discrete baker's transformation $(i, j) \rightarrow (i', j')$ is defined by

$$x = ih , \qquad y = jh ,$$
$$x' = 2x - [2x] , \quad y' = \tfrac{1}{2}(y + [2x]) , \qquad\qquad (1)$$
$$i' = x'/h , \qquad j' = y'/h .$$

One can readily check that this transformation maps two pairs (i, j) on one (i', j') if i' is odd and no pairs if i' is even. To construct the image ϕ' of ϕ under the transformation, consider for i' odd the sum s of the $\phi_{i,j}$ at the preimages of (i', j'), and set $\phi_{i',j'} = 0$ if $s = 0$, $\phi_{i',j'} = 1$ if $s \geq 1$. Then, if $s \leq 1$ set $\phi_{i'+1,j'} = 0$, if $s = 2$ set $\phi_{i'+1,j'} = 1$. This construction preserves $\int \phi dx dy$ (and indeed, $\int \phi^q dx dy$ for all q). The use of an integer- valued function ϕ is designed to eliminate any entropy increase due to smoothing. It is easily seen that, given an initial ϕ with few "holes", the first few steps of the transformation (1) approximate well the first few steps of the continuous baker's transformation. However, the sequence of discrete maps of ϕ is periodic with period $2m$.

Start with initial data ϕ with few "holes", for example, with a function ϕ which equals 1 in some $\ell h \times \ell h$ square, ℓ integer, $\ell < n$. This ϕ and its discrete baker's transforms can be expanded in Fourier or, even better, Haar series. For the first few steps, the "energy" moves to smaller and smaller scales, as in simple models of turbulent cascades. In the next m steps the energy comes back; the time average of the energy per scale is a constant independent of scale. One can readily find initial data for which the energy starts out roughly equidistributed among scales and remains so. We shall not bother to define a temperature and an entropy for this system and shall identify the "equidistribution" ensemble with a thermal equilibrium; for details, see [12]; the energy spectrum has the form $E(k) = Ak^\gamma$, with $\gamma = 0$, and its amplitude A depends on the mean amplitude of the initial data.

Now add an irreversible energy cascade to this model: Start with "smooth" initial data, for example $\phi_{ij} = 1$ for $i, j \leq I$, $I = [\sqrt{\beta} n]$, where $\beta = \Sigma \phi_{ij}/n^2 < 1$ is the fraction of sites where $\phi_{ij} = 1$. Remove energy at small scales by setting ϕ_{ij} to 0 whenever $\phi_{ij} = 1$ and

$\phi_{i\pm1,j} = 0$, $\phi_{i,j\pm1} = 0$ (i.e., whenever there is an isolated 1). Feed energy at large scales by filling in the missing 1's whenever the initial data would have been recovered if it were not for the removal, i.e., once every $2m$ steps. Define a rate of energy dissipation ε as the ratio of the number of 1's removed per q steps divided by q (the limit $q \to \infty$ is reached in $2m$ steps). It is easy to see that the form of the spectrum is unchanged, and that for small β, $A = 2\varepsilon + o(\beta)$, i.e., $E(k) = 2\varepsilon k^{-\gamma}$. (The dependence of A on ε is more interesting near $\beta = \frac{1}{2}$, but this is not the place to discuss it.) One can reverse the usual argument and say that here (and possibly in turbulence theory) there is a relation between E and ε not because energy dissipation creates the spectrum but conversely, because the more energy there is in the "equilibrium" scales, more of it can be dissipated. Note that the dissipation mechanism described for the baker's transformation is in fact a faithful representation of the effect of the scales omitted by the discretization. The recurrent behavior of the system is not without parallel in vortex dynamics [34].

Equilibria of vortex filaments

We now consider thermal equilibria of vortex filaments and their relevance in turbulence theory. The steps are

(i) analysis of the equilibrium configurations of a single vortex filament,

(ii) analysis of a many-vortex collection, including numerical results and an analytical description based on a low fugacity, i.e., sparse system theory of Kosterlitz-Thouless type, and

(iii) a brief mention of the modifications needed to accommodate an energy cascade.

The passage from a many-vortex system to a continuum is well analyzed in other contexts (for a review, see [31]) but not in the statistical context, and is probably the weakest link in what remains a speculative chain. The main points in the chain are: there exist stable equilibria that attract vortex systems embedded in a three-dimensional incompressible flow; these vortex equilibria have Kolmogorov-like spectra (and higher moments); appropriate irreversible (but not one-way) cascades live in the vicinity of these equilibria.

First, a common misapprehension should be dispelled: the vortices and vortex filaments we are dealing with can be used to represent an arbitrary incompressible flow, under wide conditions well studied in numerical analysis; they are not the coherent, organized vortices one sees in some turbulent flows. A coherent, organized vortex is a collective mode of this many-vortex system. The vortex representation is general and requires no à *priori* assumptions about the role of organized structures in turbulence. A similar situation holds in two-dimensional vortex-based analyses (reviewed e.g. in [10],[18]).

A single vortex filament

The construction in this section is amply described elsewhere, see e.g. [8],[10]. Consider a single long vortex filament, with some small, finite, constant cross-section. It is immaterial whether the filament is open (a topological cylinder) or closed (a topological doughnut). The energy of a compactly-supported vorticity field $\boldsymbol{\xi} = \boldsymbol{\xi}(\mathbf{x})$ is given by the integral

$$E = \frac{1}{8\pi} \int d\mathbf{x} \int d\mathbf{x}' \, \frac{\boldsymbol{\xi}(\mathbf{x}) \cdot \boldsymbol{\xi}(\mathbf{x}')}{|\mathbf{x} - \mathbf{x}'|} \, , \tag{2}$$

where $|\mathbf{x}|$ is the length of the vector $\mathbf{x}$. To specialize this integral to the case of a filament, suppose the filament center-line can be approximately mapped on a connected set of N bonds of a cubic lattice; then E is approximated by

$$E_N = \sum_i \sum_{j \neq i} \frac{\mathbf{t}_i \cdot \mathbf{t}_j}{|i - j|} + N\mu \, , \tag{3}$$

where $\mathbf{t}_i$ is a vector centered at the center of the lattice bond occupied by the i-th piece of the filament, parallel to that bond, pointing in the same direction as $\boldsymbol{\xi}$ in the bond, with $|\mathbf{t}_i| = $ circulation of the vortex divided by $\sqrt{8\pi}$, $|i - j|$ is the straight-line distance between segments i and j, and μ is that portion of the integral in (2) which corresponds to $\mathbf{x}, \mathbf{x}'$ in the same bond. The parameter μ can be interpreted as a chemical potential. The vortex filament has many configurations. Assign to each the probability $Z^{-1} \exp(-\beta E_N)$, where Z is the appropriate normalization factor, $\beta = 1/T$, and T is the temperature.

As in two space dimensions, T can be positive or negative; $T < 0$ is "beyond infinity" rather than below 0; negative temperatures occur whenever the maxima of the entropy and of the energy fail to coincide. In the limit $N \to \infty$, the vortex filament is in one of three states: For $\beta > 0$ it is balled up into a crumpled ball, for $\beta < 0$ it is a straight line, and for $\beta = 0$ all configurations are equally likely and one has a "polymer" — a self-avoiding equal probability random walk. The heuristics of this situation are simple: $\beta > 0$ favors low energy states, and low energy for a vortex is obtained by folding it and allowing the velocity fields induced by its several pieces to cancel; the converse holds for $\beta < 0$. In the boundary case $\beta = 0$ the energy spectra has the form $E(k) \sim k^{-\gamma}$. For an infinitely thin vortex $\gamma = D'$, the vector-vector correlation exponent for a polymer, which has been calculated in several ways; a plausible conjecture by Akao [1] yields $D' = 2 - \frac{1}{\alpha}$, where α is the Flory exponent for a polymer, $\alpha \cong .59$ and thus $\gamma \cong .3$. (A related conjecture has been offered by Shenoy [6]). For a filament with a finite, non-uniform cross-section, a heuristic analysis yields $\gamma \cong 5/3$, a Kolmogorov-like result (note that this Kolmogorov-like result is found at equilibrium).

Consider now a smooth physical vortex with some finite cross-section. It can be modeled as a finite collection of segments, at a negative β. If the vortex is imbedded in a random flow it will presumably stretch; if its energy is conserved one can see that $|\beta|$ will decrease until $\beta = 0$ is reached, assuming one can represent the evolution of the vortex as a succession of equilibrium states. Energy conservation forbids the crossing of the boundary $\beta = 0$, which is therefore attracting. We thus have a simple near-equilibrium cartoon of the effect of vortex stretching and of the reason for the appearance of a Kolmogorov spectrum. In two space dimensions, by contrast, there is no stretching, the temperature of a vortex system is invariant (assuming adiabatic walls), and there is no universal spectrum. Note that here, just as in two dimensions, $\beta = 0$ is the boundary between classical and non-classical (e.g., quantum) hydrodynamics.

A collection of vortex filaments

We now consider many filaments. The significant parameters in the problem are $\beta = 1/T$, as before, and μ, the "chemical potential", defined in equation (3). The equilibria of many-vortex systems have been studied numerically (see e.g [2],[11],22]), and they exhibit a phase transition line that is no longer on the μ axis ($\beta = 0$) in the (β, μ) plane. As before, smooth vortex filaments live on the small β side of this line (which includes the half plane $\beta < 0$, since negative temperatures are "hotter" than positive temperatures), and vortex stretching draws them to the transition line. The analysis of this attraction provides an interesting interpretation of intermittency [7],[10].

To understand the situation analytically, we proceed via a Kosterlitz-Thouless (KT) "dielectric" analysis [23],[24], which is a two-term expansion in powers of the fugacity $y = e^{-\beta\mu}$, and applies to sparse (but no longer one-filament) systems. This methodology is ambiguous [14] and to simplify the presentation here we shall pick its easiest variant [13]; the KT analysis has the great advantage of leading naturally to irreversible extensions of the equilibrium theory.

First, following [13],[27],[32], we simplify the vorticity field and represent it as a sparse collection of circular vortex loops. Recent work, in particular by Buttke [4],[5], shows that any vorticity field can be approximated by such a union of circular loops (also known as "magnets", or "elements of impulse" or "velicity"), but the representation is not unique (creating problems when entropy is calculated) and the loops are generally not independent (creating problems for the argument that follows here). The loop representation thus constitutes a substantative simplification [11].

Assume now that in some part of the (β, μ) plane, to be determined, large loops "polarize" smaller loops, i.e., orient them in a specific way, creating a "dielectric" medium [21]. In the simplest form of a Kosterlitz-Thouless argument, one assumes that there exists a dielectric constant e and that the response of the system of loops to an imposed velocity, as well as its energy, is reduced by e. (In a more detailed theory, e is a function of scale.) When $\beta > 0$, the Gibbs factor favors lower energies, and thus arranges smaller loops to oppose the flow; then $e > 1$. It is customary to work in terms of the parameter $K = \beta/e$. From general statistical mechanical principles, one can calculate the susceptibility of a vortex loop (the mean amount by which it reduces an imposed velocity $\mathbf{u}$, divided by $|\mathbf{u}|$), as well as the density of loops of each radius r, and by adding up the contribution of all loops one obtains for K the integral equation

$$K^{-1} = \beta^{-1} + \text{constant} \times \int_{\delta}^{\infty} r^6 \exp(-KE(r,\mu))dr \; , \qquad (4)$$

where $E = E(r,\mu)$ is the energy of a loop of radius r, which depends on μ. δ is the thickness of a loop, assumed small but finite. The integral equation is nonlinear even in this linear theory because the effective energy depends on K. The integral equation (4) defines a function $K = K(\beta,\mu)$, as well as a set of renormalization group transformations which do not concern us here. The (β,μ) plane is divided into a region where K is real and positive, a region where K (which must be real) is not defined, divided by a "transition curve".

A small imaginative leap identifies the region where K is well defined with the "folded" region of the one-vortex theory. Indeed, polarization and vortex folding are similar; one can observe the reduction of the effective velocity by vortex folding and bending in standard vortex computations. Similarly, the region where K is ill-defined is the region in which vortex lines are smooth and the representation by small, independent loops is inapplicable; the boundary between them is the phase transition line. In superfluid theory, the "folded region" is the region in which there is a superflow, and the "unfolded region" is the normal fluid region.

To make the theory concrete, consider the case of very thin vortex filaments, $\delta << 1$. The integral equation (4) reduces [13] to

$$K^{-1} = \beta^{-1} + c_1 \int_{\delta}^{\infty} r^6 \exp(-c_2 K\mu r)dr \; , \qquad (5)$$

where c_1, c_2 are known constants, and μ can be calculated from (2) once δ is known. Following an argument due to Hald [20], equation (5) can be rewritten in the form

$$T = \beta^{-1} = f(K) \; , \qquad f(K) = K^{-1} - c_1 \int_{\delta}^{\infty} r^6 \exp(-c_2 K\mu r)dr \; .$$
$$(6)$$

(Note that the integral can be evaluated exactly.) The function $f(K)$, plotted in Figure 1 for $c_1 = c_2 = \mu = 1, \delta = 0$, is convex in K and has a single maximum, $T_{\max}$, reached at a point $K = K_{\max}$. For $T > T_{\max}$, equation (5) has no real solution. The chemical potential μ as a function of $\beta_{\max} = 1/T_{\max}$ is plotted in Figure 2. Note that $\beta < 0$ is to the right of $\beta > 0$, being "hotter". This phase transition line is qualitatively similar to the computed phase transition line, see e.g. [2],[11],[22]. The trajectory of a vortex system undergoing stretching is also marked in Figure 2. One can deduce from (5) an inertial range spectrum and an inertial exponent, which is not exact, not surprisingly in view of the crudeness of the model.

A further conceptual difficulty has to be faced here: arguments of the type presented above are familiar in the context of superfluid vortex systems, where vortices are generated by thermal fluctuations and can reconnect by quantum mechanisms. We are applying these arguments to classical systems, where vortices are generated by vortex stretching and reconnect through either singularity formation or viscous effects. In neither case is it obvious that thermal equilibrium is reached, since potential barriers must be forded in the former case and topological constraints overcome in the latter. One must assert without proof a "generalized ergodic principle" [26], to the effect that whatever the obstacles, equilibrium is reached in a reasonable time.

One obtains from all this not only a qualitative description of equilibrium vortex distributions, but also a recipe for simplifying this description. Suppose a homogeneous vortex system near an attracting equilibrium is described (for example, on a computer) by means of a large number of vortex loops. Suppose one deletes all loops below some scale d. To compensate for this removal, one has to endow the system with a dielectric constant, $e = \beta/K$, where $K^{-1} = \beta^{-1} + c_1 \int_\delta^d r^6 \exp(-c_2 K \mu r) dr$, c_1, c_2 are known, μ can be found from (2), $\beta = 1/T_{\max}$ and $K = K_{\max}$ can be found on the transition curve by identifying the maximum of $f(K)$. This is a renormalization of the system.

Note that if $\mu \to \infty$ the system becomes sparse; as one can see from Figure 2, the transition curve is asymptotic to $\beta = 0$, and one recovers the one-vortex description of the previous section. At $\beta = 0$ the Gibbs factors are all unity, the polarizability of a loop is zero, and small loops can be removed with no penalty. One recovers the hairpin removal of reference [9].

The modification needed to handle inhomogeneous systems, as well as practical applications of these constructions, will be discussed in [16].

Irreversibility in vortex systems

The use of KT-like theory is not without problems, but it does provide an elegant transition to the study of "nearby" irreversible process, and in particular energy cascades. It is quite easy to add to the KT formalism terms that describe a finite rate of relaxation to equilibrium for vortex loops of scale r in the presence of a varying imposed velocity. The "dielectric function" acquires a delay, and as is well known, the imaginary part of the Fourier transform of the delay function represents energy loss (the Kramers-Kronig formalism, see e.g. [25]). As is already known from superfluid calculations [3],[33], the energy loss is largest near the phase transition line, as it should be from the discussion above. This remark, to be amplified elsewhere [15], connects the equilibrium vortex model with the irreversible aspects of turbulence.

Conclusions

An alternate approach to turbulence theory has been outlined. It can be viewed as a generalization to three dimensions of two-dimensional vortex equilibrium theory started by Onsager and so ably developed since. It provides a simple explanation for the universality of the Kolmogorov spectrum and for the origins of intermittency. The general premise can be restated in fairly anodyne terms: Since some kind of perturbation theory seems to be unavoidable in turbulence theory, one should start with a "ground state" that is as close as possible to the final truth. Vortex equilibria near the vortex critical line may well offer the natural starting point, as they already present many of the features of the Kolmogorov theory.

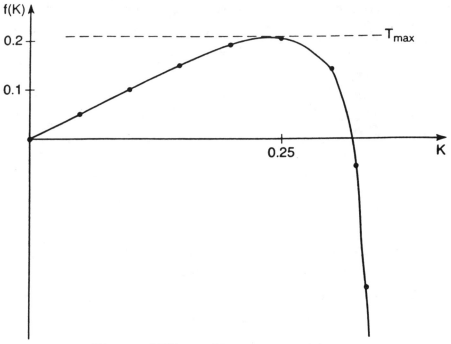

Fig. 1. $f(K)$ vs. K in equation (6).

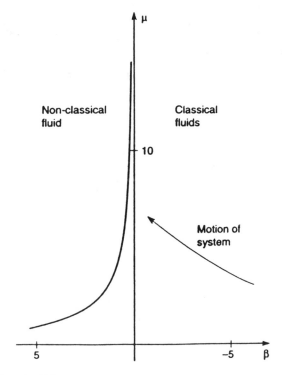

Fig. 2 Phase transition line in simple KT model.

References

[1]   J. Akao, Scaling laws for fractal objects, Manuscript, 1994.

[2]   J. Akao, Phase transitions and connectivity in three-dimensional vortex equilibria, PhD Thesis, UC Berkeley Math. Dept., 1994.

[3]   V. Ambageokar, B. Halperin, D. Nelson and E. Siggia, Dissipation in two-dimensional superfluids, *Phys. Rev. Lett.* **40**, 783–786 (1978).

[4]   T. Buttke, Lagrangian numerical methods which preserve the Hamiltonian structure of incompressible fluid flow, in *Vortex Flows and Related Numerical Methods*, J. T. Beale, G. H. Cottet and S. Huberson, editors, NATO ASI Series, vol. **395**, (Kluwer, Norwell, 1993).

[5]   T. Buttke and A. J. Chorin, Turbulence calculations in magnetization variables, *Appl. Num. Math.* **12**, 47–54 (1993).

[6]   B. Chattopadhyay, M. Mahato and S. Shenoy, Vortex loop crinkling in the three-dimensional XY model: Numerical evidence in support of an ansatz, Phys. Rev. B, **47**, 15159-15169 (1993)

[7]   A. J. Chorin, Turbulence and vortex stretching on a lattice, *Comm. Pure Appl. Math.* **39** (special issue), S47–S65, (1986).

[8]   A.J. Chorin, Equilibrium statistics of a vortex filament with applications, *Comm. Math. Phys.* **141**, 619–631 (1991).

[9]   A.J. Chorin, Hairpin removal in vortex interactions II, *J. Comput. Phys.* **107**, 1–9 (1993).

[10] A. J. Chorin, *Vorticity and Turbulence*, (Springer, 1994).

[11] A. J. Chorin, Vortex phase transitions in 2.5 dimensions, *J. Stat. Phys.* **76**, 835–856 (1994).

[12] A. J. Chorin, Turbulent cascades through equilibrium spectra, Manuscript, Dept. of Mathematics, UC Berkeley, 1995.

[13] A. J. Chorin and O. Hald, Vortex renormalization in three space dimensions, *Phys. Rev. B* **51**, 11969–11972 (1995).

[14] A. J. Chorin and O. Hald, Analysis of Kosterlitz-Thouless transition models, Manuscript, Dept. of Mathematics, UC Berkeley, 1995.

[15] A. J. Chorin and O. Hald, Turbulence and linear response, Manuscript, Dept of Mathematics, UC Berkeley, 1995.

[16]   A. J. Chorin and O. Knio, Renormalization of vortex calculations, in preparation.

[17]   R. J. Donnelly, *Quantized Vortices in Helium II*, (Cambridge, 1991).

[18]   G. L. Eyink and H. Spohn, Negative temperature states and equivalence of ensembles for the vortex model of a two-dimensional ideal fluid, *J. Stat. Phys.* **70**, 833–886 (1993).

[19]   M. Goldman, Strong turbulence of plasma waves, *Rev. Mod. Physics* **56**, 709–735 (1984).

[20]   O. Hald, Private communication.

[21]   J. D. Jackson, *Classical Electrodynamics*, (Wiley, NY), 1974.

[22]   G. Kohring and R. Shrock, Properties of generalized 3D $O(2)$ model with suppression/enhancement of vortex strings, *Nuclear Physics B* **288**, 397–418 (1987).

[23]   J. Kosterlitz, The critical properties of the two-dimensional XY model, *J. Phys. C: Solid State Phys.* **7**, 1046–1060 (1974).

[24]   J. Kosterlitz and D. J. Thouless, Order, metastability and phase transitions in two-dimensional systems, *J. Phys. C: Solid State Phys.* **6**, 1181–1203 (1973).

[25]   L. Landau and E. Lifshitz, *Statistical Physics*, 3d edition, part 1, (Pergamon, NY, 1980).

[26]   J. Langer and J. Reppy, Intrinsic critical velocities in superfluid helium, in *Progress in Low Temperature Physics*, vol. **VI**, C. Gorter, editor, (North-Holland, Amsterdam, 1970).

[27]   F. Lund, A. Reisenegger and C. Utreras, Critical properties of a dilute gas of vortex rings in three dimensions and the lambda transition in liquid helium, *Phys. Rev. B* **41**, 155–161 (1990).

[28]   D. McComb, *The Physics of Fluid Turbulence*, (Oxford University Press, Oxford, 1990).

[29]   C. Meneveau, Dual spectra and mixed energy cascade of turbulence in the wavelet representation, *Phys. Rev. Lett.*, **66**, 1450-1453 (1991)

[30]   J. Moser, *Stable and Random Motion in Dynamical Systems*, (Princeton Univ. Press, Princeton, 1973).

[31]   E. G. Puckett, A review of vortex methods, in *Incompressible*

Computational Fluid Mechanics, R. Nicolaides and M. Gunzburger, editors, (Cambridge, 1992).

[32] G. Williams, Vortex ring model of the superfluid lambda transition, *Phys. Rev. Lett.*,**59**, 1926–1929 (1987).

[33] G. Williams, Vortex dynamics and superfluid relaxation near the ^{4}He lambda transition, Manuscript, Physics Dept, UCLA, 1994.

[34] H. Zhou and A. J. Chorin, Vortices: Thick and thin, Manuscript, Dept. of Mathematics, UC Berkeley, 1995.

Lectures in Applied Mathematics
Volume 31, 1996

HOMOGENIZATION AND RENORMALIZATION:
THE MATHEMATICS OF MULTI-SCALE RANDOM MEDIA
AND TURBULENT DIFFUSION

Marco Avellaneda[1]

This lecture describes some mathematical methods developed in recent years to study the macroscopic behavior of heterogeneous media and diffusion in turbulent fluids. This area of Applied Mathematics comprises a diversity of beautiful ideas. These include the averaging of partial differential equations, or homogenization theory, probabilistic techniques using Brownian motion, and recursive scaling methods, known as renormalization group techniques, which originated in High-Energy Physics and in Statistical Physics for the study of phase-transitions. My goal is to describe these ideas "at work" on a classical problem, the advection-diffusion of a passive scalar in a multiple-scale, random velocity field, showcasing the inter-relation between these methods, as well as their scope and limitations. The applicability of homogenization and renormalization methods extends well beyond the topics of this talk to many other areas of Applied Mathematics.

I present an outline of what follows. In the first section, I briefly describe homogenization theory and its application to computing the effective conductivity of a heterogeneous medium. The second section deals with the effective diffusivity of a passive scalar undergoing advection-diffusion in an incompressible flow. In Section 3, I discuss the method of iterated homogenization — a mathematical approach to evaluate iteratively the effective properties of multi-scale, self-similar media. The rest of the paper focuses on advection-diffusion exclusively, concentrating on the phenomenon of anomalous diffusion due to "infrared-divergent" velocity fields. After formulating the renormalization problem in Section 4, I present two examples in which renormalization can be carried out in full mathematical rigor. These examples illustrate the use of scaling techniques combined with homogenization in the same problem. In particular, the mathematical relation between decimation of high-wavenumber components and homogenization is analyzed. Section 7 discusses the renormalization group. Finally, in Section 8, I propose an approximate solution to the (open) problem of characterizing the large-scale statistics of a passive scalar evolving in an isotropic, incompressible, infrared-divergent velocity field.

1. Homogenization Theory. I begin by a brief review of the classical problem of computing the *effective conductivity* of a composite material. To fix ideas, consider a heterogeneous material made from two different conductors with conductivities σ_1 and σ_2 and volume fractions f_1 and f_2, respectively. Mathematically,

---

1991 *Mathematics Subject Classification.* Primary 76F10.

[1]Courant Institute of Mathematical Science, New York University, 251 Mercer St., New York, NY 10012, USA. Lecture presented at MSRI, July, 1994 and at the International Congress of Mathematicians, Zurich, August 1994.

its "microscopic" conductivity is described by the function

$$\sigma(\mathbf{x}) = \sigma_1 \chi_1(\mathbf{x}) + \sigma_2 \chi_2(\mathbf{x}), \tag{1}$$

where $\chi_1(\mathbf{x})$ and $\chi_2(\mathbf{x})$ are the characteristic functions of the regions of space occupied by each of the two constituents. The partial differential equation satisfied by an electric potential ϕ inside this composite is

$$\nabla \cdot (\sigma(\mathbf{x})) \cdot \nabla \phi(\mathbf{x})) = 0. \tag{2}$$

The characteristic composite dimension, L, is assumed to be large with respect to the "microscale", l. Let us introduce the small parameter $\epsilon = l/L$, where L represents the dimensions of the composite. Then, defining the *macroscopic* coordinate $\mathbf{x}' = \epsilon \mathbf{x}$, we rewrite (2) as

$$\nabla \cdot (\sigma(\mathbf{x}'/\epsilon)) \nabla \phi_\epsilon(\mathbf{x}')) = 0, \tag{3}$$

with

$$\phi_\epsilon(\mathbf{x}') \equiv \epsilon \phi(\mathbf{x}'/\epsilon).$$

Let us also make an auxiliary assumption: that the conductivity $\sigma(\mathbf{x})$ is periodic in $\mathbf{R}^3$ with period 1 in each direction, so that the heterogeneous material is "microscopically periodic"[2]. The macroscopic behavior of the composite is then described by the equation satisfied by $\phi_0(\mathbf{x}') \equiv lim_{\epsilon \to 0} \phi_\epsilon(\mathbf{x}')$, which turns out to be[3]

$$\nabla \cdot (\sigma_{eff} \cdot \nabla \phi_0(\mathbf{x}')) = 0. \tag{4}$$

This equation and, in particular, the value of σ_{eff}, can be derived using the *homogenization method* or method of multiple-scale asymptotics [BLP], [SA], [B]. Let us expand formally the potential as a series

$$\phi_\epsilon(\mathbf{x}')) = \phi_0(\mathbf{x}') + \epsilon \phi_\epsilon^{(1)}(\mathbf{x}', \mathbf{x}'/\epsilon) + \epsilon^2 \phi_\epsilon^{(2)}(\mathbf{x}', \mathbf{x}'/\epsilon) + \dots \tag{5}$$

This is a perturbation series in ϵ, in which the successive terms depend on the "macroscopic" variable $\mathbf{x}'$ as well as on the "microscopic" variable $\mathbf{x} = \mathbf{x}'/\epsilon$. To calculate the unknown terms in the expansion, replace $\mathbf{x}'/\epsilon$ by $\mathbf{x}$ in (3) and (5), and the operator ∇ by $\nabla_{x'} + \epsilon^{-1} \nabla_x$. Then substitute (5) into (3) and, *treating $\mathbf{x}'$ and $\mathbf{x}$ as independent variables*, expand equation (3) in powers of ϵ. This results in a hierarchy of partial differential equations that completely determine the functions $\phi^{(k)}$ for all k ([BLP]).

It can be shown that the *effective conductivity*, σ_{eff}, is given by[4]

$$\sigma_{eff} = \langle (\mathbf{e} + \nabla \chi(\mathbf{x}))^t \cdot \sigma(\mathbf{x}) \cdot (\mathbf{e} + \nabla \chi(\mathbf{x})) \rangle. \tag{6}$$

---

[2] The validity of this assumption will be discussed below.

[3] Strictly speaking, we must supplement (3) with boundary conditions on the boundary of the conductor. It can be shown that the form of the homogenized equation is independent of these conditions.

[4] For simplicity, we assume that the composite is macroscopically isotropic, so that σ_{eff} is a scalar. In general, the effective conductivity would be a tensor $\sigma_{eff,ij}$.

where e is a unit vector and the "potential" $\chi(\mathbf{x})$ is a periodic solution of the the PDE

$$\nabla_x \cdot (\mathbf{e} + \nabla\chi(\mathbf{x})) = 0 \qquad (7)$$

The brackets $\langle \cdot \rangle$ in (6) represent averaging over the period cell of $\sigma(\mathbf{x})$. Equation (7) is known as the *cell problem* associated with the homogenization problem (3). The effective conductivity characterizes the conduction properties of the composite on a macroscopic scale.

Homogenization theory is not restricted to periodic media. Kozlov [Ko] and Papanicolaou and Varadhan [PV] established the existence of a homogenized limit for equation (3) for very general stationary ergodic random conductivities $\sigma(\mathbf{x})$ as $\epsilon \to 0$, using the *ansatz* (5). This theory requires however that the conductivity be bounded away from 0 and ∞, i.e., that

$$0 < \lambda \leq \sigma(\mathbf{x}) \leq 1/\lambda \qquad (8)$$

for some constant $\lambda \in (0,1]$, uniformly in $\mathbf{x}$. In this more general setting, the effective conductivity is obtained by solving (6) in a suitable Hilbert space and evaluating (7). (The brackets in (7) represent then statistical averaging.)

2. Advection-diffusion. Let us consider a different but related physical problem: the mixing of a passive scalar in an incompressible, steady flow. Typically, this scalar represents the volumetric concentration of a solute substance dispersing in the fluid under the action of the flow and of molecular diffusion[5].

The equations for the concentration $C(\mathbf{x}, t)$ and the velocity $V(\mathbf{x})$ are

$$\frac{\partial C(\mathbf{x}, t)}{\partial t} + V(\mathbf{x}) \cdot \nabla C(\mathbf{x}, t) = D\Delta C(\mathbf{x}, t) \qquad (9a)$$

and

$$\nabla \cdot V(\mathbf{x}) = 0. \qquad (9b)$$

For simplicity, we assume that the velocity field V is smooth and periodic in $\mathbf{R}^3$ and that it has average zero over the period: $\langle V \rangle = 0$. To compute the effective, large-scale diffusion rate, we make, as before, a change of variables $\mathbf{x}' = \epsilon\mathbf{x}$, $t' = \epsilon^2 t$ and rewrite equation (9a) as

$$\frac{\partial C_\epsilon(\mathbf{x}', t')}{\partial t'} + \frac{1}{\epsilon} V(\mathbf{x}'/\epsilon) \nabla C_\epsilon(\mathbf{x}', t') = D\Delta C_\epsilon(\mathbf{x}', t') \qquad (10)$$

where

$$C_\epsilon(\mathbf{x}', t') = \epsilon^{-3} C(\mathbf{x}'/\epsilon, t'/\epsilon^2).$$

In the case of periodic velocity fields, the homogenization of equation (10) is essentially equivalent to the problem of Section 1, with a minor modification. To

---

[5] For a background on this problem see, for instance, Monin and Yaglom [MY].

see this, notice that, since V is incompressible and has mean zero, it has a smooth, periodic vector-potential satisfying

$$\nabla \times A(\mathbf{x}) = V(\mathbf{x}).$$

Define an *antisymmetric* matrix $H_{ij}(\mathbf{x})$ by setting

$$H_{12} = A_3, \quad H_{13} = -A_2, \quad H_{23} = A_1, \quad \text{etc}.$$

Using this matrix, the rescaled equation (10) can be rewritten as a divergence-form parabolic PDE *with a tensor-valued, non-symmetric "conductivity"*. Specifically, C_ϵ satisfies

$$\frac{\partial C_\epsilon(\mathbf{x}',t)}{\partial t'} = \nabla \cdot \left[(D\mathbf{I} + \mathbf{H}(\mathbf{x}'/\epsilon)) \cdot \nabla C_\epsilon(\mathbf{x}',t') \right]$$

where $\mathbf{I}$ is the identity matrix and $\mathbf{H} = H_{ij}$. Using the homogenization method (i.e. expansion (5)), it can be shown [BLP] that $C_\epsilon(\mathbf{x}',t')$ converges to the solution of the diffusion equation

$$\frac{\partial C_0(\mathbf{x}',t')}{\partial t'} = D_{eff} \Delta C_0(\mathbf{x}',t'),$$

with *effective diffusivity*, D_{eff}, given by[6]

$$D_{eff} = D \left[1 + \left\langle |\nabla \chi(\mathbf{x})|^2 \right\rangle \right], \tag{11}$$

where $\chi(\mathbf{x})$ is a periodic solution of the "cell-problem"

$$\nabla \cdot \left[(D\mathbf{I} + \mathbf{H}(\mathbf{x})) \cdot (\nabla \chi(\mathbf{x}) + \mathbf{e}) \right] = 0. \tag{12}$$

Generalizing this result to the setting of random velocities V is important for physical applications. However, the adaptation of homogenization methods to advection-diffusion in random velocity fields is a more delicate matter. This is due to the fact that *the natural stationary field is the velocity V and not the vector potential A*. To apply the homogenization method with random velocities one must make additional hypotheses. If the vector potential is uniformly bounded and stationary – and thus a coercivity relation of type (7) holds for the "conductivity" $D\mathbf{I} + \mathbf{H}$ – the theory of Kozlov and Papanicolaou and Varadhan can be directly applied to show that the tracer obeys a macroscopic diffusion equation. A generalization of this result, obtained with Andy Majda [AM1], shows that the scalar will satisfy an effective diffusion equation provided that the velocity admits a *stationary, square-integrable* vector potential. (See also Oelschlager [O]). The significance of the result is that it is, in a sense, sharp. (The condition is also necessary.)

---

[6]We assume, for simplicity, that the system is macroscopically isotropic so that the effective diffusivity is a scalar.

A random vector field V has a stationary vector potential if and only if

$$\int_{\mathbf{R}^3} \frac{\hat{R}(\mathbf{k})}{|\mathbf{k}|^2}\, d^3\mathbf{k} < \infty, \tag{13}$$

where $\hat{R}(\mathbf{k})$ is the spectral measure of V.

Intuitively, if condition (13) is satisfied, the "correlation length" of the stationary vector potential A determines the "microscopic" scale in the problem. Homogenization theory then tells us that the system behaves diffusively at distances which are large compared to the correlation length.

However, the assumption of stationarity of the vector-potential is not always suitable from a physical viewpoint. I will show that new phenomena arise if this condition (13) fails, i.e. if A has larger and larger fluctuations as $|\mathbf{x}| \to \infty$. In the latter case, the correlation length of the vector potential is infinite and, as we shall see, the macroscopic motion is no longer diffusive. This puts the problem beyond the reach of "classical" homogenization theory and new methods must be introduced.

3. Iterated Homogenization and Rigorous Decimation Procedures. Homogenization theory can be applied to study *multiple-scale* heterogeneous media and flows. This is an important application of the theory, which I regard as a "precursor" of the physicists' renormalization theory discussed in §7. Many situations in Physics lead to the consideration of self-similar or fractal media. Thee include the analysis of phase transitions and of turbulent mixing in flows with many active energetic scales.

Suppose that the microscopic conductivity of a material can be expressed in the form

$$\sigma_\epsilon(\mathbf{x}) = \tilde{\sigma}(\mathbf{x}\epsilon, \mathbf{x}\epsilon^2, \ldots, \mathbf{x}\epsilon^N), \tag{14}$$

where ϵ is a small parameter and the function $\tilde{\sigma}$ is defined in $(\mathbf{R}^3)^N$ and is periodic of period 1 in each variable. A "medium" with this conductivity is composed of "elements" of different characteristic lengths, each being much larger or smaller than the other. Homogenization theory tells us (cf. [BLP]) that *the solution of the cell problem associated to (14) can be done recursively, by solving independent partial differential equations at each scale.* This is done as if $\mathbf{x}\epsilon^1, \mathbf{x}\epsilon^2, \ldots, \mathbf{x}\epsilon^N$ were independent variables. To illustrate this idea, let

$$\sigma_\epsilon(\mathbf{x}) = \tilde{\sigma}(\mathbf{x}\epsilon, \mathbf{x}\epsilon^2). \tag{15}$$

represent the conductivity of a "two-scale" material. Accordingly, in order to calculate its effective conductivity as $\epsilon \to 0$, we should first homogenize the shorter scale $\mathbf{x}\epsilon$, treating the longer one, $\mathbf{x}\epsilon^2$, as a parameter. The result is a "partially homogenized" conductivity

$$\sigma_\epsilon^{(1)}(\mathbf{x}) = \sigma^*(\mathbf{x}\epsilon^2),$$

where σ^* is obtained by solving a cell problem of type (6), with $\mathbf{x}\epsilon$ as the independent variable and $\mathbf{x}\epsilon^2$ in held fixed, and averaging over the corresponding period cell. In a second stage, we should homogenize $\sigma_\epsilon^{(1)}(\mathbf{x})$, using again a cell problem. The result of these two successive homogenizations is exactly the effective conductivity of the two-scale material (15) as $\epsilon \to 0$.

As an application of this idea, consider now a *self-similar* material with conductivity (14). By this, I mean a material for which the spatial fluctuations in the conductivity are *identical* from one scale to the next after dilatation by a factor of ϵ. Using iterated homogenization, we see that the effective conductivity of such a material can be obtained (in the limit $\epsilon \to 0$) by *iterating a map*: i.e.,

$$\sigma^{(N)} = \Omega \circ \Omega \circ ... \Omega(\sigma_0)$$
$$= \Omega^{(N)}(\sigma_0), \tag{16}$$

where σ_0 is an initially given conductivity and $\sigma \mapsto \Omega(\sigma)$ is the result of a "single-step" homogenization. This operation is identical at every scale due to the self-similarity. When the map Ω is known, we can calculate the effective conductivity by doing successive iterations. If the number of scales N is large or infinite, the problem reduces to the analysis of the fixed points of Ω ([Mi], [A]).

It is worthwhile to emphasize that the principle of iterated homogenization hinges on the large *separation of scales* imposed in the microstructure by assuming that ϵ is a small parameter. This assumption leads to nice mathematical results but may not always be justified in practice.

This idea can also be used to evaluate the effective diffusivity for advection-diffusion with velocities with multiple scales. For instance, consider a velocity field of the form

$$V_\epsilon(\mathbf{x}) = \sum_{n=1}^{N} \beta^n \epsilon^n V_0(\mathbf{x}\epsilon^n), \tag{17}$$

where V_0 is a given incompressible periodic vector field and ϵ and $\beta \in (0,1)$ are parameters. The effective diffusivity for (17) can be calculated exactly in the limit $\epsilon \to 0$ as follows. First, we observe that the corresponding H-matrix is given by

$$\mathbf{H}(\mathbf{x}) = \sum_{n=1}^{N} \beta^n \mathbf{H}_0(\mathbf{x}\epsilon^n), \tag{18}$$

where $\mathbf{H}_0$ is the H-matrix for the basic field V_0. The smallest spatial scale corresponds to the term with $n = 1$. Iterated homogenization allows us to "average out" the combined effects of molecular diffusion and this first advective scale *independently of the larger scales* ($n \geq 2$). Let us define the map

$$F(D) \equiv \text{homogenized diffusivity for the operator} \quad D\Delta + V_0 \cdot \nabla.$$

The effect of "decimating" the smallest spatial scale is a *modified diffusion coefficient* D_1 given by

$$D_1 = \beta F\left(\frac{D}{\beta}\right).$$

We can continue "decimating" the successive scales $n = 2, 3 \ldots$. A simple calculation shows that the diffusivity D_j, obtained after eliminating the jth scale, satisfies

$$D_j = \beta^j \, F(\frac{D_{j-1}}{\beta^j}) \, . \tag{19}$$

Supplementing this relation with the initial condition $D_0 = D$, we obtain a recursion relation from which the effective diffusivity D_N can be expressed in terms of D and the "one-step map" F (the analogue of Ω for advection-diffusion).

Consider now what happens if $N = \infty$ in (17). It is easy to show that, for $\beta < 1$, the diffusion coefficient D_N approaches a finite limit as $N \to \infty$. On the other hand, if $\beta \geq 1$, the velocity is well-defined for each value of $\epsilon < 1/\beta$, but D_N diverges as $N \to \infty$. For instance, if $\beta = 1$, the iteration scheme (19) is

$$D_N = F \circ F \circ \ldots \circ F(D)$$
$$= F^{(N)}(D) \, .$$

Recall, from (10), that the homogenized diffusivity satisfies $D_{eff} = F(D) > D$ for all D. Therefore, the map F increases the diffusivity with each step. Thus, D_N increases without bound as $N \to \infty$. This is an indication that the effective diffusion coefficient corresponding to advection-diffusion by the velocity

$$\sum_1^\infty \epsilon^n \, V_0(\mathbf{x} \, \epsilon^n)$$

is *infinite*[7]. It is noteworthy that the velocity field (17) has a uniformly bounded vector potential if $\beta < 1$, for all $\epsilon > 0$ and all N. On the other hand, if $\beta \geq 1$, the vector potential of (17) diverges as $N \to \infty$.

4. Anomalous Diffusion and the Renormalization Problem. Advection-diffusion in flows for which the large-scale behavior of the passive scalar is *superdiffusive*, i.e. for which the effective diffusion coefficient is infinite, have been the object of much investigation. Recent progress has been made in understanding this phenomenon from a rigorous standpoint, although a complete theory is still missing. Anomalous diffusion in incompressible flows can arise only when the vector-potential is non-stationary, i.e. when

$$\int_{\mathbf{R}^3} \frac{\hat{R}(\mathbf{k})}{|\mathbf{k}|^2} \, d^3\mathbf{k} = \infty \, . \tag{20}$$

The large-scale analysis of the advection-diffusion equation (9a) for velocities satisfying (20) can be formulated as follows. First, let $\delta > 0$ be a small parameter

---

[7]It is an "indication" rather than a proof, because I took the limit $\epsilon \to 0$ first and then the limit $N \to \infty$, rather than the other way around. The same conclusion and caveat apply for $\beta > 1$.

representing the ratio between the velocity correlation length and the macroscopic scale. We make the change of coordinates $\mathbf{x}' = \delta \mathbf{x}$, so that $\mathbf{x}'$ represents distance measured in the macroscopic scale. Let us also define a macroscopic time-scale t' by setting $t' = \rho^2(\delta)\, t$, where $\rho^2(\delta)$ is a yet unspecified function of δ.[8]

Unlike the cases treated before, the macroscopic time-scale is expected to be different than in homogenization theory, to account for the infinite diffusion coefficient when $\rho = \delta$. We shall see that this time-scale depends on the singularity of the function $\frac{\hat{R}(\mathbf{k})}{|\mathbf{k}|^2}$ at $|\mathbf{k}| = 0$.

Define the large-scale concentration field

$$C_\delta(\mathbf{x}'\, t') = \delta^{-2} C(\mathbf{x}'\, \delta^{-1}, t'\, \rho^{-2}). \tag{21}$$

The *renormalization problem* ([AM2]) consists in i) finding a scaling function $\rho(\delta)$ such that

$$\lim_{\delta \to 0} C_\delta(\mathbf{x}', t') \equiv C_0(\mathbf{x}', t') \tag{22}$$

defines a non-trivial function, different both from zero and from the initial datum, and ii) to characterize the limiting function $C_0(\mathbf{x}', t')$, by means of an an appropriate evolution equation.

An important difference between systems with stationary vector potentials, satisfying (13), and the present ones is that, in the former, the limiting function $C_0(\mathbf{x}', t')$ is always *non-random*, due to the "self-averaging" property of vector-potentials with finite correlation length [PV]. In contrast, for *infrared-divergent flows* satisfying (20), the limiting function C_0 is expected to be a *random process*, with a distribution determined by a macroscopic equation of motion.

5. Simple Shear Flows with Infrared Divergences. The simplest mathematical model to study anomalous diffusion is a simple shear flow [AM3]

$$V(x_1, x_2) = \big(U(x_2), 0\big),$$

where

$$U(x_2) = \bar{U} \int_{|k| \le k_0} |k|^{(1-\epsilon)/2} \cos kx_2\, N(k)\, dk. \tag{23}$$

The function $N(k)$ in the integrand represents one-dimensional Gaussian white noise. Thus, $U(x_2)$ is a Gaussian field with spectral function

$$\hat{R}(k) = \begin{cases} \bar{U}^2\, |k|^{1-\epsilon} & , |k| \le k_0 \\ 0, & |k| > k_0. \end{cases}$$

---

[8]The parameter δ plays the same role as ϵ in Sections 1 and 2; ϵ will be used to denote another quantity (cf. (23)).

The integral $\int \hat{R}(k)/|k|^2 \, dk$ converges if $\epsilon < 0$ and diverges for $\epsilon \geq 0$. The parameter ϵ varies in the range $-\infty < \epsilon < 2$.[9] It controls the degree of divergence of the integral (20) since , for $\epsilon > 0$,

$$\int\limits_{\delta \leq |k| \leq k_0} |k|^{1-\epsilon}/|k|^2 \, dk \approx \delta^{-2\epsilon} \quad \delta \ll 1,$$

and the stream-function, $\psi(x_2)$, (the 2D analogue of the vector potential) for this flow satisfies

$$\psi(x_2) \approx |x_2|^{2\epsilon} \quad |x_2| \gg 1.$$

Let us derive the macroscopic evolution equation for this system. Due to the anisotropy of the random shear flow, it is convenient to define C_δ in a slightly different way than (21), namely,

$$C_\delta(\, x_1' \, , x_2' \, , t' \,) = \delta^{-1} \rho^{-1} \, C(\, x_1'/\delta \, , x_2'/\rho \, , t' \,/\rho^2 \,).$$

Using equation (9a), we find that the PDE satisfied by C_δ is

$$\frac{\partial C_\delta(\, x_1' \, , x_2' \, , t' \,)}{\partial t'} + (\delta \,/\rho^2 \,) \cdot U(\, x_2'/\rho \,) \frac{\partial C_\delta(\, x_1' \, , x_2' \, , t \,)}{\partial x_1'} =$$

$$D \left[(\delta^2/\rho^2 \,) \frac{\partial^2 C_\delta(\, x_1' \, , x_2' \, , t' \,)}{\partial (x_1')^2} + \frac{\partial^2 C_\delta(\, x_1' \, , x_2' \, , t' \,)}{\partial (x_2')^2} \right]. \tag{24}$$

A choice must now be made for the scaling function $\rho(\delta)$. We choose

$$\rho(\delta) = \delta^{2/(2+\epsilon)} \tag{25}$$

for reasons that will become immediately clear. Using this value of ρ, we can rewrite equation (24) as

$$\frac{\partial C_\delta(\, x_1' \, , x_2' \, , t' \,)}{\partial t'} + \rho^{\epsilon/2 - 1} \, U(\, x_2'/\rho \,) \frac{\partial C_\delta(\, x_1' \, , x_2' \, , t' \,)}{\partial x_1'} =$$

$$D \left[\rho^\epsilon \frac{\partial^2 C_\delta(\, x_1' \, , x_2' \, , t' \,)}{\partial (x_1')^2} + \frac{\partial^2 C_\delta(\, x_1' \, , x_2' \, , t' \,)}{\partial (x_2)^2} \right]. \tag{26}$$

We observe that the random field

$$\rho^{\epsilon/2 - 1} \, U(\, x_2'/\rho \,)$$

converges in distribution as $\rho \to 0$ to the Gaussian field

$$U_0\,(x_2') = \bar{U} \int\limits_{-\infty}^{+\infty} |k|^{(1-\epsilon)/2} \cos k x_2' \, N(k)\, dk\,.$$

---

[9]The upper bound $\epsilon < 2$ is imposed so as to have a velocity with finite "energy" $\langle \, |V(\,\mathbf{x}\,)|^2 \,\rangle$. In [AM 3], a wider range $0 < \epsilon < 4$ is studied by introducing a natural infrared cutoff for the velocity statistics.

This is a generalized (distribution-valued) random field, in the sense of Gel'fand and Vilenkin [GV]. Therefore, formally,

$$\lim_{\delta \to 0} C_\delta \left(x_1' , x_2' , t' \right) = C_\delta \left(x_1' , x_2' , t' \right),$$

where C_0 satisfies the *stochastic PDE*

$$\frac{\partial C_0(x_1' , x_2' , t')}{\partial t'} + U_0 \left(x_2' \right) \frac{\partial C_0(x_1' , x_2' , t')}{\partial x_1'} = D \frac{\partial^2 C_0(x_1' , x_2' , t')}{\partial (x_2')^2} . \qquad (27)$$

The reader may wonder what is the meaning of a partial differential equation in which one of the coefficients is a *random distribution*! Nevertheless, it can be proved rigorously that, in spite of the presence of the singular coefficient $U_0(x_2')$, equation (27) has a well-defined solution which can be expressed as an expectation over a Wiener paths and, moreover, that this equation characterizes the limit of C_δ as $\delta \to 0$. The solvability of (27) justifies the choice of the scaling function and is really the essence of the renormalization. The key element needed to solve (27) is the fact that

$$\int_0^t U_0 \left(\beta(s) \right) ds$$

is a well-defined random variable with finite moments.

As an application of this result, let us calculate the *averaged Green function*, which describes the evolution of the statistical mean of the random function C_0. This is done by solving (27) and averaging over velocity statistics[10]. Accordingly, we find that

$$\left\langle C_0 \left(x_1' , x_2' , t' \right) \right\rangle = \int \int \bar{P}(x_1' - y_1 , x_2' - y_2 , t') \, C_{init} (y_1 , y_2) \, dy_1 \, dy_2 ,$$

where C_{init} represents the initial condition and where the Fourier transform of $\bar{P} (x_1' , x_2' , t')$ is given by

$$\hat{\bar{P}}(k_1 , k_2 , t') = \exp \{ - D (k_2)^2 t' \} \times$$

$$\mathbf{E} \left\{ \exp \left[-\frac{\bar{U}^2 (t')^{(1+\epsilon/2)} (k_1)^2}{2^{(2-\epsilon/2)} D^{(1-\epsilon/2)}} \int_0^1 \int_0^1 F_\epsilon (\beta(s) - \beta(s')) \, ds \, ds' \right] \right\}. \qquad (28)$$

In this last equation, the expectation is taken with respect to Brownian Motion, $\beta(s) , 0 \le s \le 1$, and the function F_ϵ which appears in the exponent is given by

$$F_\epsilon(y) = \int_{-\infty}^{+\infty} |k|^{1-\epsilon} \, e^{iky} \, dk$$

$$= \begin{cases} C_\epsilon \, \frac{sign(y)}{|y|^{2-\epsilon}} & \text{for } \epsilon \ne 1 \\ C_1 \, \delta(y) & \text{for } \epsilon = 1, \end{cases}$$

---

[10] The higher-order moments of C_0 can be computed in a similar fashion

where C_ϵ and C_1 are numerical constants and $\delta(y)$ is the Dirac function. The function (28) is very far from being a Gaussian propagator $\exp(-D_{eff}|\mathbf{k}|^2 t')$. It has the form

$$\hat{\bar{P}}(k_1, k_2, t') = \Phi[D^{-(1-\epsilon/2)}(k_1)^2 (t')^{1+\epsilon/2}] \cdot \exp\{-D(k_2)^2 t'\}.$$

This Green function does not correspond to a "closed" evolution equation for the average scalar. (On the other hand, the *unaveraged* scalar follows a stochastic PDE.)

Recently, Majda obtained a complete characterization of the distribution of the concentration in the case of *time-dependent* random shear flows ([Ma1]).

6. "Nearly Stratified" Flows: Rigorous Assessment of the Influence of High-Wavenumber Components on Macroscopic Scales. The previous result can be extended to a wider class of flows with infrared divergences. These "nearly stratified" flows are no longer simple shear flows, but have the same type of spectrum for $|\mathbf{k}| \ll 1$. Since this is the spectral region which "controls" the diffusive/superdiffusive behavior, we expect qualitatively the same results as for the simple shear flows[11]. However, unlike in the previous example, the high-wavenumber components of the flow also "participate" in the renormalization [AM4].

The nearly stratified flows have the form

$$V(x_1, x_2) = \left(U(x_2) + \tilde{U}_1(x_1, x_2), \tilde{U}_2(x_1, x_2)\right) \tag{29}$$

where U is defined in (23) and $\tilde{U} = (\tilde{U}_1, \tilde{U}_2)$ is an incompressible vector field which satsifies condition (13). Notice that the associated Lagrangian trajectories are non-trivial and the advection-diffusion equation cannot be solved by separation of variables. My goal in this section is to point out that the effect of the high-wavenumber component $\tilde{U}(\mathbf{x})$ on the on the macroscopic scale is simply to *renormalize the transverse diffusion coefficient.*

This can be seen intuitively by considering the Lagrangian formulation of the problem [AM4]. A Lagrangian particle moving with the flow (29), with molecular diffusivity D, will perform a *normal* random walk in the x_2–direction at large scales, and behave superdiffusively in the x_1–direction. These two phenomena occur at different time scales. In fact, the diffusive motion is achieved at distances which are large compared to the (finite) correlation length of $\tilde{U}_2(x_1, x_2)$, but still relatively short. Because of this, the superdiffusive behavior in the x_1–direction, which is determined by the infinitesimal region of the spectrum near $|\mathbf{k}| = 0$ and hence by very large wavelength fluctuations, is like in Section 5. The system behaves as if the Lagrangian particle component $x_2(t)$, was distributed as Brownian Motion independent of $U(x_2)$, with an *enhanced diffusivity.* The macroscopic time-scale $\rho(\delta)$ which characterizes superdiffusion in the x_1 direction, is unchanged.

---

[11]In the language of the renormalization group theory, the high wavenumber components are "irrelevant" insofar as the critical exponents are concerned.

The macroscopic evolution equation for the function C_0 for "nearly stratified" flows turns out to be

$$\frac{\partial C_0(x_1', x_2', t')}{\partial t'} + U_0(x_2') \frac{\partial C_0(x_1', x_2', t')}{\partial x_1'} = D_{eff}^{\perp} \frac{\partial^2 C_0(x_1', x_2', t')}{\partial (x_2')^2}, \quad (30)$$

where $D_{eff}^{\perp}$ is a homogenized transverse diffusivity (see [AM4]. Thus, the system behaves as if there was a "separation of scales" between the "infrared-divergent" component $(U(x_2), 0)$ and the "infrared-convergent" component, $(\tilde{U}_1(x_1, x_2), \tilde{U}_2(x_1, x_2))$. The situation is actually more subtle and shows a notable difference between iterated homogenization and renormalization. In fact, $D_{eff}^{\perp}$ *depends on the random velocity* $U(x_1)$ *as well.* Thus, equation (30) does not arise from an iterated homogenization as presented in §3 — the "longer" scales are still weakly coupled to the shorter ones and play a role in determining the effective diffusivity.

To be more specific about this last point, I exhibit this interaction explicitly. It is proved in [AM4] that the transverse effective diffusivity in (30) is obtained by solving the "cell problem"

$$D \Delta \chi(x_1, x_2) + \left(U(x_1) + \tilde{U}_1(x_1, x_2) \right) \frac{\partial \chi(x_1, x_2)}{\partial x_1} +$$

$$\tilde{U}_2(x_1, x_2) \frac{\partial \chi(x_1, x_2)}{\partial x_2} = -\tilde{U}_2(x_1, x_2), \quad (31)$$

and that

$$D_{eff}^{\perp} = D \cdot \left[1 + \left\langle |\nabla \chi|^2 \right\rangle \right].$$

Notice that $U(x_2)$ is present in equation (31).

The renormalization for nearly-stratified flows has been instrumental for analyzing more complicated systems. For instance, Avellaneda, Elliott and Apelian ([AEA]) used it to explain the scaling behavior of random advection with zero diffusion in the presence of a mean flow (i.e. $\langle V \rangle \neq 0$).

7. Renormalization Group Analysis: a Non-Rigorous Decimation Procedure.
In this section, I return to the renormalization problem formulated in §4 for *isotropic* infrared-divergent velocities. The degree of infrared divergence can be quantified using the exponent ϵ, as in the shear-flow models. We shall consider an incompressible, isotropic, Gaussian, d-dimensional random velocity field $V(\mathbf{x})$ with spectrum

$$\hat{R}(\mathbf{k}) = \bar{V}^2 |\mathbf{k}|^{(2-d-\epsilon)}, \qquad |\mathbf{k}| \leq k_0, \quad (32)$$

where the parameter range is $0 < \epsilon < 2$.

It is easy to see that for any number $0 < \theta < 1$, such a velocity can be represented as an infinite series

$$V(\mathbf{x}) = \sum_{0}^{\infty} \theta^{n(1-\epsilon)} V_n(\mathbf{x}\theta^n) \quad (33)$$

where the V_n are independent, identically distributed Gaussian velocities with the same band-limited spectrum

$$\hat{R}_\theta(\mathbf{k}) = \bar{V} \, |\mathbf{k}|^{(2-d-\epsilon)}, \qquad \theta \, k_0 < |\mathbf{k}| \leq k_0.$$

Notice that the velocity (33) is analogous to the ones studied in §3. Because the singularity of the ratio $\hat{R}(\mathbf{k})/|\mathbf{k}|^2$ is the key feature in this problem, *we are naturally led to considering multi-scale, self-similar flows*. However, an important difference between this problem and iterated homogenization theory is the absence of "separation of scales". Notice also that, since $\theta < 1$, the **H**-matrix associated with the velocity in (33) is a *divergent* series

$$\mathbf{H}(\mathbf{x}) = \sum_0^\infty \theta^{-n\epsilon} \, \mathbf{H}_n\left(\mathbf{x}\,\theta^n\right), \tag{34}$$

where $\mathbf{H}_n$ is the **H**-matrix of V_n.

The *renormalization group analysis* ([FNS], [R], [AN]) is a non-rigorous approach to study this problem, which is based on successive decimations. The idea is to guess the scaling exponent (or, equivalently, $\rho(\delta)$) by analyzing recursively the effects of successive scales on the evolution of C_0. I shall focus on the renormalization group analysis of the effective diffusivity D_N of the "truncated" velocity

$$\sum_0^N \theta^{n(1-\epsilon)} \, V_n\left(\mathbf{x}\,\theta^n\right). \tag{35}$$

This coefficient can be viewed (in the context of the renormalization group analysis) as describing the effect of the top N scales on the macroscopic concentration field. Renormalization group theory *makes the hypothesis* that each scale influences the longer ones primarily as a "homogenized" coefficient, i. e. through a map. This should be understood as an "approximate homogenization", because i) θ is not an infinitesimal quantity and ii) we have shown in the previous section that the renormalized diffusivity corresponding to high-wavenumber components is *not* independent of the larger scales ($n > N$). Such hypothesis leads to a recurrence relation for the successive diffusivities $D_1, D_2, ..., D_N$ etc. Let us make this relation explicit by introducing the map

$$F(D) \equiv \text{"homogenized" diffusivity of the operator} \quad D\,\Delta + V_1 \cdot \nabla. \tag{36}$$

Like in (19), the recurrence relation has the form

$$D_n = \theta^{-n\epsilon} F(D_{n-1}\,\theta^{n\epsilon}), \quad 0 \leq n \leq N \tag{37a}$$

or,

$$\lambda_n = D_n\,\theta^{n\epsilon}, \quad \lambda_n = F(\theta^\epsilon \lambda_{n-1}). \tag{37b}$$

To analyze this relation, we use the fact that $F(D)$ is a monotone increasing function satisfying $F(D)/D > 1$ and that $F(D)/D \to 1$ as $D \to \infty$, as well as the fact that $\theta < 1$. It follows that the function $\lambda \longmapsto F(\theta^\epsilon \lambda)$ has a unique fixed point, λ^*. We conclude from (37b) that

$$D_N \propto \lambda^* \theta^{-\epsilon N}, \qquad N \gg 1. \tag{38}$$

This gives an estimate of the contribution of the high-wavenumber components ($|\mathbf{k}| \geq k_0 \theta^N$) to the diffusivity, suggesting that the effective diffusivity is infinite as $N \to \infty$ and diverges in a precise way as the infrared cutoff in (35) is removed. Recalling that diffusivity has dimensions of L^2/T and that $\theta^N k_0$ has dimensions of L, (38) suggests the space-time scaling for the macroscopic concentration

$$L^2 \propto T^{1/(1-\epsilon/4)}, \qquad T \gg 1, \tag{39a}$$

or, in terms of the time-scaling function $\rho(\delta)$ of the renormalization problem,

$$\rho(\delta) = \delta^{1-\epsilon/4}. \tag{39b}$$

These scaling relations are in excellent agreement with Monte-Carlo simulations using Lagrangian particles [AG].

8. An Approximate Evolution Equation for C_0 in the case of Isotropic, Infrared-Divergent Velocity Fields.

The renormalization problem for isotropic velocity fields with spectra (32) is still a mathematically open problem. Physicists have used renormalization group analysis as well as diagrammatic methods, or renormalized perturbation theories ([Kr], [KB]), to derive approximate the evolution equations for the mean statistics $\langle C_0(\mathbf{x}', t') \rangle$[12]. These procedures are all consistent with the scaling (39b) but differ in the form of the effective equations[13]. I shall speculate on the form of the evolution equation for the *unaveraged* concentration, assuming that (39b) is the correct scaling.

The PDE satisfied by the function $C_\delta(\mathbf{x}', t')$ defined in (21) is

$$\frac{\partial C_\delta(\mathbf{x}', t')}{\partial t'} + (\delta/\rho^2) V(\delta^{-1}\mathbf{x}') \cdot \nabla C_\delta(\mathbf{x}', t') = (\delta/\rho)^2 D \Delta C_\delta(\mathbf{x}', t').$$

Substituting the scaling function $\rho(\delta)$ from (39b), we obtain the PDE

$$\frac{\partial C_\delta(\mathbf{x}', t')}{\partial t'} + \delta^{\epsilon/2-1} V(\delta^{-1}\mathbf{x}') \cdot \nabla C_\delta(\mathbf{x}', t') = \delta^{\epsilon/2} D \Delta C_\delta(\mathbf{x}', t'). \tag{40}$$

This leads to the following observations: first, the random field

$$\delta^{\epsilon/2-1} V(\delta^{-1}\mathbf{x}') \tag{41}$$

---

[12]Majda [Ma2] obtained exact solutions for the first-order moment of C_0 for time-dependent velocities with fast time-decorrelation. This is a different problem that the one for steady fields that is discussed here.

[13]Avellaneda and Majda [AM5] put to test the validity of renormalized perturbation theories and of the renormalization group by comparing their predictions in the context of exactly renormalizable models.

converges in distribution to a Gaussian random field $V_0(\mathbf{x}')$ with spectrum

$$\hat{R}_0(\mathbf{k}) = \bar{V}^2 \, |\mathbf{k}|^{(2-d-\epsilon)}, \qquad \mathbf{k} \in \mathbf{R}^d.$$

Note that since $0 < \epsilon < 2$, the function $\hat{R}(\mathbf{k})$ is not integrable and hence $V_0(\mathbf{x}')$ is a *distribution-valued* random field. Second, $\delta^{\epsilon/2} D \Delta C_\delta(\mathbf{x}',t')$ converges weakly to zero in the Hilbert space $L^2\left(\mathbf{R}^d, L^2(< \cdot >)\right)$. We are thus led to the *formal* conclusion that the large-scale concentration should satisfy the first-order stochastic PDE

$$\frac{\partial C_0(\mathbf{x}',t')}{\partial t'} + V_0(\mathbf{x}') \cdot \nabla C_0(\mathbf{x}',t') = 0. \tag{42}$$

However, this reasoning leads to serious difficulties: first, due to the absence of any smoothing in the right-hand side of (42), the meaning of advection by a distribution-valued random function is unclear[14]. The vector field V_0 is far from satisfying the usual requirements for well-posedness of the initial-value problem (classically, Lipschitz continuity is required). Thus, equation (42) is essentially meaningless[15]. Furthermore, the passage to the limit in (40) requires evaluating the limit of the product of the singular field $\delta^{\epsilon/2-1} V(\delta^{-1}\mathbf{x}')$ with $\nabla C_\delta(\mathbf{x}',t')$. The regularity of the gradient of the solution is not expected to be sufficient to justify the above procedure. A more sophisticated renormalization is needed to make sense of the limit of (40) as $\delta \to 0$. Since numerical experiments leave little doubts as to the validity of the scaling (39b), I will use heuristic ideas to guess the form of the effective evolution equation.

As indicated by the examples of Sections 5 and 6, the large-scale evolution is determined by dissipation from the high-wavenumber components (with $|\mathbf{k}| \geq$ constant) in conjunction with the advection-dominated transport from the infrared portion of the spectrum ($|\mathbf{k}| \ll 1$). The renormalization group analysis suggests that the diffusivity "contributed" by Fourier modes with $|\mathbf{k}| \geq \kappa$ should be

$$D(\kappa) \propto \kappa^{\epsilon/2}.$$

Naively, we could implement this idea by introducing a fractional Laplacian[16], $-|\mathbf{k}|^{2-\epsilon/2} = -\left(-\Delta\right)^{1-\epsilon/4}$, on the right-hand side of (42). The corresponding evolution equation would then be

$$\frac{\partial C_0(\mathbf{x}',t')}{\partial t'} + V_0(\mathbf{x}') \cdot \nabla C_0(\mathbf{x}',t') = -\gamma(k_0) \left(-\Delta\right)^{1-\epsilon/4} C_0(\mathbf{x}',t'), \tag{43}$$

where $\gamma(k_0)$ is a non-universal constant depending on the high-wavenumber cutoff k_0 in (32). Although equation (43) is consistent with the symmetry and scaling of the problem (it is invariant under the scaling transformation associated with

---

[14]Herein lies the difference between the renormalization for isotropic velocities and the examples of Sections 5 and 6.

[15]This applies to ϵ is the range $0 < \epsilon < 2$. In Avellaneda and Majda [AM2], it is shown that (42) is the correct renormalization for the regime $2 < \epsilon < 4$, which I don't discuss here.

[16]The fractional Laplacian is the infinitesimal generator of the so-called *Levy processes* in Probability Theory.

$\rho(\delta) = \delta^{1-\epsilon/4}$), it presents a serious drawback. In fact, it can be shown that (43) leads to *infinite second-order moments* for Lagrangian mean-square displacements due to the "fat tails" of the Lévy distribution: under (43), we have

$$\int_{\mathbf{R}^d} |\mathbf{x}'|^2 \, \langle C_0(\mathbf{x}',t') \rangle \, \mathbf{d}^d \mathbf{x}' = \infty .$$

This divergence is inconsistent with numerical simulations. (Some authors have proposed heuristic renormalizations associated with "Levy flights" corresponding to modified Levy operators [SK]).

An approximation that does not lead to this last problem can be obtained by making an analogy with the effective equation for nearly stratified flows (30). Accordingly, consider the equation

$$\frac{\partial \tilde{C}_0(\mathbf{x}',\mathbf{y}',t')}{\partial t'} + V_0(\mathbf{y}') \cdot \nabla_{\mathbf{x}'} \, \tilde{C}_0(\mathbf{x}',\mathbf{y}',t') = -\gamma(k_0) \left(-\Delta_{\mathbf{y}'}\right)^{1-\epsilon/4} \tilde{C}_0(\mathbf{x}',\mathbf{y}',t')$$

$$(44)$$

with initial condition

$$\tilde{C}_0(\mathbf{x}',\mathbf{y}',t'=0) = C_{init}(\mathbf{x}') \quad \mathbf{x}' \in \mathbf{R}^d , \, \mathbf{y}' \in \mathbf{R}^d .$$

In this equation, I use an additional variable, $\mathbf{y}'$, to model the action of the high-wavenumber components on the large-scale concentration. This represents a "weak-coupling" approximation for the interaction between advection and dissipation at large scales. The "macroscopic concentration" is defined as

$$C_0(\mathbf{x}',t') = \tilde{C}_0(\mathbf{x}',\mathbf{0},t') .$$

$$(45)$$

Equation (44) mimics the equation for the concentration in nearly-stratified flows (30), with the effective transverse diffusivity $D_{eff}^{\perp}$ replaced by the *operator* $-\gamma(k_0) |\mathbf{k}_y|^{-\epsilon/2}$. It is not hard to show that this approximation produces finite mean-square displacements for Lagrangian particles. I do not claim that equations (44) and (45) represent an exact solution. They can be viewed as a "closure" for calculating the statistics of the large-scale concetration. As other approximate solutions, this one has only heuristic value. For instance, in this approximation, the Fourier transform of the averaged Green function is

$$\hat{P}(\mathbf{k},t') = \mathbf{E}\left\{ \exp\left[-\bar{V}^2 \, t'^{1/(1-\epsilon/4)} \, |\mathbf{k}|^2 \int_0^1 \int_0^1 F_\epsilon\left(\mathbf{Y}(s) - \mathbf{Y}(s')\right) ds \, ds' \right] \right\},$$

$$(46)$$

where $\mathbf{E}$ represents expectation value over over the d-dimensional Lévy process $\mathbf{Y}(s)$ and $F_\epsilon(\mathbf{y}')$ is a homogeneous function. The higher-order moments of $C_0(\mathbf{x}',t')$ can be calculated easily as well. The implications if this approximation will be investigated elsewhere.

In conclusion, I have attempted to show the relationship between homogenization theory, a rigorous method for averaging partial differential equations, and the renormalization group, a heuristic method designed to investigate scaling behavior in self-similar systems. There are important connections between these two approaches. However, the assumption of scale separation invoked in homogenization is not adequate for treating the most general problems of transport and diffusion in self-similar random media. The problem of advection-diffusion in a random velocity field showcases the difficulties which arise when the velocity has self-similar characteristics and a singular spectrum. Approximate effective equations based on exact solutions for model problems can be derived. However, the systematic understanding of anomalous scaling phenomena remains an open problem in Applied Mathematics.

Acknowledgements. I am grateful to Robert Kohn, Andrew Majda, Graeme Milton and George Papanicolaou for sharing with me their ideas on heterogeneous media and turbulent diffusion over the past years. This work was prepared under support from the National Science Foundation (NSF-DMS-92-07085 and NSF-DMS-94-02763) and the U.S. Army (ARO-DAAL-03-92-G0011).

References

[AN]   J. A. Aronovitz and D. R. Nelson, *Phys. Rev. A, 30, p. 1948 (1984)*

[A]   M. Avellaneda, *Commun. Pure Appl. Math., 40, 527-554 (1987)*

[AEA] M. Avellaneda, F. Elliott, Jr. and C. Apelian, *Journ. Stat. Phys. 72, 1227-1304 (1993)*

[AG]   M. Avellaneda and Y. Gonorovski, *unpublished numerical simulations, Courant Institute, 1993.*

[AM1] M. Avellaneda and A. J. Majda, *Commun. Math. Phys. 138, 339-391 (1991)*

[AM2] M. Avellaneda and A.J. Majda, *Phys. Rev. Lett. 68, 3028-3031 (1992)*

[AM3] M. Avellaneda and A.J. Majda, *Commun. Math. Phys. 131, 381-429 (1990) and Commun. Math. Phys. 146, 139-204 (1992)*

[AM4] M. Avellaneda and A.J. Majda, *J. Stat. Phys. 69, 689-729 (1992)*

[AM5] M. Avellaneda and A.J. Majda, *Physics of Fluids A 4, 41-57 (1992)*

[B]   N. Bakhvalov and G. Panasenko, *Homogenisation: Averaging processes in Periodic Media, Kluwer, Dordrecht (1989)*

[BLP] A. Bensoussan, J.L. Lions and G. C. Papanicolaou, *Asymptotic Analysis of Periodic Structures, North Holland, Amsterdam (1979)*

[FNS] D. Forster, D. Nelson and M.Stephen, *Phys. Rev. A 16, 732-749 (1977)*

[KB]   D. Koch and J. Brady, *Phys. Fluids A, 1, 47-51 (1989)*

[Ko]   S. Kozlov, *Russian Math. Surveys 40*, p.73 (1985)

[Kr]   R. Kraichnan, in *The Padé Approximant in Theoretical Physics* p. 127 ,
          G. A. Baker Jr. and J. Gammel editors, Academic Press, New York (1970)

[Ma1]   A. J. Majda, *Journ. Stat. Phys. 73*, 515-542 (1993)

[Ma2]   A. J. Majda *Journ. Stat. Phys.*, preprint, *to appear in vol 76 (1/2)* (1994)

[Mi]   G. W. Milton, in *Physics and Chemistry of Porous Media, D. L. Johnson,
          and P. N Sen, editors, AIP Conference 107, American Institute of Physics,
          New York, 1984*

[MY]   A.S. Monin and A. M. Yaglom, *Statistical Fluid Mechanics: Mechanics of
          Turbulence, Vol 2, MIT Press, Cambridge, Mass.,* (1975)

[O]   K. Oelschlager, *Annals of Prob. 16*, 1084-1126 (1988)

[PV]   G. C. Papanicolaou and S. R. S. Varadhan, in *Random Fields, Coll. Math.
          Soc. Janos Bolyai 27, Fritz, Lebowitz and Szasz editors, 835-873, North-
          Holland, Amsterdam* (1982)

[R]   H. Rose, *J. Fluid Mech. 81*, 719-734 (1977)

[S]   E. Sanchez Palencia, *Non-Homogeneous Media and Vibration Theory, Lec-
          ture Notes in Physics, 127, Springer-Verlag* (1980)

[SK]   M. F. Schlesinger and J. Klafter, *Phys. Rev Lett. 54*, 2551 (1985)

Other Titles in This Series